THE CORRESPONDENCE BETWEEN SIR GEORGE GABRIEL STOKES AND SIR WILLIAM THOMSON, BARON KELVIN OF LARGS

VOLUME I 1846–1869

THE CORRESPONDENCE BETWEEN SIR GEORGE GABRIEL STOKES AND SIR WILLIAM THOMSON, BARON KELVIN OF LARGS

VOLUME I 1846–1869

EDITED WITH AN INTRODUCTION
BY DAVID B. WILSON

The right of the University of Cambridge to print and sell all manner of books was granted by Henry VIII in 1534. The University has printed and published continuously since 1584.

CAMBRIDGE UNIVERSITY PRESS

CAMBRIDGE
NEW YORK PORT CHESTER
MELBOURNE SYDNEY

Published by the Press Syndicate of the University of Cambridge
The Pitt Building, Trumpington Street, Cambridge CB2 1RP
40 West 20th Street, New York, NY 10011, USA
10 Stamford Road, Oakleigh, Melbourne 3166, Australia

First published 1990

Printed in Great Britain at the University Press, Cambridge

British Library cataloguing in publication data

Stokes, Sir, George Gabriel
The correspondence between Sir George Gabriel Stokes
and Sir William Thomson, Baron Kelvin of Largs.
1. Physics. Kelvin, William Thomson, Baron, 1824–1907
I. Title
II. Kelvin, William Thomson, Baron, 1824–1907
530'.092'2

Library of Congress cataloguing in publication data

Stokes, George Gabriel, Sir, 1819–1903.
The correspondence between Sir George Gabriel Stokes and
Sir William Thomson, Baron Kelvin of Largs / edited with
an introduction by David B. Wilson.
p. cm.
Contents: v. 2. 1870–1901.
0 521 32831 4
1. Physics – History.
2. Mathematics – History.
3. Stokes, George Gabriel, Sir, 1819–1903 – Correspondence.
4. Kelvin, William Thomson, Baron, 1824–1907 – Correspondence.
5. Physicists – Great Britain – Correspondence.
6. Mathematicians – Great Britain – Correspondence.
I. Kelvin, William Thomson, Baron, 1824–1907.
II. Wilson, David B.
III. Title.
QC7.S79 1990
530'.092'2–dc20 89–17463 CIP

ISBN 0 521 32831 4 Set of two vols.

FOR

JULIE AND NATASHA

HOWARD AND MARY SUE

HARRY AND COLETTA

When Kelvin was speaking, Stokes would remain silent until Kelvin seemed at any rate to pause. On the other hand, when Stokes was speaking, Kelvin would butt in after almost every sentence with some idea which had just occurred to him, and which he could not suppress.

J. J. THOMSON, *Recollections and Reflections* (London, 1936), 50.

The existing letters from Lord Kelvin, with the complementary letters from Sir George Stokes which Lord Kelvin has promptly supplied, suffice to form a collection by themselves, and a memorial of the lifelong friendship and collaboration of the writers.

JOSEPH LARMOR, 'Preface' to *Memoir and Scientific Correspondence of the Late Sir George Gabriel Stokes*, ed. J. Larmor, 2 vols. (Cambridge, 1907), I, viii.

While Thomson was daring in speculation, moving swiftly, almost erratically, to some intuitive result, Stokes was methodical, conservative, cautious. . . . Whenever the letters between Stokes and Thomson come to be published, it will be seen how invaluable to both was the friendship so formed.

S. P. THOMPSON, *The Life of William Thomson, Baron Kelvin of Largs*, 2 vols. (London, 1910), I, 299.

I shall always remember Lord Kelvin, as he stood at the open grave, almost overcome by his emotion, saying in a low voice: 'Stokes is gone and I shall never return to Cambridge again'.

ARTHUR SCHUSTER, *Biographical Fragments* (London, 1932), 242.

CONTENTS

VOLUME 1

VOLUME 2

PREFACE

Some decades ago when most of the letters in this volume were in his care, Sir Joseph Larmor evidently planned an edition of correspondence between Lord Kelvin and Sir George Gabriel Stokes. He arranged the letters chronologically, pencilling editorial notes on many in preparation for printing. I do not know why the work was not published. At any rate, regret at not having such an edition available all these years may be somewhat offset by the knowledge that Larmor apparently intended to omit some letters altogether and to delete diagrams and 'embarrassing' passages from others. Moreover, he probably would not have included the several letters printed here from collections not in his care and was apparently unaware of a few letters between Kelvin and Stokes in the material he did have charge of. Nevertheless, Larmor's edition would have been highly useful and would undoubtedly have impelled scholarship on Kelvin and Stokes along much faster than it has in fact proceeded. In any case, Kelvin and Stokes were on the whole well served by Larmor as editor and biographer, and, so far as I know, the present volume completes the projects envisioned by him.

Plans for the current edition grew out of my employment at the Cambridge University Library in the early 1970s, when I arranged and catalogued the Library's collections of Stokes and Kelvin manuscripts. In organizing the collections, I broke up Larmor's arrangement of the Stokes–Kelvin correspondence (which constituted a large part of Add. MS 7618), putting the letters from Stokes to Kelvin in the Kelvin Collection (Add. MS 7342) and those from Kelvin to Stokes in the Stokes Collection (Add. MS 7656). The *Catalogue of Manuscript Collections of Sir George Gabriel Stokes and Sir William Thomson, Baron Kelvin of Largs, in Cambridge University Library* was published by the Cambridge University Library Press in 1976. It was decided that an edition of the correspondence between Stokes and Kelvin would be a valuable sequel to the *Catalogue* of their collections, and I thus began preparing it during my final year in Cambridge.

Although the great majority of the letters printed here are from the Cambridge University Library's Stokes and Kelvin Collections, forty-four are not. Letter 118 is from Stokes's *Mathematical and Physical Papers*, letter 515 is from the H. F. Stewart Collection in the Cambridge University Library, letters 170–1, 173, 209, and 236 are from the Kelvin Papers in the Glasgow University Library, letters 360–2 are from the private collection of Mr Francis Bottomley, and thirty-four letters are from the archives of the

Royal Society of London. The latter come mainly from the Royal Society's collection of referees' reports on papers submitted to the Society for publication in its journals. Often, there was no clear distinction in Kelvin's mind between a personal letter to Stokes and a referee's report sent to Stokes as secretary of the Royal Society. Generally, I have included those reports by Kelvin in the Royal Society's archives which are written as letters to Stokes (beginning 'Dear Stokes') and which also go beyond the report itself. Conversely, there is one item in the Stokes Collection at Cambridge which I have not included because it is a straightforward referee's report by Kelvin, not even beginning 'Dear Stokes'. This is a report on a paper by William Hopkins, K99 in the Stokes Collection.

For permission to publish these letters, I am grateful, respectively, to the Syndics of the Cambridge University Library, the Court of Glasgow University, Mr Francis Bottomley, and the President and Fellows of the Royal Society of London. I am also grateful to Harold Sharlin for calling my attention to Mr Bottomley's collection.

I gladly acknowledge financial assistance from various sources. A grant from IBM United Kingdom to Cambridge University made my employment by the Cambridge University Library possible. A research grant from Iowa State University allowed me to work on the project full time for one summer. A grant from the National Science Foundation allowed me to do the same for a second and enabled me to make a return trip to Cambridge for a final examination of the original letters.

In preparing both the *Catalogue* and the present edition, I have worked under the general supervision of the Head of the Manuscripts Department of the Cambridge University Library, A. E. B. Owen, who has given welcome advice on archival and editorial matters. I am indebted for many things to my former colleagues in the Manuscripts Department, Peter Gautrey, Jayne Ringrose, and Margaret Pamplin. D. J. Hall, Administrative Services Officer for the Library, has assisted me at several points along the way.

The typescript for the edition was prepared mainly by Susie Ulrickson and also by Laurie Helmers, typists for the Iowa State University Department of History.

Finally, I should especially like to acknowledge the support of the late E. B. Ceadel, Cambridge University Librarian from 1967 until his death on 1 June 1979. Ultimately, it was his commitment to publication of this edition which made the project possible. He had received from me the typescript of the letters and had worked out publication arrangements with the Cambridge University Press only shortly before his death.

D. B. W. Cambridge, July 1980

When financial strictures halted production of this book in 1981, the Cambridge University Library Press had nearly completed its composition. Fortunately, the Cambridge University Press has been able to use this prior work. For their invaluable assistance in resurrecting the project, I wish to thank several people: Arthur Bergles, former chairman of the Department of Mechanical Engineering at Iowa State University, D. J. Hall of the Cambridge University Library, Roger Smith and J. Graham Smith, successive editors of the Monograph Series of the British Society for the History of Science, S. Millard and Richard Ziemacki of the Cambridge University Press, Crosbie Smith, P. M. Harman, and M. Norton Wise. Audrey Burton typed the indexes. I have further developed parts of the introduction in various publications since 1980. See my 'Experimentalists among the mathematicians: physics in the Cambridge Natural Sciences Tripos, 1851–1900', *Historical Studies in the Physical Sciences*, **12** (1982), 325–71; 'The educational matrix: physics education at early-Victorian Cambridge, Edinburgh and Glasgow Universities', in P. M. Harman (ed.), *Wranglers and Physicists: Cambridge Mathematical Physics in the Nineteenth Century* (Manchester, 1985), 12–48; and *Kelvin and Stokes: A Comparative Study in Victorian Physics* (Bristol, 1987).

D. B. W. Ames, July 1987

EXPLANATORY NOTES

The heading for each letter gives the location of the letter and cites any previous publication of it. Not cited is the recent biography of Kelvin, *Lord Kelvin: The Dynamic Victorian* (University Park and London: The Pennsylvania State University Press, 1979), by Harold Issadore Sharlin in collaboration with Tiby Sharlin, which cites or quotes from nearly seventy Stokes–Kelvin letters. Also not cited is Crosbie Smith, 'Engineering the universe: William Thomson and Fleeming Jenkin on the nature of matter', *Annals of Science*, **37** (1980), 387–412, which quotes from letters 161–2 and 164–5.

Somewhat more than the first half of the correspondence is almost entirely in the hands of Kelvin and Stokes themselves. In 1878, much to the relief of his correspondents, Stokes began using the first of three typewriters. It had all capital letters and he used it from 1878 to 1885. In 1886 he began using the second, which also had all capital letters, though larger than those of the first typewriter. The third typewriter, with both capital and small letters, was used from 1890 to the end of the correspondence. Only fifteen letters from 1878 to 1901 are in his own hand. Most of Kelvin's later letters are in his own hand, but some are in Lady Kelvin's hand, in the hand of one of his assistants, or typed. I have noted the form of the letter when relevant to the reading of the letter. I have also noted when letters appear to be copies or drafts of letters, rather than the letters themselves.

The order of the letters in this book is sometimes different from that in my *Catalogue* of their collections in the Cambridge University Library. This is because I have now been able to give at least approximate dates to all the undated letters and have discovered that a few letters were misdated by Kelvin or Stokes. Also, letters S559 and S560 in the Kelvin Collection are not included here because I now think that neither is a letter to Kelvin. Each is an undated fragment of a letter consisting of one sheet signed by Stokes. Written in Stokes's hand, S559 appears to be a letter to W. J. M. Rankine from 1857. Typed on the second of Stokes's three typewriters, S560 dates from the late 1880s and is possibly to William Abney.

Notes to the letters include ones by me, ones by Kelvin and Stokes to their own letters, and a few by Larmor, which he wrote on the letters themselves. The notes explain which are by whom.

I have referred to Kelvin as Kelvin throughout, even though he was not raised to the peerage until 1892. I have referred to his first wife as Lady Thomson, even though she enjoyed that title only from 1866 until her death in

1870. I have referred to Kelvin's second wife as Lady Kelvin, even though she was Lady Thomson from 1874 to 1892 and Lady Kelvin only from 1892.

I have silently modified punctuation in very few cases, adding full stops and an occasional apostrophe. On the other hand, until 1852 Kelvin usually wrote 'your's' (especially when signing letters), and I have left that uncorrected and not followed it with *sic*.

Unless otherwise noted, square brackets [ ] enclose material added by me. Pointed brackets ⟨ ⟩ enclose material written by Kelvin or Stokes and then deleted.

Stokes's and Kelvin's articles are cited in the notes by numbers, the number of each corresponding to its number in the Index of their publications. The numbering of the articles in the Index is based on that in the *Catalogue of Scientific Papers*, compiled by the Royal Society of London, 19 vols., vols. I–XII (London, 1867–1902), vols. XIII–XIX (Cambridge, 1914–25).

Abbreviations used by me in the notes are:

Kelvin

Baltimore Lectures (1884)	*Notes of Lectures on Molecular Dynamics and the Wave Theory of Light* (Baltimore, 1884)
Baltimore Lectures (1904)	*Baltimore Lectures on Molecular Dynamics and the Wave Theory of Light* (London, 1904)
MPP	*Mathematical and Physical Papers*, 6 vols. vols. IV–VI edited by Joseph Larmor (Cambridge, 1882–1911)
PEM	*Reprint of Papers on Electrostatics and Magnetism*, 2nd edn (London, 1884)
PLA	*Popular Lectures and Addresses*, 3 vols. (London, 1889, 1894, 1891)
TNP	W. Thomson and P. G. Tait, *Treatise on Natural Philosophy* (Oxford, 1867)
Thompson	Silvanus P. Thompson, *The Life of William Thomson, Baron Kelvin of Largs*, 2 vols. (London, 1910)

Stokes

Burnett Lectures	*Burnett Lectures. On Light*, 3 vols. (London, 1884–87)
Memoir	*Memoir and Scientific Correspondence of the Late Sir George Gabriel Stokes*, ed. Joseph Larmor, 2 vols. (Cambridge, 1907)

MPP	*Mathematical and Physical Papers*, 5 vols., vols. IV and V edited by Joseph Larmor (Cambridge, 1880–1905)

Journals

Brit. Assoc. Rep.	*Report of the British Association for the Advancement of Science*
Camb. Dub. Math. J.	*Cambridge and Dublin Mathematical Journal*
Camb. Math. J.	*Cambridge Mathematical Journal*
Comptes Rendus	*Comptes rendus hebdomadaires des séances de l'Académie des Sciences*
Phil. Mag.	*The London, Edinburgh, and Dublin Philosophical Magazine and Journal of Science*
Phil. Trans.	*Philosophical Transactions of the Royal Society of London*
Proc. Camb. Phil. Soc.	*Proceedings of the Cambridge Philosophical Society*
Proc. Roy. Inst.	*Proceedings of the Royal Institution of Great Britain*
Proc. Roy. Soc. Edinb.	*Proceedings of the Royal Society of Edinburgh*
Proc. Roy. Soc.	*Proceedings of the Royal Society of London*
Trans. Camb. Phil. Soc.	*Transactions of the Cambridge Philosophical Society*
Trans. Roy. Soc. Edinb.	*Transactions of the Royal Society of Edinburgh*

INTRODUCTION

I

Stokes took his degree from Cambridge University in 1841, the same year Kelvin matriculated there. Stokes remained after graduation as a fellow of his college, and it was thus in Cambridge in the early-to-mid 1840s that they formed the lifelong association documented by the letters in this book.[1] Early friendship, longevity, similar research interests, high professional status, and geographical separation all combined to produce an abundant and rich correspondence, easily the most extensive extant between two major Victorian physicists.

George Gabriel Stokes (1819–1903) grew up in County Sligo, Ireland, son of an Anglican priest who died in 1834. The year after his father's death, Stokes entered Bristol College, run by a friend of Stokes's older brother, William. William had broken with the family's traditional ties with Trinity College, Dublin, and had gone instead to Cambridge University, graduating as sixteenth wrangler in the mathematical tripos of 1828 and becoming a fellow of Caius College. George followed William to Cambridge in 1837. Upon graduating as senior wrangler and first Smith's prizeman in 1841,[2] he was elected to a fellowship in his college, Pembroke. In October, 1849, a few months after declining to apply for the chair of mathematics at Glasgow because of the religious tests there (letters 34–43), Stokes was elected Lucasian professor at Cambridge. He informed Kelvin that he had been the only candidate (letter 51). In November, Kelvin was 'very glad you are to lecture on Hydrostatics & Optics!' (letter 57), which Stokes did for the rest of the century. At least partially for financial reasons, he also accepted a chair at the new Government School of Mines in London, which he held during the 1850s, and, in 1854, a secretaryship of the Royal Society of London, which he filled until 1885 when he became president for five years. His most notable scientific research included his investigation of friction in fluids, his efforts to determine the properties of the luminiferous ether, his explanation of fluorescence, and his dynamical theory of diffraction. His most powerful work came before the mid-1850s, and he sank most of his later efforts into his Royal Society duties, made especially time consuming by his own supreme conscientiousness. As a fellow secretary remarked, 'It has been painful to see how his energy has been wasted in this way'.[3] As early as 1859 Kelvin tried to persuade Stokes to take the Glasgow professorship of astronomy by arguing 'the importance to science of getting you out of London & Cambridge, those great Juggernauts under which so much potential energy for original investigation is crushed'. (Letter 176.) In 1884 Kelvin urged Stokes to secure more time for research by applying for the Cavendish professorship at Cambridge. 'I suppose', Kelvin wrote, 'the income of the Exp^{l} Physics chair is decidedly more than you have in the Lucasian and I thought possibly the difference might amount to even a money compensation for giving up the R[oyal] S[ociety] work'. (Letter 468.) Stokes remained as Lucasian professor, however, and through his vast correspondence, occasional publications, and high offices retained an eminent position in Victorian science.[4]

William Thomson, Baron Kelvin of Largs (1824–1907), was born in Belfast, where his father was professor of mathematics at the Royal Academical Institution, but lived most of his life in Glasgow. The Thomson family moved there in 1832 when Professor Thomson took the chair of mathematics at Glasgow, and Kelvin entered the university as a student two years later at the age of ten. He went to Cambridge in 1841, and while he was an undergraduate, the Glasgow professor of natural philosophy, William Meikleham, fell seriously ill, prompting Kelvin and his father to plan an assault on the soon-to-be-vacated chair. After graduating as second wrangler and first Smith's prizeman in 1845, Kelvin spent much of the summer in Henri Regnault's laboratory in Paris doing experimental work in preparation for applying for the Glasgow position. He lectured as a fellow of Peterhouse during the 1845/6 academic year and, after Meikleham's death, did gain the Glasgow professorship in 1846. With his move to Glasgow, the correspondence between Stokes and Kelvin began, Kelvin writing in November, for example, 'Yesterday I gave the introductory lecture, w^{h} was rather a failure as I had it all written, and I read it very fast'. (Letter 3.) By this time Kelvin had already published some twenty papers, the first being an anonymous defence of Fourier analysis against erroneous charges made by the professor of mathematics at the University of Edinburgh.[5] He was sixteen when he wrote it, and his father sent it to the *Cambridge Mathematical Journal*, which published it and most of Kelvin's other articles in the 1840s. His early researches focused on mathematical theories of heat, electricity, and magnetism, exploring analogies between them and reconciling apparently contradictory previous research – that of Coulomb and Faraday in electricity and that of Carnot and Joule in heat. The latter led to his formulation of the two laws of thermodynamics, his most fundamental contribution to physical theory.[6] In the 1850s he turned to the practical work on submarine telegraphy that eventually made the Atlantic Cable successful and Kelvin famous. In the 1860s he worked with P. G. Tait (senior wrangler 1852), professor of natural philosophy at Edinburgh, on their highly influential *Treatise on Natural*

Philosophy, a physics textbook published in 1867. His later research sought, in Silvanus Thompson's words, a 'great comprehensive theory' – a dynamical theory of the motions of matter which underlay the observed phenomena of heat, electricity, magnetism, and light.[7] One can trace the search through his theory of vortex atoms of the late 1860s, his remarkable *Baltimore Lectures* of 1884, his elaborate models of the ether in the late 1880s, and in papers like 'On the generation of longitudinal waves in ether' (1896)[8] and 'On the duties of ether for electricity and magnetism' (1900),[9] leading to a second, expanded edition of the *Baltimore Lectures* in 1904. He succeeded Stokes as president of the Royal Society in 1890 and was raised to the peerage in 1892. With his immense scientific accomplishments and widespread popular renown, Kelvin was *the* eminent Victorian among physical scientists.[10]

Hence, while Stokes quietly managed the business of Victorian science, Kelvin irrepressibly pursued nature through one theory after another, and the correspondence between them reflects such differences. For example, many of Kelvin's letters to Stokes – most notably those refereeing papers submitted to the *Philosophical Transactions of the Royal Society* – were sent to Stokes in his capacity as secretary of the Royal Society. More significantly, the correspondence as a whole represents Kelvin's research more than Stokes's. Stokes often served as Kelvin's critic and helped Kelvin keep abreast of research that had already been done in particular areas. In this regard, the Stokes–Kelvin correspondence is simply the most important of many surviving groups of letters between Stokes and other Victorian scientists. He corresponded extensively over the years, for example, with the talented experimental physicist and chemist, William Crookes, whose lack of mathematics 'was often masked in later years through his friendship with Stokes, who privately solved many mathematical problems in physics for him'.[11] His intensive correspondence in the 1890s with Arthur Smithells on the latter's research on flame spectra constituted, said Smithells, 'one of the great experiences of my life'.[12] At the century's end, grateful British scientists widely recognized that, unlike Kelvin's, Stokes's mind had for decades been exercised not primarily in his own research programme, but in response to the research of others. As Smithells wrote, 'What Stokes did for his generation can hardly be estimated'.[13] Kelvin would have agreed; 'I always consulted my great authority Stokes whenever I got a chance',[14] he said, and noted further that Stokes 'gave generously and freely of his treasures to all who were fortunate enough to have opportunity of receiving from him'.[15]

Though science naturally has its international aspects, the context of the Kelvin–Stokes correspondence was very much a British setting. For example, of the major correspondents of Kelvin and Stokes, only Helmholtz was not British.[16] Of the more than 200 papers which Kelvin and Stokes refereed for the Royal Society, less than five were by foreigners.[17] Such intensively localized communication between scientists helps explain, of course, the development of national divisions in scientific thought or style and even of what resembles national jealousies. In the 1850s, for instance, Stokes and Kelvin wondered at French excitement over experiments by Regnault and Foucault, whose results followed naturally from the Englishman Joule's much earlier work. (Letters 110–11, 134–7, and 260.) Kelvin observed that Regnault 'does not understand English at all' (letter 111) and remarked that 'Joule in 1843 was rather ahead of France in 1855' (letter 135). In the 1890s, they were amazed that German physicists could regard cathode rays as ethereal waves instead of agreeing with British physicists that they were particles of matter. (Letters 543, 545, and 546.) Kelvin noted the 'partisan spirit' (letter 546) with which the Germans pursued their 'perversity' (letter 543). Awkward phraseology in a French paper on the reflection of light drew from Kelvin a comment on 'French logic, which generally strikes me as peculiar'. (Letter 460.)

Therefore, it should be no surprise that historians of science have long recognized a nineteenth-century *British* school of physics with its characteristic methods and conclusions,[18] and it should be even less surprising that *British* universities and scientific societies constituted the primary institutional framework for Stokes's and Kelvin's careers. Moreover, just as British society at large expanded and improved in many ways during the century, so did the ideas and institutions associated with Kelvin and Stokes. The staggering progress in physics during their lifetimes resulted largely from British accomplishments. The education and organization of British scientists went from amateur to professional during the century. The remainder of this introduction, then, considers the place of Kelvin and Stokes in such institutional and intellectual developments, attempting thereby to provide both a context for the letters and a partial guide to them.

II

Something of an institutional basis for British science did exist before the nineteenth century, of course. The Royal Society of London had been established in the seventeenth century, and some local societies (like the Manchester Literary and Philosophical Society) and specialized societies (like the Linnean Society) had emerged during the eighteenth. One could even find a smattering of science taught at some universities around 1800. But from the vantage point of 1900, the situation in 1800 would have looked meagre indeed.[19] During the nineteenth century, the old universities increasingly emphasized science education, which was also an important aspect of the many new universities founded during the period. Societies devoted to geology, astronomy, chemistry, zoology, and eventually physics appeared. The growing numbers of local societies provided a supportive matrix for the British Association for the Advancement of Science.[20] It was founded in 1831 and met annually in different cities, allowing an intermingling of the well-known men of British science with local 'cultivators of science'. The Royal Society itself passed through a turmoil of reform to regain a respected status.[21]

Stokes and Kelvin were involved with many of these institutions. They both attended meetings of the British Association and served it as officers, each being president for a year. Kelvin was active in the Glasgow Philosophical Society, was president of the Physical Society of London during 1880/1, and presented many of his most significant papers[22] to the Royal Society of Edinburgh, of which he was president for more than twenty years. However, the three institutions I shall discuss here – the ones most closely tied to Stokes's and/or Kelvin's careers – are Cambridge University, Glasgow University, and the Royal Society of London.

CAMBRIDGE UNIVERSITY

Depending on one's perspective, one can view the Cambridge University of the late 1830s and early 1840s – the period of Stokes's and Kelvin's undergraduate years – in various ways.[23] One interested in subjects other than mathematics and classics would find no tripos covering them. One interested in classics would discover that the classical tripos did not count for an honours degree and that it had to be taken after the mathematical tripos, which did. One interested in the general study of mathematics would see that its teaching by college lecturers and university professors was often weak or non-existent and that extra-collegiate coaches – hired for extra fees – often merely crammed mathematics into their students. One interested in students overall would find that most did not study for honours anyway. Yet, if we look at the very best mathematical students, we see a high-level, nearly professional *milieu*, involving an educational system, a scientific society, two scientific journals, and positions relating to mathematics and natural philosophy.

For students like Stokes and Kelvin, life was ordered by the mathematical tripos. Huge quantities of time went into preparation for it and results could heavily influence careers. The tripos had been modernized in the years just before 1820, when three graduates of 1813 – John Herschel, Charles Babbage, and George Peacock – and their younger colleague William Whewell (second wrangler 1816) became involved in administering the examination and introduced the Continental version of the calculus into it.[24] Whewell and the first great product of the improved tripos, George Biddell Airy (senior wrangler 1823), wrote widely used textbooks on mechanics and astronomy incorporating the changes.[25] Airy introduced the new wave theory of light into the second edition of his *Mathematical Tracts* in 1831. The tripos of the 1830s and 1840s, then, covered not only pure mathematics, but also its applications in such areas as statics, dynamics, optics, astronomy, hydrostatics, and hydrodynamics. Though many students prepared with coaches who apparently advocated cramming and memorization, strong students usually worked with more enlightened coaches. Stokes and Kelvin both read with William Hopkins (seventh wrangler 1827), most illustrious of the coaches of this period.[26] In 1862, Kelvin hesitated to criticize a portion of Hopkins's research, 'having been one of his pupils, and having experienced the greatest possible benefit from his teaching'. (Letter 216.) Therefore, one should not regard the tripos as having been simply a race to be won by weak students with memorized answers and speedy pens. To be sure, memory and speed were important and future mathematicians and physicists of real creativity – like Kelvin – did not always finish first; but they frequently did, and from 1830 to 1850 they always ranked within the first five.[27]

Developed within the climate surrounding the mathematical tripos, *The Cambridge Mathematical Journal* first appeared in the autumn of 1837. Edited by D. F. Gregory (fifth wrangler 1837) until 1843 and then by Robert Leslie Ellis (senior wrangler 1840) until Kelvin replaced him in 1845, the journal focused on the subjects of the tripos and contained papers mainly by Cambridge wranglers. From its beginning it also attracted a few papers by non-Cambridge authors, including George Boole. To mark his strengthening of the journal's non-Cambridge ties when he became editor, Kelvin changed the title to *The Cambridge and Dublin Mathematical Journal*. As an absentee editor, Kelvin relied on friends in Cambridge to help with editorial chores, including the evaluation of submitted papers. (See letters 5, 6, 21, 50, 67–8, 73, 76–8, and 80.) As he wrote to Stokes in an unsuccessful attempt to persuade him to take over the editorship in 1851, 'My mathematical friends have always been most kind in helping me, by giving me reports on papers, and I have never once found it necessary to wade through a paper on a subject that was at all out of my way, as you, and others have formed a sufficient *council* to enable me always without difficulty to find a willing referee'. (Letter 80.) Written and edited chiefly by former Cambridge students, the journal was intended in part for current students. Hence, Kelvin and Stokes's six 'Notes on hydrodynamics'[28] were written 'for the *men* not for the *philosophers*' (letter 43), and Stokes published two papers on Clairaut's theorem, one in the *Cambridge and Dublin Mathematical Journal*[29] 'for the sake of the men' (letter 45), the second in the *Transactions of the Cambridge Philosophical Society*.[30]

Founded in 1819 mainly through the efforts of Adam Sedgwick (fifth wrangler 1808), professor of geology, and John Henslow (sixteenth wrangler

1818), later to be professor of botany, the Cambridge Philosophical Society met bi-weekly during full term, providing a continuing forum for Cambridge men interested in the natural sciences. The Society's first volume of *Transactions*, covering the years from 1819 to 1822, included papers by Sedgwick and Henslow on geology, Whewell on astronomy and mineralogy, Herschel on optics, and Babbage on mathematics. Airy presented numerous optical papers to the Society between 1822 and 1836, which appeared in the second to sixth volumes of the *Transactions*.

The seventh volume contained Stokes's first paper[31] and a good deal of other original research that had been read to the Society during the years from 1836 to 1842. Philip Kelland (senior wrangler 1834) contributed an investigation of molecular forces in 1838 and, after he had become professor of mathematics at Edinburgh, a paper on the wave theory of light in 1840. In a paper which influenced Kelvin's early researches, Samuel Earnshaw (senior wrangler 1831) examined the molecular forces operating between particles of ether.[32] The volume included two papers on experimental optics by the professor of mineralogy, W. H. Miller (fifth wrangler 1826), and two on hydrodynamics by the Plumian professor, James Challis (senior wrangler 1825). Matthew O'Brien (third wrangler 1838) presented a study of optical dispersion. James Power (eighth wrangler 1841) tried to prove a hydrodynamical theorem which Stokes later treated more fully.[33] Above all, volume seven of the *Transactions of the Cambridge Philosophical Society* contained two monumental articles on physical optics by George Green (fourth wrangler 1837) which remained relevant to physical theory throughout the century, being discussed, for example, in several letters between Stokes and Kelvin in the 1880s when Kelvin turned to this area of research. (Letters 449–51, 456–71, and 492–3.)

Hence, although Sedgwick and Henslow undoubtedly had at first envisioned an institution to foster natural sciences like botany and geology, as well as those already entrenched at Cambridge due to the mathematical tripos, apparently the tripos's pervasive presence had shaped the Society's interests differently. Mathematics and physics dominated the seventh and eighth volumes of the Society's *Transactions* (which contain papers read between 1836 and 1849), and in these areas, the Society's members had heard a number of significant papers over the years, thus fulfilling the founders' hope that 'by the co-operation of minds thus accustomed to investigation, some services, not otherwise to be expected, might eventually be rendered to the cause of science'.[34]

Finally, although most wranglers became clergymen or lawyers, an appreciable number of high wranglers found academic employment in the general area of their tripos studies. The six Cambridge professorships relating to mathematics, astronomy, natural philosophy, mineralogy, and chemistry were all filled by tripos graduates. Moreover, some two dozen of the graduates between 1830 and 1850 found professorships of mathematics or natural philosophy in other British universities, colleges, or military academies or in universities in Australia, Canada, or America. For many more, success in the tripos competition brought election as fellows of their colleges where they often lectured on tripos subjects. Men already mentioned, such as Green, Gregory, Ellis, and Power, who never became professors, held college fellowships. For some, college fellowships served until a professorship came open. John Couch Adams (senior wrangler 1843) relied on his fellowship until he became Lowndean professor at Cambridge in 1859. As we have seen, Stokes's and Kelvin's careers fit in with these established patterns, with each holding a college fellowship until he obtained a university professorship.

Consequently, for students like Stokes and Kelvin, Cambridge during the 1830s and 1840s exhibited a significant degree of professionalism. They read for a severely competitive examination, a high finish in which promised successful employment and suggested future accomplishment in mathematics or physics. Reading for the examination occurred within an atmosphere of research activity – which supported two leading scientific journals – meaning that one's own research would be influenced not only by Continental figures like Fourier, Poisson, Fresnel, and Gauss and by other British scientists like Joule and Faraday, but to some extent also by the Powers, Earnshaws, and O'Briens. Around mid-century this Cambridge scene underwent the first of various transformations which would make science education at Cambridge in 1900 much stronger than it had been in 1850. Stokes and Kelvin, though not prime movers of these changes, were certainly caught up in them.

The mid-century developments in science education involved the introduction of a new tripos and the change of an old one.[35] The new natural sciences tripos, held first in 1851, encompassed several sciences and eventually provided the foundation for Cambridge 'schools' of physiology and geology. Physics was largely excluded from the natural sciences tripos until the early 1870s, when the Cavendish Laboratory opened and James Clerk Maxwell (second wrangler 1854) became the first Cavendish professor of experimental physics. By the 1890s the natural sciences tripos was graduating physicists of the first rank, including two Nobel prizewinners. The mid-century reorganization of the mathematical tripos was mainly procedural, but it did involve a formalization of the examination's content. However, the major change regarding content came in the early 1870s when heat, electricity, and magnetism were added to the old physical subjects, making the mathematical tripos – like the natural sciences tripos – representative of contemporary physics.

Stokes and Kelvin were primarily involved with the mathematical tripos, about their only connection with the natural sciences tripos being their respective refusals to become the Cavendish professor (letters 468–70),[36] whose activities were germane to both triposes. Stokes, an examiner in the first reorganized mathematical tripos in 1848, was complimented by Kelvin for the problems he had set (letter 22) and informed Kelvin that the new system allowed examiners to 'pretty well avoid setting the questions so as to be answered by crammed men' (letter 23). A year later Stokes reported on discussions within the Board of Mathematical Studies as to whether the figure of the earth should be retained as an examination topic and explained how his own work on Clairaut's theorem stemmed from the discussions. (Letter 45.) After becoming Lucasian professor, Stokes no longer examined for the tripos, but he was one of the Smith's prize examiners for the next four decades (see letter 76, note 2) and was a member of the committee which advocated the introduction of heat, electricity, and magnetism into the tripos. For his part, Kelvin declined to examine in the tripos in 1851 because of his obligations in Glasgow (letter 73) but lent his support to the major reforms of the early 1870s by agreeing to be the 'additional examiner' in 1874 in charge of the new subjects.[37] He also examined once for the Smith's prizes, in 1883, (letters 436–45) commenting that 'I am disappointed on the whole with the performances. I did not expect *very* much but the results are not so good as I expected'. (Letter 445.)

GLASGOW UNIVERSITY

Scottish educational ideals contrasted sharply with English, and Glasgow *circa* 1840 differed greatly from Cambridge. Students entered university at a younger age from a wider spectrum of the population in Scotland and, once there, usually decided by their own vote which of their fellow students deserved recognition for academic performance. Although the Scottish curriculum included natural philosophy, classics, and mathematics, philosophy was its centrepiece. Attention focused on the logic and moral philosophy courses, with other subjects being treated largely from philosophical and historical perspectives. As chronicled – and deplored – in G. E. Davie's *The Democratic Intellect*,[38] the development of university education in Victorian Scotland featured more-or-less continuous strife between adherents to Scottish university traditions and proponents of the English values of specialization, with Anglicization – after initial rejection and later delay – finally coming in the 1890s.

The status of mathematics and natural philosophy at Glasgow during the 1830s would seem to fit the overall picture of instruction at a popular level. Neither the professor of natural philosophy (Meikleham) nor the professors of mathematics (James Millar until 1832, James Thomson from 1832 to 1849) embodied the spirit of original research. J. P. Nichol, the professor of astronomy who substituted for Meikleham in 1839/40, wrote popular books nicely reflecting the Scottish historical-philosophical approach to astronomy and natural philosophy.[39] David Thomson, the ailing Meikleham's substitute in the early 1840s, had graduated from Cambridge in 1839 as only a senior optime. Furthermore, the mathematical attainments of Glasgow students were much lower than the professors would have liked. Such lament, for example, ran through Millar's and Meikleham's testimony to the Royal Commissioners in 1827. On January 8th Meikleham answered questions:

Do you generally find that your students are sufficiently prepared with Mathematics? – I certainly do not. They are beginning to be better prepared.

Is there great deficiency in that respect? – There is very considerable deficiency.

Is it such as materially to interfere with the improvement they would otherwise get? – It is, certainly.[40]

The next day Millar read and commented upon the university regulation requiring students 'to have studied Mathematics for one session' prior to 'being enrolled in the Natural Philosophy or Physic class'. Although the first course in mathematics only included algebra, geometry, and plane trigonometry (conic sections, spherical trigonometry, and calculus being reserved for the second course), Millar complained that many students had not mastered even that level.[41]

Undoubtedly, faculty at both Cambridge and Glasgow complained about students, but that should not hide the differences between the universities. For better or worse, in the 1830s and 1840s Glasgow did not possess the same orientation towards research in mathematics and natural philosophy as Cambridge, nor did it produce a Cambridge-like stream of graduates in these areas. Indeed, many promising students – like Kelvin – went on from Glasgow to Cambridge for their degrees.

Thus, when one reads Kelvin's statements that his Glasgow professors extolled French mathematical physicists like Laplace, Fresnel, and Fourier, one should not conclude that the Glasgow curriculum included these men in all their mathematical splendour.[42] Kelvin's notes for the natural philosophy course he took from Meikleham and Nichol, for example, contained little mathematics and only a bit of elementary calculus.[43] Indeed, as Kelvin himself pointed out, Nichol, in recommending Fourier, only claimed to know that Fourier was important, not to have *understood* him.[44] Although he greatly valued his early Glasgow days for turning him towards the French mathematical physicists, the fact that Kelvin read their works at a technically proficient, original level was nearly unique for Glasgow in the 1830s – either for an undergraduate or a professor. Moreover, the Glasgow tradition of popular lectures in natural philosophy was, of course, what led Professor Thomson repeatedly to urge his son to develop the art of lecturing at an elementary level with experimental demonstrations before he applied for the Glasgow chair of natural philosophy in 1846. The existence of the tradition also causes one to wonder whether the application of the strong mathematician, Stokes, for the Glasgow mathematical chair (letters 34–43) would have been so highly regarded by other Glasgow professors as by Kelvin, and whether Hopkins's recommendation of the mathematically weaker Hugh Blackburn (fifth wrangler 1845) for the position as 'one who has received his early education in your own institutions' (letter 34, note 5) was so faint praise as it might now seem.

In any case, although Kelvin's influence on the teaching of natural philosophy at Glasgow has yet to be studied in great detail, his tenure brought many changes. In the first place, whereas his predecessor, Meikleham, had responded 'I never do' to the Royal Commission's question, 'Do you make them perform experiments?',[45] Kelvin greatly expanded the experimental side of the natural philosophy course. In his first couple of years he persuaded the university to spend large sums on new apparatus for experimental demonstrations[46] and, shortly afterwards, turned an abandoned wine cellar into a research laboratory for himself and his students. Although not an official part of the course, the laboratory work attracted a number of students. As Kelvin explained to Stokes in 1860, in order to justify spending government-grant funds on his laboratory, 'I generally find several among them efficient for original investigation, to whom I give ⟨some of the⟩ work under some of the heads specified in my applications to the Grant Committee', enabling him 'to do some thing more in original investigation than I could accomplish without such assistance'. (Letter 181.) Having received in 1862 additional room for his laboratory from the university and additional funds thanks to a Royal Commission, Kelvin gave Stokes his address as 'Natural Philosophy Laboratory' and wrote 'I have at last, for the first time got a students' working laboratory. It will be ready for the beginning of this session & will accommodate about 24 students'. (Letter 225; see also letter 226, note 1.) Preceding the establishment of similar laboratories at other universities, Kelvin's laboratory continued to expand and to blend his teaching with his research.[47] Moreover, in addition to his emphasis on experimentation, Kelvin introduced advanced lectures on mathematical physics, sharing the teaching chores with his assistant and nephew, James Thomson Bottomley.[48] (In letter 485 – a testimonial for Bottomley – Kelvin outlined Bottomley's duties.)

Hence, by the late Victorian period, although great differences in staff and resources still separated Glasgow and Cambridge, physics education at the two institutions was more nearly parallel than before.

By the time Stokes and Kelvin became closely associated with it, the Royal Society of London had already passed through the tumultuous phase of its nineteenth-century life. In the 1820s, reform-minded members of the Society saw its current weakness as largely the legacy of Sir Joseph Banks's unenlightened, and nearly unending, reign as president from 1778 to 1820. During that time, the Society's standards had declined, its membership had soared, and it even had opposed the founding of the Geological Society in 1809 and the Astronomical Society about a decade later.[49] In the late 1820s, the reformers mounted an internal attempt to transform the Society, ending in their failure in 1830 to elect as president the Society's premier man of science, John Herschel.[50] Having been much criticized already, the Society received a slashing, public attack in 1830 in a well-known book by one of the reformers, Charles Babbage. In his *Reflections on the Decline of Science in England and on Some of Its Causes*, Babbage portrayed the Society as an irresponsible gentlemen's club with its too-large membership publishing much too little.[51] Of the 714 members, he reported, only 109 had contributed even one paper to the Society's *Philosophical Transactions*. Babbage wanted to reduce the president's power of selecting members of the Society's governing council, to increase the Society's vigilance over the quality of papers it published, and, above all, to effect membership reforms. He endorsed, for example, the reformers' suggestion that only four scientific fellows be elected per year until the Society's bloated membership declined to a more respectable number of 400.

Such reforms did come, but not all of them immediately. In the 1830s, the Society published two volumes of accumulated *Abstracts of the Papers Printed in the Philosophical Transactions of the Royal Society of London, from 1800 to 1830 Inclusive*. The third volume, entitled the *Proceedings of the Royal Society*, chronicled the Society's meetings from 1830 to 1837. In subsequent volumes, this new journal became a receptacle not only of abstracts, but also of publishable papers too short or limited for the *Transactions*, one intention presumably being to help increase the quality of the *Transactions*. Moreover, in the 1830s the Society's committee on papers began taking more care in deciding which papers to publish, sometimes asking for reports from referees. The refereeing process apparently became systematic around mid-century, resulting in the several volumes of carefully preserved and fully indexed referees' reports now in the Society's archives. In 1846, under the urging of Sir W. R. Grove, the Society passed regulations allowing only fifteen new members per year. As a consequence, by 1860 the scientific fellows were finally in the majority and by 1900 the membership had fallen to about 450.[52] Furthermore, in 1850 the Society began administering the dispersal of £1000 per year (increased in 1876 to £5000) in government funds as grants to support individual scientific research. They awarded ninety grants in the 1850s, the number increasing to 473 in the 1890s.[53] Hence, with its restrictive membership and encouragement of sound research, the Royal Society at the end of the century differed markedly from what it had been in the 1820s.

Elected fellows in 1851 under the new regulations, Stokes and Kelvin came to be intimately involved with the reformed Royal Society, especially since Stokes was a principal administrator of the new procedures for so long, first as secretary and then as president. The Stokes Collection in Cambridge contains a group of more than 2400 letters concerning the business of the Royal Society,[54] and the Royal Society possesses some 100 referee's reports from each of the two. In addition, around mid-century, after several years of publishing primarily in the two Cambridge-based journals, Stokes and Kelvin began publishing more elsewhere, including the Royal Society's journals. Stokes published six papers in the *Transactions* (the first in 1848) and nearly thirty in the *Proceedings* (the first in 1849). Kelvin had eight in the *Transactions* (the first in 1851), nearly fifty in the *Proceedings* (the first in 1854). Moreover, Kelvin enjoyed a good deal of support from the government-grant fund – so much so, in fact, that Stokes had to write him in 1879 that the Society's president, William Spottiswoode, 'had heard [f]rom more than one remarks as to the number of grants made to [yo]u, who are not in the same condition as many others as to [de]pendence upon these grants for carrying on the investigations you have in your mind'. (Letter 347.) Those aspects of the Royal Society most frequently represented in the correspondence, then, are Kelvin's papers for the *Transactions*, his reports on others' papers, and, to a lesser extent, his applications for grants.[55]

The earliest of Kelvin's six *Transactions* papers discussed by Stokes and Kelvin in their correspondence was his Bakerian Lecture, 'On the electrodynamic qualities of metals', delivered 28 February 1856.[56] A detailed experimental investigation of thermo-electric properties of metals, the paper reported research done at Kelvin's newly established laboratory in Glasgow. In many of the letters from 6 November 1855 to 17 June 1857 (letters 133–62), the correspondence traces the lecture's administrative history from proposal to publication. In December, Stokes informed Kelvin of his official appointment as Bakerian Lecturer (letter 144), and, later, as Kelvin was very slowly sending the final version to the printers, instalment by instalment, a mildly exasperated Stokes wrote:

> Verily I must make a rule not to send a paper to the press till it is complete. You know these things are laid on the Secretary, and the worst of it is that you are taking away the character of my wife to be; for people will be saying that she has been keeping me from my work, and lo it is all your fault. (Letter 159.)

For much of the time they were discussing the Bakerian Lecture, they also discussed Kelvin's 'Elements of a mathematical theory of elasticity'.[57] (Most of the letters from 145 to 155.) This is the paper which Kelvin based on his modifications of Rankine's terms *stress* and *strain* and which Silvanus Thompson regarded as 'the foundation for all that has since been written on the subject'.[58] In one letter, Kelvin pointed out that 'there is very little direct connection' between the views in his paper and those in a paper by Rankine on elasticity which Kelvin had been refereeing, 'although', Kelvin wrote, 'I believe it will be important to investigate the relations between them'. (Letter 145.)

Kelvin's next two papers in the *Transactions* (for 1863) combined his interests in elasticity with his continuing research on the physical properties of the earth. As discussed by Burchfield, Kelvin maintained a sixty-year interest in the earth, using various physical considerations to study its internal state (solid or fluid), its age (indefinitely old or relatively young), and its physical changes over time.[59] With alterations in detail, Kelvin consistently argued that a relatively young earth had undergone significant changes, including its almost total solidification from an initially molten globe. The first of the two 1863 papers investigated the rigidity of the earth (read to the Society in May 1862)[60] and depended upon general conclusions presented in the second, 'Dynamical problems regarding elastic spheroidal shells and spheroids of incompressible liquid' (read the following November).[61] Stokes and Kelvin discussed the papers in many of their letters between 25 October 1861 and 20 April 1864. (Letters 204–44.) Kelvin began by asking Stokes whether 'Poisson or any one else has solved the problem of finding the whole effect produced through a uniform elastic sphere by any arbitrarily given stresses applied to its surface or by any arbitrarily given displacements of all the points of its surface' (letter 205), and Stokes answered that he felt 'pretty sure that the solution of the problem has not hitherto been effected' (letter 206). To Kelvin's conclusion that the earth 'must be far more rigid than glass otherwise the solid would yield so much to tidal influence of sun & moon as to leave no sensible tides of water relative to solid land' (letter 205), Stokes replied that he found the result 'very remarkable, and not at all what I should have expected' (letter 206).

Continuations of Kelvin's earlier Bakerian Lecture, his final two papers in the *Transactions* investigated the effects of stress on the magnetization of wires composed of various metals. They were published, respectively, in 1876[62] and 1879[63] and were mentioned in several letters between 13 April 1874 and 27 August 1879 (letters 306–66). The first paper discussed experiments made on wires of steel and iron and reported certain anomalous results with soft iron. Within a week or so after presenting the first paper to the Royal Society in May 1875, Kelvin had obtained important results concerning the anomalies.

In July, he wrote to Stokes about publishing a short paper[64] which stated the new results and which would also 'serve as a stepping-stone to the long continuation for the Transactions which I am preparing'. (Letter 311, note 1.) Kelvin briefly described the experiments contained in the 'long continuation' – the paper published in 1879 – in this same letter (letter 311) and in another of 10 March 1877 (letter 334).

Kelvin's many referee's reports included in this edition range from the brief to the substantial. Most recommended publication of the paper in question. Indeed of the some 100 papers refereed by Kelvin altogether about three-quarters were published in the *Transactions* and most of the rest were published in the *Proceedings*. Kelvin commented extensively on Samuel Earnshaw's 'On the mathematical theory of sound' in *seven* letters (letters 174, 182–3, 186–8, and 190), initially concluding 'very decidedly that it should not be published in the Transactions in its present form' (letter 174). Earnshaw's paper eventually was published in the *Transactions*,[65] though, and Kelvin's letters help to explain why it was. Kelvin could be unsparing in his criticism: writing in 1861, 'I am considerably bored by a paper of Sir W. S. Harris' which you have sent me. It is so bad, like all he has done ...' (letter 203); in 1870, 'Mr Barlow's paper on Flexure.... is wrong from beginning to end, and shows the author to be altogether ignorant of the terms, grammar, and principles of the mathematical theory of elasticity ...' (letter 269); and in 1879, 'Can it be that [Osborne] R[eynolds] does not understand the principle he is dealing with?' (Letter 367.)

Brief discussions of government grants appear frequently in the correspondence, and three of the letters hinted at controversy regarding the administration of these funds. Two of the three have already been cited; they concerned Kelvin's defence of spending such funds on his students' laboratory (letter 181) and complaints that the well-to-do Kelvin was receiving too many grants (letter 347). Even more interesting is Kelvin's letter of 21 March 1877. Clearly upset with the Society's refusal to support the research of Fleeming Jenkin and J. A. Ewing, Kelvin wrote of widespread criticism of the Society:

> I find there is a great deal of dissatisfaction in respect to actions taken by the Council of the Royal Society regarding the action of the Committee for the Government £4000.... I find there is a great deal of feeling in the R[oyal] S[ociety of] E[dinburgh], and in the general public on the matter.... (Letter 335.)

SUMMATION

Though controversy, disappointment, and dissatisfaction accompanied the process at times, Stokes and Kelvin participated in or witnessed numerous dramatic changes in Britain's scientific institutions. At the end of the century, their respective universities gave physics education – including both its mathematical and experimental aspects – a respected place it had not enjoyed before. Kelvin, of course, had shaped the Glasgow system himself, and both he and Stokes had supported the basic reforms at Cambridge. As successive presidents of the Royal Society from 1885 to 1895, the two presided over an organization which was much more scientific, professional, and sympathetic towards original research than it had been half a century earlier. Thus, it was within a changing and improving institutional framework that Stokes and Kelvin sought to change and improve the science of physics.

III

The concepts of ether and energy were basic to the nineteenth-century British understanding of physical nature.[66] Moreover, early in their careers, Stokes and Kelvin put their respective stamps on these areas of physics – Stokes with a series of articles on the ether in the late 1840s, Kelvin with his formulation of the laws of thermodynamics in 1851. The physics of energy, though, was in a sense more settled than that of the ether during the second half of the century. Physicists tended to view the laws of thermodynamics as providing a context within which *other* physical problems could be solved.[67] The luminiferous ether, on the other hand, was largely a question requiring answers, as physicists tried to determine how the ether and ordinary matter interacted with each other and to decide how the properties of the ether compared and contrasted with those of matter. Such investigations were also among those which followed from the British concept of energy, the ultimate goal being a comprehensive, dynamical theory which would explain how interconversions of different kinds of energy were, in reality, simply transitions from one state of ether and matter to another.

Hence, it is probably natural that Stokes and Kelvin would correspond more about ether and matter than about energy. In any case, I shall concentrate here on their discussions of ether and matter as the best subject offered by the correspondence for understanding their physical ideas. Their letters will help us sharply to revise some of our established views of Kelvin's thought and to see that Stokes disagreed with Kelvin in important ways.

Stokes and Kelvin approached physical issues with a methodology characteristic of mid-to-late nineteenth-century British physicists.[68] Developed largely in reaction to prevailing French methodologies, it was a highly flexible, cautious realism utilizing mechanical models. In contrast to the positivism associated with Fourier and Auguste Comte, Stokes and Kelvin affirmed the possibility of genuine knowledge of actually existing unobservable entities. Hence, from at least as early as the late 1840s to the end of his life, Kelvin proclaimed that to explain light waves 'it is absolutely necessary to suppose a medium'.[69] In contrast to the 'astronomical' view associated with Laplace and Poisson, Stokes and Kelvin were cautious in incorporating unobservables into their theories. Stokes wrote, for example, that 'Duhamel according to the French fashion works on the hypothesis of intra molecular radiation, which for my own part I don't much like making the results depend upon'. (Letter 74.) The ability to know unobservables seemed to demand extra responsibility in postulating their existence. Thus, when one lacked sufficient evidence, one emphasized a phenomenological approach or turned to mechanical models to help probe the unseen. Because both macroscopic and microscopic realms behaved in accordance with Newtonian mechanics, macroscopic models yielded, by analogy, insights into nature's microscopic workings, though without necessarily providing detailed pictures of them. Hence, in the letters we find references to numerous analogies: an apparatus made of wooden bars, steel wire, and treacle used to represent the phenomena of phosphorescence (letter 440), a row of ivory balls to represent a cathode-ray phenomenon (letter 545), jelly to represent ether (letter 6), smoke rings, atoms (letter 250), piano strings, the production of spectra (letter 277), and a bag of marbles, the internal friction in fluids (letter 495). Although most of these examples occur in Stokes's letters to Kelvin, Kelvin was the more accomplished model builder of the two and more frequently commented on the method, as in his *Baltimore Lectures* of 1884. There, in several isolated passages, he indicated that, though a model was a 'rude mechanical illustration',[70] the ability to make such a model was also 'the test of "Do we or not understand a particular subject in physics?"'[71] Moreover, he said,

> that something exists in the luminiferous ether and acts upon it in the manner that is faultily illustrated by our mechanical model, I absolutely believe. I have no more doubt that something of the kind is true, than I have of my own existence.[72]

Consequently, though we must be careful ourselves not to conclude too readily that any particular imagery proposed by Stokes or Kelvin was meant to correspond precisely with nature's hidden realm, it was just that realm that they were trying to get at.

In his papers on the ether in the late 1840s, Stokes tried to resolve the question of the ether's apparently contradictory properties.[73] Like a solid, the ether supported transverse waves, but, like a tenuous fluid, it seemed to allow planets to move through it with little or no resistance. Calling upon the law of continuity – 'a law which seems to pervade nature'[74] – Stokes envisioned an ether with both solid and fluid characteristics. He likened the ether to glue water, which, when it had been greatly diluted, would still – by the law of continuity – retain vestiges of the solid properties it had had in concentrated form.[75] Such an ether, Stokes suggested, would behave like a fluid for large-scale, relatively slow planetary motions, but like a solid for the small, rapid vibrations of light. Stokes's glue-water analogy possibly emerged from conversations with Kelvin, for in an 1847 letter Stokes mentioned 'the jelly-like fluid that we once spoke of'. (Letter 6.)[76] Indeed, looking ahead to Kelvin's views, we should note that Stokes also acknowledged that the law of continuity meant 'that air, like the luminiferous ether, ought to admit of transversal vibrations', although they would be 'utterly insensible, unless we had organs with a delicacy equal to that of the retina adapted to receive them'.[77] Kelvin was to emphasize the similar, but much bolder, continuity of the existence of longitudinal waves in ether as well as in air.

In fact, for nearly a decade at least, from the early 1850s to the early 1860s, Kelvin actually thought ether and air were probably the same thing. We can find this idea in 1854 both in a published paper and in correspondence with Stokes (letters 100–5) and, again, in 1862 in correspondence with P. G. Tait.

Kelvin's main concern in his remarkable little paper read to the Royal Society of Edinburgh on 1 May 1854 was to calculate the density of the ether.[78] He had written to Stokes two months earlier that recent measurements of the energy of solar radiation gave a value for the mechanical energy of the vibrating ether transmitting sunlight. This mechanical energy was, in turn, a function of the ether's density and of the maximum velocity of the portions of the ether undulating to and fro. Hence, a reasonable estimate of the velocity would give an approximate value for the density. Kelvin wrote that if the velocity were 1/100 of the speed of light, 'you will be astounded with the greatness of the density of the luminiferous medium required to produce the mech^{l} eff^{ts}'. (Letter 100.) His value was 32.2×10^{-23} pounds per cubic foot. Although Stokes had done no calculations, he told Kelvin that he, too, had 'contemplated' determining the density of the ether in this manner, and he found Kelvin's estimate plausible, responding to Kelvin that 'I confess I do not think there is anything unreasonable in it'. (Letter 101.) Similar figures appeared in Kelvin's published account.

The remarkable part of Kelvin's paper of 1854 concerned the connection between ether and matter. He, unlike Stokes, thought that the earth's atmosphere extended into interplanetary space and that air and ether were probably one and the same. He opened his paper with the statement:

> That there must be a medium forming a continuous material communication throughout space to the remotest visible body is a fundamental assumption in the undulatory Theory of Light. Whether or not this medium is (*as appears to me most probable*) a continuation of our own atmosphere, its existence is a fact that cannot be questioned....[79] [My italics.]

Kelvin's correspondence with Stokes during the several weeks preceding the paper's reading amply confirms that Kelvin did seriously think – in spite of

Stokes's objections – that the earth's atmosphere extended into space. Kelvin first broached the subject on 2 March 1854. (Letter 100.) After Stokes wrote that he was 'altogether sceptical about the existence of air in the planetary spaces' (letter 101), Kelvin asked, 'How can you think the air stops?' (Letter 103.) Stokes explained how he could in some detail (letter 104), but Kelvin responded, 'I still believe in the continuity of atmosphere through space.... I am much disposed to go back to the Vortices, differing only from DesCartes' in being dragged round by the planets instead of drag[g]ing them round'. (Letter 105.) Moreover, the correspondence suggests that Kelvin may have been assuming for some time that the ether of space was a continuation of the earth's atmosphere. His value for the ether's density was surprising because it was so much greater than the density the atmosphere should have possessed at that distance from the earth. As he wrote to Stokes in his first letter on the subject:

Now the density of the air in interpl. space, *if the temperature were uniform from the surf. of the earth upwards*, [Kelvin's italics] would be only some $1/10^{230}$ of the dens[ity] at the surface of the earth. *What is the lum^s medium then?* [My italics.] ⟨I suppose⟩ There must be matter in interplanetary space perhaps 10^{200} times as dense as the air w^d be on that hypoth[esis]. (Letter 100.)

However, such discrepancies did not prevent his identifying the interplanetary atmosphere with the luminiferous ether in his paper, which he presented shortly after his exchange of letters with Stokes.

Kelvin's communication with Tait on this issue in 1862 is contained in a notebook which the two posted back and forth to obtain one another's comments as they worked out plans for their *Treatise on Natural Philosophy* published five years later.[80] Kelvin sketched some of his thoughts on Newton's laws of motion, writing on one page of the 'resistance of interstellar air' to the motions of the sun and the earth. In red ink, Tait encircled the word *air* and wrote on the facing page:

That is one of the great stumbling-blocks between us as joint authors. I can't rightly appretiate [*sic*] your idea of an unlimited atmosphere. I have seen *hints* of it in your papers, but *no reasoning*. Why not say matter?

Adjacent to Tait's remark, Kelvin scrawled, 'Matter or medium if you please. But air seems to me simpler'.

Given Kelvin's inclination to regard ether and air as the same thing, we can better understand a series of letters he wrote Stokes between May and December 1857 (letters 161–2 and 164–5) and a notebook entry of January 1858.[81] With the model presented there, he was especially trying to explain the Faraday effect which he had discussed in an 1856 paper.[82] In the Faraday effect, the plane of transverse vibrations of light is rotated when the light is transmitted through heavy glass within a magnetic field, and Kelvin had argued in his paper that the magnetic field must give rise to some kind of rotations of matter which exist independently of the presence of light and which themselves cause the plane of vibrations of light to rotate. Speculating on exactly what these rotations might be, he described a provisional model, which – though 'it does not seem probable that a complete theory of physical science can be founded on such a hypothesis'[83] – was intended as a step towards a comprehensive theory of the varieties of physical phenomena. He envisioned a perfect, frictionless fluid containing solid 'motes', and he hoped to correlate observed physical phenomena with motions of the fluid and solids. Concerning the Faraday effect, for example, he wrote Stokes that 'if there was a preponderance of rotation [of the motes] in one direction about parallel axes, waves of transverse vibrations ⟨about⟩ in planes perpendicular to these axes would have Faraday's optical property of heavy glass . . .'. (Letter 161.) Moreover, he thought translational motion of a solid mote with respect to the fluid would tend to become rotational motion and that such a rotating mote 'will experience a highly intens⟨ive⟩e repulsive action when brought near either fixed boundaries of other solids rotating or not'. (Letter 162.) Not only did he think these microscopic processes could be shown to correspond to the observed properties of ordinary solids, fluids, and gases, he also wrote that 'an infinite number of such motes all rotating with great angular velocities will repel one another & keep up the kind of stability & relative stiffness, *required for luminiferous vibrations*'. (Letter 162. My italics.)[84] Unsurprisingly, then – given his equating of air and ether – Kelvin thought the 'stiffness' imparted by rotating motes would account for properties of everyday, observable materials as well as for the ability of air–ether to transmit transverse vibrations of light. As we shall see, even when he abandoned the oneness of air and ether, he evidently retained the idea that they had similar structures.

Kelvin gave up the identity of ether and air, he tells us, because of the kinetic theory of gases,[85] which was being developed by Clausius and Maxwell during the late 1850s and the 1860s. Although Kelvin did not explain *why* the kinetic theory had this effect on his views, I think we can see why it did in a short note Kelvin published in *Nature* in 1870 entitled 'The size of atoms' and in his presidential address to the British Association the next year. He began the article in *Nature* declaring that because of difficulties in conceptualizing the atom and its properties 'chemists and many other reasonable naturalists of modern times [including Kelvin himself, if my interpretation is correct], losing all patience with it, have dismissed it to the realm of metaphysics, and made it smaller than "anything we can conceive"'.[86] In his address he made the point by saying, 'Chemists and other naturalists had been in the habit of evading questions as to the hardness or indivisibility of atoms by virtually assuming them to be infinitely small and infinitely numerous'.[87] But now, he explained in *Nature*, the kinetic theory of gases (plus three other areas of science) indicated that atoms were much larger than previously thought, and, to help his readers 'form some conception of the degree of coarse-grainedness indicated by this conclusion', he pointed out how large the molecules in a drop of water would be if the drop were the size of the earth.[88] Whether Kelvin was correct or not concerning the views of other naturalists, the conceptual transition from matter containing 'infinitely small and infinitely numerous' atoms to matter as something much more coarse-grained did, I would suggest, occur in his own thinking, causing him to distinguish between ether and matter as he had not before. In 1884 he declared:

It seems probable that the molecular theory of matter may be so far advanced sometime or other that we can understand an excessively fine-grained structure and understand the luminiferous ether as differing from glass and water and metals in being very much more finely grained in its structure.[89]

Hence, it would seem that whereas Kelvin thought it plausible that air composed of infinitely closely packed particles (something like the rotating motes) might be able to transmit the very rapid and very short waves of light, the 'coarse' air of the kinetic theory could not do so. Presumably from around the mid-1860s, therefore, Kelvin regarded ether and air as different, but similar, entities, one fine-grained, the other coarse-grained.

It was also during this period that Kelvin published his 1867 paper,[90] in which, as Knudsen has observed, the solid motes of the late 1850s were 'replaced by the famous vortex atoms'.[91] Drawing upon Helmholtz's theoretical hydrodynamics and P. G. Tait's demonstrations with smoke rings, Kelvin transferred the rotational motion from the solid motes to portions of the fluid itself, thus picturing atoms as smoke-ring-like whirlpools of ether. Moreover, in continuation of his earlier conception, Kelvin apparently thought these microscopic vortices would not only constitute atoms of ordinary matter, they also would be distributed throughout the luminiferous ether, providing the required stiffness formerly provided by rotating motes. As he wrote in introducing an 1887 paper:

I have found something seemingly towards a solution (many times tried for within the last twenty years) of the problem to construct, by giving vortex motion to an incompressible inviscid fluid, a medium which shall transmit waves of laminar motion as the luminiferous aether transmits waves of light.[92]

Towards the end of his life, he recalled his much earlier conception that 'ether is a fluid presenting appearances of elasticity due to motion, as in collisions between Helmholtz vortex rings.[93]

The correspondence touches on the vortex-atom theory briefly in 1867 and more extensively in 1872 and 1873. In a now-incomplete letter, probably written in early 1867, Kelvin asked Stokes about the possible divisibility of 'a continuous bulk, V, of a perfect fluid' and disclosed that Helmholtz's and Tait's accomplishments 'have set me to a very promising atomic theory'. (Letter 250.) Before Stokes could reply, Kelvin had partially convinced himself of the fluid's indivisibility. (Letter 252.) In 1872 and 1873, Kelvin and Stokes discussed vortex motions at some length, Kelvin writing, for example, that 'void-cored ring vortices, with varieties of knots, serve well for atoms, *being expansible* yet rigourously permanent & stable'. (Letter 295; see also letters 286, 293–4, and 296–9.)

Kelvin's basic concerns continued into the 1880s. He still looked to rotatory motion, for example, as the way to explain elasticity, as indicated by the title of his 1881 lecture, 'Elasticity viewed as possibly a mode of motion'.[94] Moreover, especially in his *Baltimore Lectures* of 1884, he focused his general concerns specifically on optics in attempting to address difficulties facing the wave theory of light. He wrote to Stokes in January 1883 that 'I want to do away with elasticity of solid entirely and with nothing but rotating liquid & perfectly rigid solid containing vessels & links, to get undulations proper for undulatory theory of light. (Letter 443.) The *Baltimore Lectures* themselves, delivered in October 1884, discussed problems like dispersion, reflection, and refraction, none of which was satisfactorily explained by existing versions of the wave theory. Broadly speaking, Kelvin was impressed with George Green's optical theory, but tried to go beyond it by taking into account the continual interaction between the ether and the material particles constituting optical bodies. This deficiency in optical theories had been noted by Stokes in 1862 in his report to the British Association on double refraction.[95] In correspondence with Kelvin immediately after Kelvin's return from Baltimore, Stokes noted certain weak points in Green's theory, if considered 'as a physical, as didtinguished [*sic*] from a merely mathematical, theory'. (Letter 464.) He also remarked 'that the supposition that the ponderable molecules have done all their work when they have modified the ether, and you may then throw them overboard and suppose you have got two uninterrupted homogeneous isotropic media to deal with, is most likely wide of the actual truth'. (Letter 463.)

Though in 1884 Kelvin apparently still hoped for the eventual success of the vortex-atom theory, in his *Baltimore Lectures* he instead used 'rude' mechanical models, especially a series of hollow, concentric shells connected by springs, which represented material atoms. As we have seen, he thought by this time that air and ether were different from one another, and he delivered his lectures largely in the language of ether and matter as quite different, but interacting, entities. Nevertheless, as we have seen, his statement in the first lecture about the fine-grained structure of ether indicated a deeper view of ether and matter as essentially similar to one another, as differing in degree, not kind.[96]The idea reappeared frequently in the lectures, usually in comparisons of ether and air. Noting in his second lecture, for example, that 'a solid mass must act relatively to the luminiferous ether as an elastic body imbedded in it of enormous mass compared with the mass of the luminiferous ether that it displaces', he thought that, whereas such an interaction was 'infinitely difficult to understand' in the case of glass or water, 'the luminiferous ether in air is very easily understood. We just think of the molecules of oxygen and nitrogen as if they were groups of jelly relative to the luminiferous ether . . .'.[97] Later, in introducing a possible mode of producing longitudinal waves, Kelvin asked his audience to imagine the generating body 'at any place in air, or in luminiferous ether (I cannot distinguish now between the two ideas) . . .'.[98] In contrast to the usual notion that ether differed from ordinary elastic solids by not transmitting longitudinal waves, Kelvin declared, 'It seems to me that there are exceedingly strong probabilities that there will be waves of condensation and rarefaction of the luminiferous ether'.[99] He thought such waves would be shown to explain the propagation of electrostatic forces,[100] as suggested by a paper of his own from 1847.[101] Finally, he stated that 'we have not the slightest reason to believe the luminiferous ether to be imponderable; it is just as likely to be atttracted to the sun as air is'.[102]

Often cryptic, Kelvin's statements on ether and matter from the 1850s to the 1880s are not easily deciphered. But we can speculate on his speculations, trying to fathom the conceptual motives behind them.

It would appear that Kelvin's philosophy of nature was driven by the overlapping notions of continuity, unity, and simplicity. It was as though he had fashioned a physical principle from the advice of John Herschel, who wrote of

> that general law which seems to pervade all nature – the law, as it is termed, of continuity, and which is expressed in the well known sentence, 'Natura non agit per saltum'. The pursuit of this law into cases where its application is not at first sight obvious, has proved a fertile source of physical discovery, and led us to the knowledge of an analogy and intimate connection of phenomena between which at first we should never have expected to find any.[103]

In his British Association address of 1871, Kelvin referred favourably to the 'doctrine of continuity'[104] as set out by W. R. Grove, who spoke of continuity as 'a law of nature, the true expression of the action of Almighty Power'.[105] The underlying unity of superficially distinct areas of physical nature would have been supported by the many experimentally verified interconnections between different areas, by similar mathematical structure found in the various areas,[106] and, above all, by the principle of the conservation of energy. Kelvin may also have had a theological basis for such unity, for he declared in 1860 that research was progressing towards 'a stage of knowledge ... in which unity of plan through an inexhaustibly varied execution will be recognised as a universally manifested result of creative wisdom'.[107] Perfectly in harmony with such thinking, equating air with ether was, as Kelvin had after all answered Tait, 'simpler'. Indeed, that he could hold this view in spite of obvious contrasts between sound waves and light waves – and by implication between the respective media which supported them – would seem to attest mainly to the prominence of the ideas of continuity and unity in his philosophy of nature.

Nevertheless, nature could be simple, continuous, and unified in more than one way, and conceptions of nature – no matter how appealing – had to square with mathematically developed and experimentally verified physical theories. Thus, as we have seen, the new kinetic theory of gases caused Kelvin to give up his early conception of ether and air. Yet, in Kelvin's later view that ether and air differed only in degree, the initial idea survived in modified form, a form which preserved the view of physical nature consisting of one *kind* of matter, not two or more. However, it was a view, as we shall now see, unacceptable to Stokes.

Indeed, Stokes's reaction to Kelvin's ideas were frequently unfavourable. Although in the 1840s he had allowed for similarities between ether and air, he had also viewed ether as 'incomparably rarer than air'.[108] For him there *was* a dichotomy between ether and air, and, as we have seen, he showed no sympathy for Kelvin's interplanetary atmosphere. Moreover, in two letters from February 1858 (letters 166–7), he criticized Kelvin's conception of a perfect fluid with solid motes. Not only was he doubtful about the hydrodynamical stability of motions within the system, he even called into question Kelvin's basic view of the Faraday effect which had led Kelvin into this area of speculation in the first place. Stokes wrote that 'I certainly am by no means clear that magnetic rotation must be due to motions going on independently of luminiferous vibrations'. (Letter 166.) In their discussion of vortex motions in the early 1870s, Stokes – once again in cautious counterpoint to Kelvin's optimism – declared, 'I confess I am sceptical about the stability of many of the motions which you appear to contemplate'. (Letter 298.) Their differences persisted into the 1880s. While Kelvin was lecturing his physicists in Baltimore, Stokes was in the midst of a three-year series of talks to a general audience in Aberdeen entitled *On Light*. The tone and substance of their respective remarks on the ether differed considerably. Kelvin confidently proclaimed that 'we know more about it [ether] than we do about air or water, glass or iron', while Stokes warned that 'we must beware of applying to the mysterious ether the gross notions we get from the study of ponderable matter'.[109] In contrast to Kelvin's lectures, Stokes's consistently presented ether and ponderable matter as two entirely different things. There were no longitudinal ethereal waves in Stokes's book, and he observed, for example, 'that the elasticity of the ether is of an altogether different nature from that of air'.[110] Moreover, Stokes remained sceptical about specific theories of ether–matter interactions, declaring at one point, 'It is easy to frame plausible hypotheses which would account for the result, but it is quite another matter to establish a theory which will admit of, and which will sustain, cross-questioning in such a variety of ways that we become convinced of its truth'.[111] Again – this time regarding aspects of the Faraday effect – Stokes stated – almost as though responding directly to Kelvin's proposals – 'all these are questions concerning the true answers to which we can affirm nothing, though plausible conjectures may in many cases be framed'.[112]

In the 1890s, the last full decade of their correspondence, we can perhaps see both change and constancy in Kelvin's and Stokes's attitudes. Although Kelvin was still vastly more active than Stokes in scientific research, his views of the limits of current physical knowledge approached those of Stokes more closely. It was apparently around 1890 or just before that Kelvin abandoned his vortex-atom theory, primarily because he decided that such motion would not be stable after all.[113] With the vortex atom went his best hope for a comprehensive dynamical theory. Hence, in his presidential address to the Royal Society in 1893, after so many decades of trying to comprehend ether and matter, Kelvin could comment on the importance of cathode-ray research 'if a *first* step towards understanding the relation between ether and ponderable matter is to be made . . .'.[114] (My italics.) He expressed the same sentiment three years later in an oft-cited passage from his address during the celebration of his Jubilee as professor of natural philosophy at Glasgow: 'I know no more of electric and magnetic force, or of the relation between ether, electricity, and ponderable matter than I knew and tried to teach to my students of natural philosophy fifty years ago in my first session as Professor'.[115]

Yet Kelvin by no means abandoned efforts to unravel the mysteries of ether and matter; nor had he abandoned all his differences with Stokes, as we can see in their letters on X-rays, the major topic in the correspondence from the 1890s (letters 543–79). As before, Kelvin expected ether, like ordinary matter, to possess longitudinal waves. Not long after Röntgen's discovery, Kelvin informed Stokes, 'I feel strongly disposed to Röntgen's own supposition of condensational-rarefactional waves, but still I see tremendous difficulties'. (Letter 546.) Stokes, on the other hand, wrote Kelvin that 'normal vibrations in the ether did not fit in with my own speculations [on X-rays]'. Continuing, he explained:

If the ether be compressible, the only examples of compressibility that we know anything about would lead us to attribute discontinuity to the ether, to suppose that it too is made up of discrete things, molecules shall I call them? But how do these 'things' act on each other? Is it by a sort of impact and rebound? That would lead us to attribute again structure and molecular constitution to these supposed molecules of ether. Is it by action at a distance? that leads us to the old alternative of action across a perfect void, or an ether of the second order to account for the supposed action of one on another of two molecules of the ether of the first order. Either supposition reminds me of the rhyme

> ... and dogs have fleas that bite 'em,
> And big fleas have little fleas, and so ad infinitum.

My mind inclines me to scratch off the fleas, and rest in the idea of a plenum of incompressible ether, which our present knowledge does not authorise us in going beyond. (Letter 551.)

Within only a month or so of the publication of Röntgen's paper in English, experimental results had convinced both Kelvin and Stokes that X-rays could not be longitudinal waves. At the end of February 1896, Kelvin announced to Stokes his conversion to the view that X-rays were transverse waves. (Letter 548.) But their initial responses to Röntgen's paper had reflected the contrasting theoretical views they had developed over the years, Kelvin once more seeking longitudinal ethereal waves, Stokes once more noting the differences between ether and matter in a statement that recalled the limits of 'present knowledge'.

Consequently, even in so brief a discussion as this, the letters between Kelvin and Stokes help us understand important aspects of their thought. Though they shared a methodology of cautious realism, they certainly emphasized different aspects of it. While Kelvin enthusiastically tried almost to will nature into a coherent dynamical system, Stokes cautiously pointed out the imperfections of current scientific knowledge. Moreover, so far as I am aware, it has not been generally recognized that Kelvin for a decade or so regarded air and ether as the same thing, that for at least three decades he thought the ether ultimately was not continuous and was like ordinary matter, that the vortex-atom theory was not his first attempt to remove the dichotomy between ether and matter, nor that the vortex-atom theory concerned properties of ether as well as properties of matter.[116] With important modifications, Kelvin's basic conviction that there was no dichotomy between ether and matter apparently had a lifetime at least from the early 1850s to the late 1880s, giving way at some point to the view of the 1904 edition of the *Baltimore Lectures*, in which matter and ether, if not so different from one another as Stokes would have had them, were at least classed as ponderable and imponderable.[117] Stokes's ideas during this period were more predictable than Kelvin's, but even so, it is useful to have them placed in contrast to Kelvin's. For one thing, his thought during the latter part of his career has been little discussed, but, more importantly, in many respects he represented the orthodoxy Kelvin was trying to move beyond.

IV

In editing this book and in focusing its Introduction on institutional and often-speculative intellectual matters, I have seen the book as reflecting current perspectives in the modern discipline of the history of science. Historians of science have realized that recounting success stories of those scientific ideas which have wound up in modern science textbooks is not the only, nor the most illuminating, approach to the history of science. We have learned that the ideas scientists have puzzled most over have often not been the ones we associate with them from reading such textbooks. Indeed, we have learned that such 'familiar' ideas frequently are much-changed versions of the past scientists' own ideas. And we have learned that scientists have lived not simply in a world of experiments and theories, but within networks of people and institutions, of jealousies and friendships, in which the experiments they have chosen to perform and the theories they have decided to support have usually depended in part on where they have lived, whom they have known, and what language they have spoken.

In the case of nineteenth-century British physical science, there are now a number of interesting studies sensitive to such considerations. Yet, until very recently, and perhaps even now, such scholarship has concentrated unduly on the thought of Faraday and Maxwell whose ideas 'point' towards modern physical theories. Hence, there is room for a great deal of scholarship from many different viewpoints, and experience tells us that such scholarship will usually be the better for having encountered relevant manuscript materials.

Broadly speaking, if we are to obtain and to impart a more and more mature understanding of the history of science and, therefore, of science itself, we require selected volumes of unpublished materials. It is towards such a general understanding that I should hope this specialized book would contribute.

NOTES TO THE INTRODUCTION

1 They had certainly met by early 1845, when Kelvin discussed his Smith's prize examination with Stokes. (Robert John Strutt, Fourth Baron Rayleigh, *Life of John William Strutt. Third Baron Rayleigh*, an augmented edition with annotations by the author and foreword by John N. Howard (Madison, 1968), 242.) On 6 July 1846 Stokes, 'as a personal acquaintance of his', wrote a testimonial supporting Kelvin's application for the chair of natural

philosophy in the University of Glasgow. (Kelvin Collection, Tm23; printed in Thompson, I, 176.)

2 For explanations of these Cambridge honours, see letter 22, note 1, and letter 76, note 2.

3 Michael Foster to Lady Rayleigh, 1884, cited in Rayleigh, *Rayleigh*, 168.

4 For further biographical information, see the sketch by his daughter in Stokes's *Memoir*, I, 1–49; Larmor's article on Stokes in the *Dictionary of National Biography;* E. M. Parkinson's article on Stokes in the *Dictionary of Scientific Biography*, vol. XIII (New York, 1976), 74–9; and the obituary notices by Kelvin, (375L), in Kelvin's *MPP*, VI, 339–44, and by Rayleigh in Stokes's *MPP*, VI, pp. ix–xxv.

5 Kelvin (1).

6 See Crosbie W. Smith, 'William Thomson and the creation of thermodynamics: 1840–1855', *Archive for History of Exact Sciences*, **16** (1976), 231–88.

7 This is the title of Chapter XXIV in Thompson II, 1012–85, but Kelvin also used the phrase. See, for example, Kelvin (307), in *MPP*, III, 510.

8 Kelvin (352).

9 Kelvin (373).

10 For further biographical information in addition to Thompson, see Joseph Larmor's obituary of Kelvin in *Proc. Roy. Soc.*, series A, **81** (1908), pp. iii–lxxvi, and Jed Buchwald's article on Kelvin in the *Dictionary of Scientific Biography*, vol. XIII (New York, 1976), 374–88.

11 William Brock's article on Crookes in *Dictionary of Scientific Biography*, vol. III (New York, 1971), 474.

12 Smithells to Joseph Larmor, 6 July 1905, in Stokes's *Memoir*, I, 266.

13 *Ibid.* On the collaboration between Stokes and Smithells, see Robert DeKosky's article in *Ambix*, **27** (1980), 103–23.

14 Kelvin, *Baltimore Lectures* (1884), 158.

15 Kelvin (375L), in *MPP*, VI, 344.

16 See David B. Wilson (ed.), *Catalogue of the Manuscript Collections of Sir George Gabriel Stokes and Sir William Thomson, Baron Kelvin of Largs, in Cambridge University Library* (Cambridge, 1976).

17 This information comes from the Royal Society's card catalogue for their manuscript holdings.

18 One thinks of the well-known discussion in Pierre Duhem, *The Aim and Structure of Physical Theory*, trans. Philip P. Wiener (New York, 1962, from the 2nd French edn of 1914). I do not mean to preclude the existence of regional differences within a country.

19 For an overall discussion, see D. S. L. Cardwell, *The Organisation of Science in England*, 2nd edn (London, 1972).

20 For a recent treatment that discusses other research on the founding of the British Association, see Susan Faye Cannon, *Science in Culture: The Early Victorian Period* (New York, 1978).

21 See Henry Lyons, *The Royal Society, 1660–1940: A History of its Administration under its Charters* (Cambridge, 1944).

22 Kelvin (42), (47), (59), (62), (63), and (64) are Kelvin's earliest papers in the publications of the Royal Society of Edinburgh.

23 See D. A. Winstanley, *Early Victorian Cambridge* (Cambridge, 1955); Sheldon Rothblatt, *The Revolution of the Dons: Cambridge and Society in Victorian England* (New York, 1968); and W. W. Rouse Ball, *Cambridge Papers* (London, 1918).

24 See J. M. Dubbey, 'The introduction of the differential notation to Great Britain', *Annals of Science*, **19** (1963), 37–48; Elaine Koppelman, 'The calculus of operations and the rise of abstract algebra', *Archive for the History of Exact Sciences*, **8** (1971), 155–242; Crosbie Smith, '"Mechanical Philosophy" and the emergence of physics in Britain: 1800–1850',

Annals of Science, **33** (1976), 3–29; and H.W. Becher's unpublished dissertation, 'William Whewell and Cambridge mathematics', (University of Missouri-Columbia, 1971).

25 W. Whewell, *An Elementary Treatise on Mechanics* (Cambridge, 1819); W.Whewell, *A Treatise on Dynamics* (Cambridge, 1823); and G. B. Airy, *Mathematical Tracts* (Cambridge, 1826).

26 See, for example, Rothblatt, *Revolution of the Dons*, 191–2n.

27 The tripos lists are in J.R. Tanner (ed.), *The Historical Register of the University of Cambridge* (Cambridge, 1917).

28 Stokes (16), (19), and (28), and Kelvin (29.1), (29.2), and (40).

29 Stokes (27).

30 Stokes (31).

31 Stokes (1).

32 See Jed Z. Buchwald, 'William Thomson and the mathematization of Faraday's electrostatics', *Historical Studies in the Physical Sciences*, **8** (1977), 101–36.

33 Stokes (11).

34 *Trans. Camb. Phil. Soc.*, **1** (1819–22), pp. iv–v.

35 My statements about science education at Cambridge during the second half of the century depend largely on unpublished papers by William Bynum, Roy Porter, and myself, dealing with aspects of the natural sciences tripos. See also Roy Porter, 'Gentlemen and geology: the emergence of a scientific career, 1660–1920', *The Historical Journal*, **21** (1978), 809–36; and Gerry Roberts's article on chemistry in the natural sciences tripos in *Historical Studies in the Physical Sciences*, **11** (1980), 157–83.

36 See also Thompson, II, 694–5 and 840–2.

37 Maxwell had been the first additional examiner the year before.

38 George Elder Davie, *The Democratic Intellect: Scotland and Her Universities in the Nineteenth Century* (Edinburgh, 1961).

39 See J. P. Nichol, *Views of the Architecture of the Heavens* (Edinburgh and London, 1837) and *The Phenomena and Order of the Solar System* (Edinburgh and London, 1838).

40 'Evidence, Oral and Documentary, taken and received by the Commissioners Appointed ... for visiting the Universities of Scotland, vol. II: University of Glasgow', *Parliamentary Papers*, **36** (1837), 120.

41 *Ibid.*, 143–5.

42 See Kelvin (375P), and the excerpt from Kelvin's inaugural address as Chancellor of Glasgow University in 1904 in Thompson, I, 13n. Thompson is one who implies a higher level of mathematical expertise in the Glasgow natural philosophy class than seems actually to have existed. (Thompson I, 12.)

43 Kelvin's notes from the natural philosophy course are in the Kelvin Collection, NB9 and NB10.

44 Thompson, I, 14.

45 'Evidence, Oral and Documentary', 119.

46 Thompson, I, 193–6.

47 There are various nineteenth-century accounts of Kelvin's laboratory. See Kelvin (267E); J.T. Bottomley, 'Physical science in Glasgow University', *Nature*, **6** (9 May 1872), 29–32; and A. Gray, 'Lord Kelvin's laboratory in the University of Glasgow', *Nature*, **55** (25 March 1897), 486–92. See also Romualdas Sviedrys, 'The rise of physics laboratories in Britain', *Historical Studies in the Physical Sciences*, **7** (1976), 405–36.

48 The announcement of Kelvin's and Bottomley's lectures for the session of 1885/6 is printed in Thompson, II, 851. One source of anecdotal remembrances of Kelvin's style in the classroom is Andrew Gray, *Lord Kelvin: An Account of His Scientific Work* (London, 1908), 88–98 and 279–98.

49 See M. J. S. Rudwick, 'The foundation of the Geological Society of London: its scheme for co-operative research and its struggle for independence', *The British Journal for the History of Science*, **1** (1963), 325–55.

50 See L. Pearce Williams, 'The Royal Society and the founding of the British Association for the Advancement of Science', *Notes and Records of the Royal Society*, **16** (1961), 221–33.

51 Charles Babbage, *Reflections on the Decline of Science in England and on Some of Its Causes* (London, 1830).

52 See Lyons, *The Royal Society*, 341–4.

53 See Roy M. MacLeod, 'The Royal Society and the government grant: notes on the administration of scientific research, 1849–1914', *Minerva*, **14** (1971), 323–58.

54 See Wilson, *Catalogue*, part I, p. ii.

55 But see his attack on John Tyndall in 1859 when Tyndall was apparently in the running for a Royal Medal from the Royal Society. (Letter 173.) He also mentioned at least four men for possible election as fellows: E. O. W. Whitehouse (letter 163), Fleeming Jenkin (letters 239 and 244), Lewis Gordon (letter 305), and Latimer Clark (letter 309).

56 Kelvin (92).

57 Kelvin (91).

58 Thompson, I, 319.

59 Joe D. Burchfield, *Lord Kelvin and the Age of the Earth* (New York, 1975).

60 Kelvin (130).

61 Kelvin (133).

62 Kelvin (207.1).

63 Kelvin (207.2).

64 Kelvin (209).

65 S. Earnshaw, 'On the mathematical theory of sound', *Phil. Trans.* (1860), 133–48.

66 See, for example, my 'Concepts of physical nature: John Herschel to Karl Person', in U. C. Knoepflmacher and G. B. Tennyson (eds.), *Nature and the Victorian Imagination* (Berkeley, 1977), 201–15. Regarding the content of this section of the Introduction, I have had helpful discussions with Don Moyer, Joe Burchfield, and Crosbie Smith.

67 See Crosbie Smith, 'A new chart for British natural philosophy: the development of energy physics in the nineteenth century', *History of Science*, **16** (1978), 231–79.

68 There is now a good deal of literature on this general point, and I have discussed it briefly in my 'The kinetic atom', *American Journal of Physics*, **49** (1981), 217–22. For a study emphasizing a British physicist's reaction to French views, see Robert Kargon, 'Model and analogy in Victorian science: Maxwell's critique of the French physicists', *Journal of the History of Ideas*, **30** (1969), 423–36.

69 William Smith, Notes taken from William Thomson's natural philosophy lectures during the 1849/50 session at Glasgow College, MS. Gen. 142, Glasgow University Library, as quoted in Crosbie Smith, 'Engineering the universe: William Thomson and Fleeming Jenkin on the nature of matter', *Annals of Science*, **37** (1980), 387–412. For what I would regard as a non-realist view of the ether, see Karl Pearson's criticism of Kelvin, in which he states that 'the "luminiferous ether" is only an intellectual mode of briefly summarizing certain wide groups of sensations'. (Isaac Todhunter, *A History of the Theory of Elasticity and of the Strength of Materials from Galilei to the Present Time*, edited and completed by Karl Pearson, vol. II: *Saint-Venant to Lord Kelvin*, part II (Cambridge, 1893), 453.)

70 *Baltimore Lectures* (1884), 280; see also *ibid.*, 105 and 166.

71 *Ibid.*, 132; see also 198–9 and 270–1.

72 *Ibid.*, 146.

73 For a fuller discussion, see my 'George Gabriel Stokes on stellar aberration and the luminiferous ether', *The British Journal for the History of Science*, **6** (1972), 57–72.

74 Stokes (11), in *MPP*, I, 78.
75 Stokes (20), in *MPP*, II, 12–13.
76 This letter was written two years after Stokes first presented his idea of an ether with both fluid and solid properties, but the year *before* he published his glue–water analogy.
77 Stokes (11), in *MPP*, I, 127–8.
78 Kelvin (72).
79 *Ibid.*, in *MPP*, II, 28.
80 The notebook is in the Cambridge University Library, Kelvin Collection, Add. MS 7342, NB47. It still has two address labels on the cover, one addressed to Kelvin and postmarked Edinburgh, 5 February 1862, the other addressed to Tait and postmarked Glasgow, 19 November 1862, and Edinburgh, 19 November 1862.
81 The entry is printed in Ole Knudsen, 'From Lord Kelvin's notebook: ether speculations', *Centaurus*, **16** (1972), 41–53. See also Knudsen, 'The Faraday effect and physical theory, 1845–1873', *Archive for History of Exact Sciences*, **15** (1976), 235–81.
82 Kelvin (94).
83 Knudsen, 'Kelvin's Notebook', 47.
84 In letter 161 Kelvin wrote that if he could work out the model satisfactorily, 'I think I should have a very good medium for a mechanical illustration of light'. In the notebook entry, he referred to 'the kind of rigidity required for undulations of light', (Knudsen, 'Kelvin's Notebook', 49.)
85 In 1884 Kelvin commented on his views of 1854: 'I then thought that the medium must be a continuation of our atmosphere. I could not say any thing like that now'. (*Baltimore Lectures* (1884), 200). His comment in the second edition of the *Baltimore Lectures* was in a footnote dated 13 October 1899: 'Not so now. I did not in 1854 know the kinetic theory of gases'. (*Baltimore Lectures* (1904), 260n.)
86 Kelvin (172), in *Nature*, *1* (31 March 1870), 551.
87 Kelvin (173), in *PLA*, II, 167.
88 Kelvin (172), in *Nature*, *1* (31 March 1870), 553.
89 *Baltimore Lectures* (1884), 10.
90 Kelvin (161).
91 Knudsen, 'Kelvin's Notebook', 44.
92 Kelvin (274), in *MPP*, IV, 308. He was still not completely successful, concluding 'that the most favourable verdict I can ask ... is the Scottish verdict of *not proven*'. (*Ibid.*, 320.)
93 Kelvin (375JJ) in *MPP*, VI, 236. He says there that he had abandoned the idea more than thirty years earlier (*i.e.* before 1877), thus raising a possible problem of chronology, for he was favourably inclined towards the idea of elasticity resulting from vortex motion in papers published after 1877. Perhaps he meant *twenty* years ago. However, whatever the exact chronology, his statement supports my main contention that he initially intended vortices to account for the elasticity of the ether as well as to explain properties of ordinary matter.
94 Kelvin (257).
95 Stokes (72), in *MPP*, IV, 177.
96 See note 89.
97 *Baltimore Lectures* (1884), 28.
98 *Ibid.*, 41.
99 *Ibid.*, 43.
100 *Ibid.*, 42, 143.
101 *Ibid.*, 43; Kelvin (25).
102 *Baltimore Lectures* (1884), 207.
103 John F. W. Herschel, *A Preliminary Discourse on the Study of Natural Philosophy* (New York and London, 1966 reprint of 1832 edn), section 199.

104 Kelvin (173), in *PLA*, II, 203.

105 W. R. Grove, 'Address', in *Brit. Assoc. Rep.* (1866), p. lxxix. Grove thought that the law 'pervades all physical phenomena' and that it encompassed the conservation of energy as well as his own version of conservation. *The Correlation of Physical Forces* (London, 1846).

106 On this point see M. Norton Wise, 'William Thomson's mathematical route to energy conservation: a case study of the role of mathematics in concept formation', *Historical Studies in the Physical Sciences*, **10** (1979), 49–83.

107 Kelvin (108), in *PEM*, 225. I have argued that we should take Kelvin's theological statements seriously in trying to understand his concepts of nature in my 'Kelvin's scientific realism: the theological context', *The Philosophical Journal*, **11** (1974), 41–60, and Crosbie Smith has emphasized the importance of Kelvin's theology in his 'Natural philosophy and thermodynamics: William Thomson and "the dynamical theory of heat"', *The British Journal for the History of Science*, **9** (1976), 293–319.

108 Stokes (20), in *MPP*, II, 12.

109 Kelvin, *Baltimore Lectures* (1884), 10; Stokes, *Burnett Lectures*, I, 80. See also Stokes, *ibid.*, I, 23–4.

110 *Ibid.*, I, 76; see also *ibid.*, I, 79–80, and II, 1–2.

111 *Ibid.*, I, 81.

112 *Ibid.*, II, 33. For Stokes's summaries of what he did think was reliably known, see *ibid.*, I, 85–8, and II, 32–3.

113 See Robert Silliman, 'William Thomson: smoke rings and nineteenth-century atomism', *Isis*, **54** (1963), 461–74, and Thompson, II, 1046–9.

114 Kelvin (338), in *Proc. Roy. Soc.*, **54** (1893), 389.

115 Quoted in Thompson, II, 1072–3.

116 My interpretation differs, for example, from that in the most recent, extensive discussion of these matters, Barbara Gusti Doran, 'Origins and consolidation of field theory in nineteenth-century Britain: from the mechanical to the electromagnetic view of nature', *Historical Studies in the Physical Sciences*, **6** (1975), 133–260. Doran concludes that both Stokes and Kelvin were among those British physicists who by the 1850s had 'established the conception of the aether as a continuous entity ontologically different from ordinary matter'. (*Ibid.*, pp. 149–50.)

117 *Baltimore Lectures* (1904), 266, in a statement dated 17 November 1899.

LIST OF THE LETTERS

THE CORRESPONDENCE
(1846–1869)

1 STOKES to KELVIN, 15 October 1846
Kelvin Collection, S321

DEAR THOMSON,[1]

I spoke to Hopkins,[2] whom I met the other day at dinner, about the composition to the [Cambridge] philosophical Soc^{y}. He said he thought that a composition paid now would be considered to include the payment for the *present* year, but that Dr Paget[3] our treasurer could say for certain. I accordingly asked Dr Paget, who said it would. So by paying £10.10 within the present year you can be a member for life. I take for granted that you will have to pay the £1.6 for the present year for the reading room, a payment which life members who are in residence have to make. If you have already paid your subscription i.e. £1.1 for soc^{y} and £1.6 for r room, I suppose that £9.9 is all that will be required.

I suppose that you have hardly begun your lectures yet. I hope you will like your work.

Yours very truly
G. G. STOKES

Pembroke College, Cambridge
Oct 15th 1846

1 Kelvin has written at the top of the letter, 'Glasgow Oct 17, 1846 (Stokes Shedden)'. Kelvin frequently noted the place and date of a letter's reception in this fashion. Thomas Shedden (1824–1906) was an undergraduate friend of Kelvin's, and there are over thirty letters from him to Kelvin in the Kelvin Collection, dating mainly from the 1840s. Kelvin also received a letter from Shedden on 17 October 1846 (Shedden to Kelvin, 11 September 1846, Kelvin Collection, S113).

2 William Hopkins (1793–1866) was seventh wrangler in 1827 and went on to become the most renowned of the coaches for the mathematical tripos during the second quarter of the century. He taught both Kelvin and Stokes, and the notes each took while reading with Hopkins survive in the Kelvin collection, PA11 to PA17, and in the Stokes Collection, PA2 to PA24.

3 George Edward Paget (1809–92) was eighth wrangler in 1831 and, after studying medicine in Paris and London, was appointed physician to Addenbrooke's Hospital in Cambridge in 1839, a position he held for forty-five years. In 1872 he was appointed Regius professor of physic at Cambridge, and he was president of the Cambridge Philosophical Society in 1855/6.

2 KELVIN to STOKES, 25 October 1846

Stokes Collection, K14

College, Glasgow
Oct 25, 1846

MY DEAR STOKES,

I am much obliged to you for your letter w^h I received a few days ago. After I saw you last, I settled the question ⟨by⟩ about the philosophical society by paying Mr Crouch[1] £11.16 (not having previously paid my subscription for the present year) as he explained the terms to me.

I hope some of you at the philosophical society will see that the foreign periodicals, especially the Com[p]tes Rendus, are ⟨taken in⟩ procured regularly. The recent proceedings about Oceanus, or Neptune, or Le Verrier, seem rather to indicate that Cambridge is behind the rest of the world in information on scientific subjects. Prof. Challis[2] told me, a few days before I left, that he had not been able to see either paper of Le Verrier's as the Phil. Society had not received the N^{os} containing them.[3] It is a pity that Challis did not look at the results of his observations, as it appears from his letters[4] that the discovery of the planet would have been entirely secured between Adams[5] and himself.

Have you made any gun-cotton yet? It is very easy, as the usual receipt w^h I suppose you have seen in the newspapers, is quite correct, and is sure to succeed if the acid be strong enough. About a tenth part of concentrated sulphuric acid makes the success more certain. I am going to try it tomorrow, as I was told about the process by a person who had made it himself.

Your's very truly
WILLIAM THOMSON

1 John Crouch was the curator of the Cambridge Philosophical Society.

2 James Challis (1803–82) was senior wrangler in 1825 and in 1836 became Plumian professor and director of the Cambridge Observatory, holding the professorship until his death and the directorship until 1861. In the directorship, he was preceded by Airy and succeeded by Adams. His physical studies were rather lightly regarded by others; see, for example, Kelvin's comments in letter 93.

3 Urbain Jean Joseph Le Verrier read a series of papers to the Académie des Sciences in 1845 and 1846: 'Premier mémoire sur la théorie d'Uranus', *Comptes Rendus*, **21** (10 November 1845), 1050–5; 'Recherches sur les mouvements d'Uranus', *ibid.*, **22** (1 June 1846), 907–18; 'Sur la planète qui produit les anomalies observées dans le mouvement d'Uranus', *ibid.*, **23** (31 August 1846, 5 October 1846), 428–38, 657–62; and 'Comparison des observations de la nouvelle planète avec la théorie déduite des perturbations d'Uranus', *ibid.*, **23** (19 October 1846), 741. Le Verrier (1811–77) was a graduate of the École Polytechnique, did work in chemistry and meteorology as well as astronomy, and in 1854 became director of the Paris Observatory.

4 There are no letters of this date from Challis to Kelvin in either the Kelvin Collection or the Kelvin Papers at Glasgow. Kelvin may have been referring to the series of letters involving

Challis, Airy, Adams, and others which were soon to be published in *Monthly Notices of the Royal Astronomical Society*, 7 (13 November 1846), 123–44.

5 John Couch Adams (1819–92) was senior wrangler in 1843 and, like Le Verrier, predicted the existence of Neptune. Although Le Verrier's prediction was confirmed observationally before Adams's, Adams's efforts earned him an offer of a knighthood in 1847, which he declined. He became a fellow of Stokes's college, Pembroke, in 1853, and the two were close friends. He was appointed Lowndean professor at Cambridge in 1859 and director of the Cambridge Observatory in 1861.

3 Kelvin to Stokes 5 November 1846

Stokes Collection, K15

College, Glasgow
Nov. 5, 1846

My dear Stokes

I should feel excessively obliged if you would let me have another copy of your last paper (on elastic solids &c).[1] I have lent my copy to Prof. Forbes,[2] and now I am in want of a copy, for references to some particulars which I am giving in my lectures about the properties of solids, &c. If you can easily spare me a second copy, I should be very glad to have one, and then I should allow Prof. Forbes to keep the copy I lent him.

Macmillan[3] would send a copy here for me, if you would give it to them.

I have got over two lectures besides a viva voce examination. Yesterday I gave the introductory lecture, w[h] was rather a failure as I had it all written, and I read it very fast.

Your's very truly
William Thomson

1 Stokes (11).

2 James David Forbes (1809–68) was diverted from early studies in law to natural philosophy and became professor of natural philosophy at the University of Edinburgh in 1833. He was a friend of the Thomson family and was regarded as a likely candidate for the chair of natural philosophy at Glasgow in 1846. However, he decided not to stand and, instead, wrote a testimonial supporting Kelvin's candidacy. (See Thompson, I, 174, for the letter.)

3 Daniel Macmillan (1813–57), after opening bookshops in Cambridge and London in 1843, formed Macmillan and Co. in 1844 to begin publishing. Daniel stayed in Cambridge, which was the centre of the business, while his brother, Alexander, took care of the London interests. Macmillan was the Cambridge publisher for the *Cambridge and Dublin Mathematical Journal*, which Kelvin edited from 1846 to 1853. It was published in London by George Bell and in Dublin by Hodges and Smith.

4 Stokes to Kelvin, 10 November 1846
Kelvin Collection, S322

Pembroke College[1]
Cambridge
Nov. 10th 1846

Dear Thomson,

I have just folded up a copy of my paper[2] on the friction of fluids &c which I shall leave today with McMillan. I fear I shall not be able to let you have another copy, as I gave away my spare copies at the Southampton meeting [of the British Association for the Advancement of Science]; as to whether you will leave your second copy in the hands of Profr Frobes or of Liouville.[3] I am sorry I did not get more copies when I was about it. However the part [of the *Transactions of the Cambridge Philosophical Society*] in which it is to appear will very soon be published.

We had our first meeting last night at the Philosophical, of which I am glad to find you have become a member for life. The subject was the new planet. Profr Challis gave an account of the methods which he employed in hunting for it, as well as of the dates and nature of the communications received from Adams and Le Verrier. It is most unfortunate that Adams's results were not published at the time. He gave his results to Airy[4] as long ago as September 1845. As it is, Le Verrier is the name which must be associated with the planet as to its theoretical discovery. Nevertheless scientific men, who look into the history of the discovery, will perceive from the evidence that I have no doubt Airy will bring forward that the planet was discovered from pure theory independently by two persons, and in fact that in point of time Adams had the priority.

I suppose you have given many lectures by this time. When once your first course is completed you will not find it such hard work.

Believe me dear Thomson
Yours very truly
G. G. Stokes

1 Kelvin has written at the top of the letter, 'Nov 12'.

2 Stokes (11).

3 Joseph Liouville (1809–82) graduated in 1830 from the École des Ponts et Chaussées and held the chair of analysis and mechanics at the École Polytechnique from 1838 to 1851. He was one of the French scientists Kelvin met during his sojourn in Paris in 1845. Indeed, Kelvin gave to Liouville one of the two copies of Green's *An Essay on the Application of Mathematical Analysis to the Theories of Electricity and Magnetism* (Nottingham, 1828), which Hopkins had given to him just before he left for Paris. Liouville founded his *Journal de mathématiques pures et appliquées* in 1836 and edited it through thirty-nine volumes.

4 George Biddell Airy (1801–92) was senior wrangler in 1823 and, after holding the Lucasian and then the Plumian professorship at Cambridge, became Astronomer Royal in 1835. He

is especially noted for instilling order and efficiency into the operations of the Royal Observatory during his forty-six year reign. An enormous correspondence between Stokes and Airy survives in the Stokes Collection and at the Royal Observatory.

5 Kelvin to Stokes, 4 December 1846
Stokes Collection, K16

College, Glasgow
Dec. 4, 1846

My dear Stokes

I should be very much obliged, if you would look over the proof sheets of a paper of my own which I have sent to the printers, ⟨for⟩ and see if the paper seems to be all right.[1] I am getting proof sheets also, and shall attend to the errors of printing. All I wish you to do therefore is to glance over the ⟨substa⟩ [?] paper, and, if you find that there are any great mistakes, to let me know. I wrote it in a great hurry, as I had not time to spend, and thought that if I were to delay I should not get it published till this time next year.

Your's very truly
(in great haste)
William Thomson

P.S. The Macmillans are anxious to have the Jany N^o [of the *Cambridge and Dublin Mathematical Journal*] published before the end of term, so that I shall have to return the proof sheets, right or wrong, without much delay.

If you are busy, or are not inclined, do not take any trouble.

Is Ap. – 1845 the proper date, for the reading of your paper?[2]

1 Kelvin (25), which appeared in the January 1847 number of the *Cambridge and Dublin Mathematical Journal*.
2 Stokes (11) was read to the Cambridge Philosophical Society on 14 April 1845.

6 Stokes to Kelvin, 13 March 1847
Kelvin Collection, S323

Pembroke College[1]
Cambridge
March 13, 1847

My dear Thomson,

Fischer has at last made up his mind to try for the Professorship at St Andrews.[2] He has Sir David Brewster[3] in his favour. I should think he has a very good chance of it if it be not that he is too late in the field. I showed your testimonial from Liouville[4] to my brother [W. H. Stokes] today, to let him see what Liouville said both of you and of his own college man Green.[5] I could not leave him the copy as it belonged to Fischer. He asked me if I could get him any copies, and I told him I would write to you and ask you to send me one or two if you had them to spare.

I suppose you have received Mr Haughton's memoir on elastic solids.[6] He sent me a copy but I have not yet had time to read it.

I forget whether I wrote to you since you asked me to look at the proof sheet of your article[7] in the *[Cambridge and Dublin] Math^l Journal* in which you referred to my paper.[8] I had no alteration to make in it. You seem to be engaged in important speculations. Perhaps the jelly-like fluid that we once spoke of may be made in your hands to explain the law of the mutual action of electric currents and the phenomena of the induction of these currents.

I suppose you have not very much time for speculations of this kind. After you get a little used to your work you will prob^{ly} find that you will have more time.

Believe me
Yours very truly
G. G. Stokes

1 Kelvin has written at the top of the letter, 'March 16'.

2 Frederick William Lewis (or Wilhelm Ferdinand Ludwig) Fischer (1814–90) was fourth wrangler in 1845 – the year Kelvin was second – and was professor of natural and experimental philosophy at St Andrews from 1847 to 1859 and of mathematics from 1859 to 1879. Kelvin wrote a testimonial for Fischer for the St Andrews professorship, and a copy of it – as well as of thirteen other testimonials for Fischer – is in the Kelvin Collection, F87g.

3 David Brewster (1781–1868) enjoyed considerable fame for his work in experimental optics and considerable notoriety for his continued support of the particle theory of light long after most physicists had accepted the wave theory. In 1838 he became principal of the United Colleges of St Salvator and St Leonard, St Andrews, and in 1859 was elected principal of the University of Edinburgh.

4 Liouville's testimonials for Kelvin are printed in Thompson, 1, 179–82. Liouville wrote two letters: one dated 24 July 1846 to James Armitage (1818–52), the second dated 25 July 1846 to Kelvin. Armitage was a fellow of Trinity College, Cambridge, who had visited Liouville in Paris. In his letter to Armitage, Liouville refers to Kelvin as 'notre ami commun'.

5 George Green (1793–1841) was a self-educated mathematical physicist whose *An Essay on the Application of Mathematical Analysis to the Theories of Electricity and Magnetism* (Nottingham, 1828) became influential after 1845 when Kelvin's support began making it known to a wide audience. In 1833, at the age of forty, Green began formal training at Cambridge for the mathematical tripos in which he was fourth wrangler in 1837. He and W. H. Stokes were both at Gonville and Caius college.

6 Samuel Haughton (1821–97), 'On the laws of equilibrium and motion of solid and fluid bodies', *Camb. Dub. Math. J.*, **1** (1846), 173–82; **2** (March 1847), 100–8. Haughton took his B.A. at Trinity College, Dublin, in 1844 and was professor of geology at Dublin University from 1851 to 1881.

7 Kelvin (25).

8 Stokes (11).

7 KELVIN to STOKES, 25 March 1847
Stokes Collection, K17

College, Glasgow,
March 28, 1847

MY DEAR STOKES

I enclose you a few copies of the document for which you asked. You will excuse my not answering your letter sooner, as I have been very busy, and I did not know that there were any copies of the paper to be had, till my father found some a few days ago.

I hope Fischer will succeed in his present application. His chances seem to be very good, and I am sure he is far better than any other candidate who has as yet come forward. You know of course by this time about how the Clare Hall matter[1] has been settled, w^h I suppose was fixed yesterday.

I intend visiting Cambridge in May and remaining till after the examinations, and I shall have a great deal to speak about, with you.

Your's very truly
WILLIAM THOMSON

1 See the next letter.

8 KELVIN to STOKES, 30 March 1847
Stokes Collection, K18

College, Glasgow
March 30, 1847

MY DEAR STOKES,

It has just occurred to me this evening that you may possibly be able to give me some information that will help me out of a difficulty which has been puzzling me for a considerable time.

Do you know whether this problem is possible?

Water is ⟨poured⟩ made to move through a space bounded by a closed surface S, ⟨in such⟩ a given quantity running across each element of S in a unit of time; it is required to find a function ϕ, which will represent by its diff^l coeff^ts the motion of the water within S. Of course the ⟨p⟩ solution will contain an arbitrary part depending on the motion of the water which may take place within S, without any flowing across it. It is with reference to magnetism that my difficulty has occurred, but I have stated what corresponds in your favourite subject. This has just suggested to me the following analogy, which will hold whether $udx + vdy + wdz$ be a complete diff^l or not.

Let q be the velocity, and (l, m, n) the direction of motion of water at any point P in the interior of a closed vessel, S. Then if we take a piece of steel bounded by a surface similar to S, and magnetize it in such a manner that q is the intensity, and (l, m, n) the direction of magnetization at P, this mass ⟨of w⟩ will exert no force on an external point.

An obvious case of this is when S is a closed tube. The piece of steel will then be a uniformly magnetized bar (or ⟨one wa[?]⟩ if the section be variable, inversely as the section) bent with its ends together, which of course exerts no force on an external point.

Your's very truly
WILLIAM THOMSON

P.S. I hear today that you have lost Fischer, from Pembroke.[1]

1 Fischer migrated from Pembroke College to Clare Hall (now College). This is no doubt the 'Clare Hall matter' mentioned in the previous letter.

9 STOKES to KELVIN, 1 April 1847

Kelvin Collection, S324

Pembroke College,[1]
Cambridge
April 1, 1847

MY DEAR THOMSON,

I have just received your second note. In the first place I have to thank you for the copies of Liouville's testimonial. I have given two copies to my brother [W. H. Stokes] and kept one myself. I hope you did not exhaust your stock in sending them to me.

Now as to your question. I conceive it to be of this nature. Water is flowing ⟨in any manner⟩ within a closed surface S, and is also flowing in any manner, (consistent with its incompressibility,) across the surface of S. The motion of the water within S is not of the most general kind, but of that particular kind for which $udx + vdy + wdz$ is an exact differential $d\phi$. It is required to find the most general form of ϕ.

First as to the manner in which the fluid flows out of, and into the interior of S. I do not know whether you regard the fluid in S as a mass by itself, using the expression flowing in and flowing out merely as a short way of expressing the conditions which must hold at a given instant at the surface of S, or whether you regard the mass within S as part of a continuous and continuously moving mass within S', S' containing S.

In the first, we may conceive the velocity normal to the surface S given arbitrarily, (subject to the condition of incompressibility), leaving the fluid to determine its own tangential velocity.

If the second, we must consider the fluid S a part of the fluid S', which drives us back to the first case, at least if we include in this case that in which S is infinite.

Let then v be the ⟨normal⟩ velocity normal to S at a given instant. Conceive the fluid within S to have been at rest, and the physical surface to to [*sic*] have been then moved suddenly with a normal vel[y] v. You know how to determine or conceive determined the value of ϕ for this case. The problem is just the same as that of determining the permanent temperature of a homogeneous solid S when the flux of heat at each point of the surface is given. Let ϕ_1 be the part of ϕ due to v, ϕ the whole, and let $\phi = \phi_1 + \phi_2$. It is required to consider the most general value of ϕ_2. Of course as far as regards the motion expressed by ϕ_2 the fluid does not flow across the mathematical surface S.

I have not much considered the motion expressed by ϕ_2. It is not so interesting in the theory of fluid motion as that expressed by ϕ_1. I am inclined to think that the motion expressed by ϕ_2 is more particular than that expressed by ϕ_1. The motion expressed by ϕ_2 could not be generated in a fluid previously at rest, or destroyed when existing, by means of the motion of the bounding surface, or by means of forces X, Y, Z of any nature.

I would however remark, what I do not know whether you have observed, that ϕ_2 must *necessarily be discontinuous*. If it were not, the permanent temperature of a solid of the same form as S might vary from point to point without there being any flux of heat across the bounding surface; or a fluid at rest within a closed rigid surface might set itself in motion.

Conceive a closed rigid surface S filled with fluid at rest. Conceive the fluid divided into two by an infinitely thin surface σ capable of sustaining pressure. Let the several points of σ be moved impulsively in any manner consistent with the incompressibility of the fluid, and then let σ be annihilated. Then the fluid will go on moving, and there will *in general* be a surface of discontinuity corresponding to σ, in passing across which the velocity and direction of motion will alter abruptly. In particular cases however ⟨the⟩ it may happen that there will be no abrupt alteration of velocity or direction of motion, ⟨although ϕ⟩ so that the motion will be perfectly continuous although ϕ is discontinuous. It may be proved that the motion will be steady. In this case the same motion would have been produced by any surface σ drawn so as to cut all the lines of motion, and properly moved. For example let S be the space compressed [?] between two concentric circles, and let σ be [a] portion of a

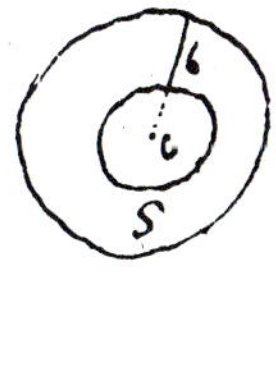

radius, and let the vel^y of the several points of $\sigma \propto \frac{1}{r}$, supposing the fluid referred to polar coordinates r, θ, r being measured from C. In this case $\phi = C(\theta - \alpha)$, α being an arbitrary constant, and C the constant which depends on the absolute velocity. I think you once set this or something like it at Peterhouse.

I do not know whether I have seized the point of your question. I should like to have a line to say whether this letter has been of any use to you. As bearing on this point you may consult my paper On some cases[2] [the rest of the letter is missing].

1 Kelvin has written at the top of the letter, 'Ap. 3, 1847'.
2 Stokes (13).

10 STOKES to KELVIN, 3 April 1847
Kelvin Collection, S325

Pembroke College[1]
Cambridge
Ap. 3^d 1847

MY DEAR THOMSON,

A thought has just struck me relative to the function ϕ_n (of my last letter) which may be of use to you.

I shall call, with Prof^r Challis, the ⟨system⟩ surfaces whose common equation is $\phi_2 = C$ surfaces of displacement. Conceive an infinite number of surfaces of displacement drawn corresponding to infinitely close equidistant values of ϕ_2. Then the motion will be everywhere normal to the surface, and the vel^y inversely proportional to the distance between consecutive surfaces. Consequently if the motion be continuous and the vel^y finite, no two surfaces can either touch or intersect.

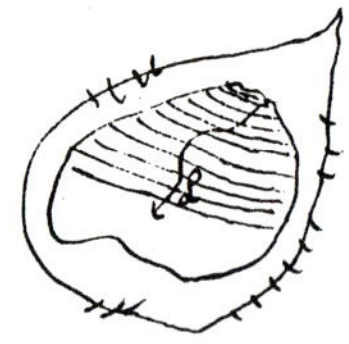

Now the 'thought' is this. Conceive a line L drawn cutting the surfaces, whether at r^t angles or not is no matter. Since $d\phi_2{}^2$ is continuous, although ϕ_2 is discontinuous, the only way in which ϕ_2 can be discontinuous is by L's getting back to the surface it started from. Hence the motion corresponding to ϕ_2 is not possible within a closed surface S without a core and therefore ϕ_2 must = 0 or a constant. For it is evident that L cannot get back to the surface it started from. Neither is it possible within a closed surface *S* with a core (Like the space inclosed between two spheres). For when *L* has got to a surface not cutting the core, or inner surface, we pass to the first case. See note (1) at end. Hence the motion can only be possible (in the case of finite surfaces) when the space considered forms a closed canal.

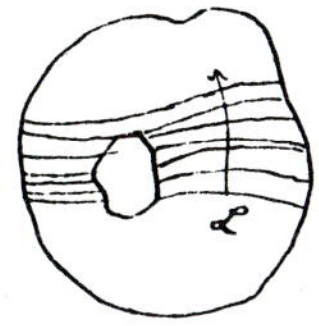

In the case of an infinite space *S* the motion is, (for aught proved to the contrary,) possible

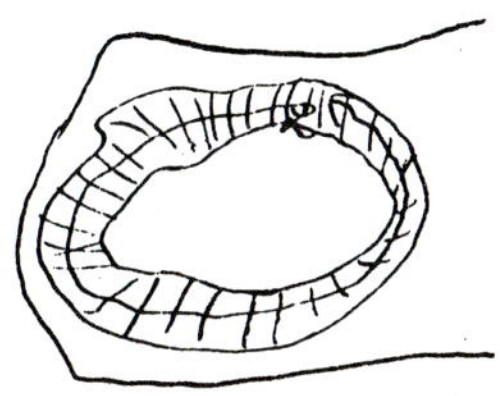

(1) if the canal instead of returning into itself be infinite in length

(2) if the outer surface of *S* move of[f] indefinitely, the core remaining of finite breadth, at least about the point first considered. In this case the form of the space taken ⟨length⟩ crossways would be

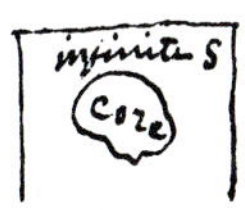

and taken lengthways would be

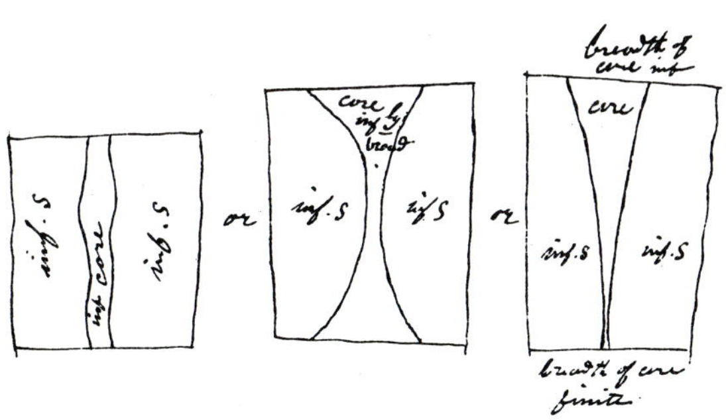

By the bye I have been using *S* for the *space* considered instead of the bounding surface, as in my last letter. Of course the surfaces of displacement must abut the bounding surfac*es* of *S* (or bounding surfac*e* if *S* be inf[inite] or have but one) every where at right angles, for the fluid is supposed not to penetrate the surfaces of S.

(3) the core may be a closed canal and ⟨the rest of S may⟩ S may be the rest of infinite space. To explain what I mean suppose the core to be a circular ring, represented in section by two circles; then the surfaces of displacement may be as in the figure. The direction of motion will be thus represented. While I was drawing the figures at the top of the last page [i.e., the three figures side by side] I was never forgetting this case. (Note (2))

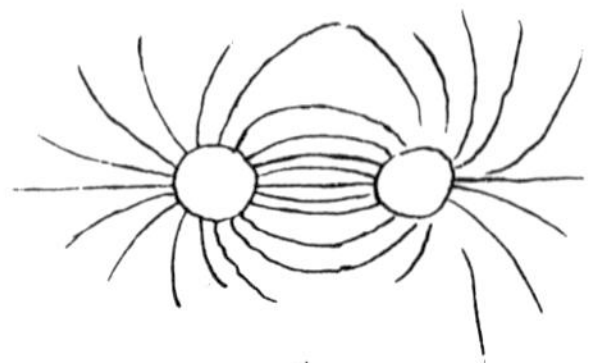

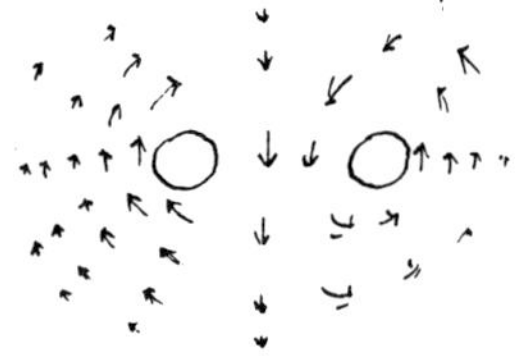

One thing leads on to another. While I was writing the above ⟨I thought⟩ a thought struck me which shows that in the case of two dimensions ϕ_2 is exactly as general as ϕ_1, and moreover is to be determined in just the same way. Let $udx + vdy$ be an exact diff^{fl} $d\phi_2$. Then ϕ_2 satisfies the eq^{n}

$$\frac{d^2\phi_2}{dx^2}+\frac{d^2\phi_2}{dy^2}=0 \tag{1}$$

The diff^{l} equation to a line of motion is $udy - vdx = 0$. Now $udy - vdx$ will be an exact diff^{l} if

$$\frac{du}{dx}=-\frac{dv}{dy}$$

or if (1) is satisfied. Let then $udy - vdx = d\psi$ so that

$$d\phi_2=\frac{d\psi}{dy}dx-\frac{d\psi}{dx}dy$$

which having to be an exact diff^{l}

$$\frac{d^2\psi}{dx^2}+\frac{d^2\psi}{dy^2}=0 \tag{2}$$

Now the bounding lines of S are lines of motion, and therefore ψ is constant for both. We have then, to determine ψ, the general eq^{n} (2) with the particular conditions

$$\psi = a \text{ at the inner surface of } S$$
$$\psi = b \text{ [at the] outer [surface of } S]$$

where a and b are two constants whose difference ⟨determines⟩ fixes the scale of the motion. Having found ψ we shall have

$$\phi_2 = \int \left(\frac{d\psi}{dy} dx - \frac{d\psi}{dx} dy \right)$$

Of course the systems of ⟨surfaces⟩ lines whose equations are ϕ_2 = cons^t, ψ = cons. are, the latter the system of isothermal ⟨surfaces⟩ lines when ⟨w⟩ the temperatures of the bounding lines are supposed to be uniform and different, and the distribution of heat permanent, and the former the system of orthogonal trajectories, which are also isothermal. This is of some interest in the theory of fluid motion as being the generalization of the case $\phi = C\theta + C'$ given by Euler, which has hitherto, so far as I am aware, been regarded as a very particular case.

In the case of three dimensions the lines of motion cannot cut the surfaces of S at finite angles. I have already mentioned that the motion must be steady, and therefore a line traced at a given instant from point to point in the direction of the motion is also that in which the particles through which it passes move. Conceive drawn the two systems of surfaces which with the system ϕ_2 = cons^t form three systems of orthogonal surfaces ⟨drawn⟩. I am almost ashamed to say I have not read the proof of the possibility of this. It is I think by Dupin.[3] If you have occasion to write perhaps you will refer me to the memoir where it is to be found. I shall call the system ϕ_2 = cons^t system 1, and the others systems 2 and 3. The system of lines formed by the intersections of systems 2 and 3 will be the system of lines of motion. Hence these lines will be closed curves, as might have been expected, and from the nature of the case they cannot cut or touch each other. Let σ be one of the bounding surfaces of S. The lines of motion will form on σ a system of closed curves which will everywhere cut at right angles the curves in which σ is cut by the surfaces of system 1. The latter system of curves may be either closed or indefinite, but the curves belonging to it cannot meet each other.

The surface or surfaces of S may or may not belong to one of the systems 2, 3. This is evident. For let a point P move in a surface of displacement and trace out in it any closed curve γ. Then the lines of motion passing through the several points of γ form a surface which may be taken for the bounding surface of the fluid. But since the form of γ is arbitrary this evidently may or may not be a surface of one of the systems 2, 3.

Note (1). If the thing were possible in the case of a closed surface with a core, each surface of displacement would have to abut both the outer surface

Another way has occurred to me of proving the propositions I gave in my last. But first I would just remark that the ϕ_1 motion and ϕ_2 motion may in some cases be the same, that is in some of the cases in which the space S occupied by fluid extends to infinity. Let AB be a canal with pistons at a, b, and let ab be filled with fluid. {I have all along been considering a homogeneous, incompressible fluid.} Let a be pushed, pushing b before it: then the ϕ_1 motion will take place in ab. Let $A'B'C'$ be a closed canal having the part $A'B'$ similar and equal to AB, and let the ϕ_2 motion take place in $A'B'C'$. Now let A, B and A', B' move off indefinitely. Then we pass to an indefinite canal in which there is no distinction between the circumstances of ϕ_1 motion and ϕ_2 motion. In this case indeed in the limit $\phi_1 = \phi_2$, at least when the breadth of the canal remains finite as we proceed indef^{y} in either direc^{n}.

Again let the fluid be contained within an envelope E which contains an annular core C. C and E being fixed we may have the circulating motion expressed by ϕ_2. Now suppose the fluid at rest, and then suppose C moved with any motion of translation only, E remaining fixed. Then we shall have the $\phi_{\langle 2\rangle 1}$ motion. Suppose now a vel^{y} equal and opposite to that of C impressed both on the fluid and on the core. We shall still have a motion of the $\phi_{\langle 2\rangle 1}$ class. Now let E move off indefinitely. We shall have two limiting cases corresponding to motions of the ϕ_2 and ϕ_1 classes respectively, and we may indeed have any combination of the two motions. In the case of the ϕ_1 motion the vel^{y} remains finite at an infinite distance.

Generally, so long as S is finite there is a precise and well-marked distinction between the ϕ_1 and ϕ_2 motions. But in order to keep up this distinction when S becomes infinite it seems that we must regard the infinite S as a limiting case of a finite S ⟨within⟩ in which either the ϕ_1 motion or the ϕ_2 motion prevails. In considering general propositions then relative to the ϕ_2 motion we may suppose S to be finite. Now the mode of considering the ϕ_2 motion to which I have alluded is this. I have already observed that if we exclude infinite velocities we cannot have the consecutive surfaces of displacement cutting or meeting one another. Let ab, cd be two consecutive surfaces of displacement.

Since ϕ_2 is constant for ab and likewise for cd, ϕ_2 must increase or decrease throughout in passing from any p^{t} in ab to the corresponding point in cd; and therefore the motion throughout ab must be either towards cd or from cd. I do not here consider a max^{m} or min^{m} value of ϕ_2, across the surface corresponding to which there could be no motion.

Now the surfaces of displacement must abut the bounding surfaces of S; for they evidently cannot be closed surfaces, on account of the incompressibility of the fluid. Hence the ϕ_2 motion is not possible unless surfaces can be drawn abutting the bounding surfaces of S without dividing the fluid completely into two. This is evident from the incompressibility of the fluid. If it be objected that the reasoning does not apply to a surface of disp[lacemen]t across which there ⟨should be⟩ [?] is no motion the answer is simple. Such a surface might be supposed rigid, so as to divide S into two other S's to each of which the reasoning would apply.

It follows at once that the motion is impossible in a closed space S without a core, or in a closed space S with one, two, or any n° of closed cores. But the motion may take place in the case of a closed canal, containing cores or not, or in the case of a closed space containing an annular core, with or without other cores annular or not.

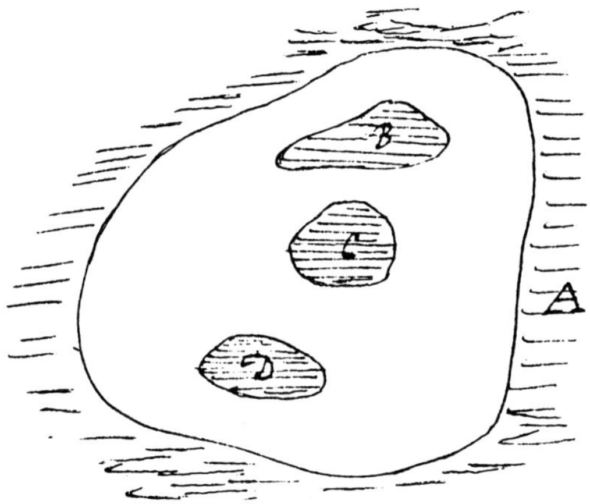

We may of course have cases of a good deal of complexity. Thus in the case of motion in two dimensions represented in the figure, where the unshaded parts represent the fluid, if we suppose the exterior space A, and the cores B, C, D kept up to constant uniform temperatures α, β, γ, δ the isothermal lines will be lines of motion, and by altering the relative magnitudes of α, β, γ, δ we may greatly alter the lines of motion.

It may be observed that in the general case of 3 dimensions ϕ_2 cannot be a max^{m} or min^{m}. For if it were the surface for which it was would be a surface of rest, and might be supposed a bounding surface. But evidently a bounding surface cannot be one of the system of surfaces of displacement.

I have not obtained a solution for the case of a closed canal, even without a core, for the general case of 3 dimensions. I am not able to prove that in the general case the solution is either impossible, possible and unique, or possible and indeterminate. Something may be done for the case of motion which is symmetrical with respect to an axis, and takes place in planes passing through the axis.

Let the axis of z be the axis of symmetry, r the coordinate perpendicular to this axis i.e. the perpendicular distance of any particle from it, w the velocity parallel to z, s that parallel to r. The diff^{l} equation to a line of motion is $sdz - wdr = 0$.

Now by virtue of the eq^{n} of continuity

$$\frac{dw}{dz} + \frac{ds}{dr} + \frac{s}{r} = 0$$

or

$$\frac{d\cdot rw}{dz} + \frac{d\cdot rs}{dr} = 0$$

$rsdz - rwdr$ is an exact diff^{l} $d\psi$

and

$$s = \frac{1}{r}\frac{d\psi}{dz}, \quad w = -\frac{1}{r}\frac{d\psi}{dr}$$

$$d\phi_2 = sdr + wdz$$

$$= \frac{1}{r}\frac{d\psi}{dz}dr - \frac{1}{r}\frac{d\psi}{dr}dz \qquad (1)$$

which having to be an exact diff^{l}

$$\frac{d}{dz}\frac{1}{r}\frac{d\psi}{dz} + \frac{d}{dr}\frac{1}{r}\frac{d\psi}{dr} = 0$$

or

$$\frac{d^2\psi}{dz^2} + \frac{d^2\psi}{dr^2} - \frac{1}{r}\frac{d\psi}{dr} = 0 \qquad (2)$$

Now if $u = C$ be the eq^{n} to an isothermal line,[2] or section of an isothermal surface by a plane passing through the axis of symmetry,

$$\frac{d^2u}{dz^2} + \frac{d^2u}{dr^2} + \frac{1}{r}\frac{du}{dr} = 0 \qquad (3)$$

Now if the iso-ψ lines were necessarily isothermals, ψ and u would have to be constants together, and therefore

$$\psi = \varpi(u)$$

Therefore from (2)

$$\varpi'(u)\left\{\frac{d^2u}{dz^2}+\frac{d^2u}{dr^2}\right\}+\varpi''(u)\left\{\left(\frac{du}{dz}\right)^2+\left(\frac{du}{dr}\right)^2\right\}-\frac{1}{r}\varpi'(u)\frac{du}{dr}=0$$

which becomes by (3)

$$\frac{\varpi''(u)}{\varpi'(u)}=\frac{\dfrac{z}{r}\dfrac{du}{dr}}{\left(\dfrac{du}{dr}\right)^2+\left(\dfrac{du}{dz}\right)^2}$$

which having to be satisfied by the most general value of u the equation

$$\frac{\dfrac{z}{r}\dfrac{du}{dr}}{\left(\dfrac{du}{dr}\right)^2+\left(\dfrac{du}{dz}\right)^2}=\zeta(u),$$

ζ being an arbitrary function, would have to be a first integral of (3), which is not. Hence the iso-ψ lines are not in general isothermal. Since however eq^{ns} (2) and (3) differ only by the sign of their last terms, it would seem that the functions u and ψ are of precisely the same degree of generality, and consequently that in this case the problem is possible and determinate, or at least only indeterminate in so far as corresponds to the arbitrary constants answering to α, β, γ, δ.

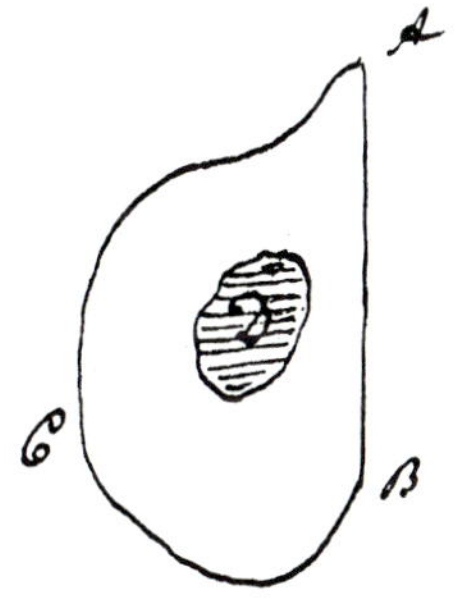

We may of course suppose r positive, or at least not negative, and ψ must be constant for the axis of z. Let AB be the axis of z, ACB the line which generates the outer surface, D the area which generates by revolution an annular core. Let ψ be determined by the general eq^{n} (2) with the conditions

$$\psi = \alpha \text{ at the perimeter } ABCA$$
$$\psi = \beta \text{ at the perimeter of } D$$

then the iso-ψ lines will be the lines of motion, and ϕ_2 will be determined from (1).

The function ψ is what I have denoted by U in a paper On the steady motion of incompressible fluids.[3]

It would be possible to find a closed canal in which the motion should be of 3 dimensions, and in which an assignable value of ϕ_2 would give a possible motion. But this would not determine, even in that particular case, whether for an assigned form of surface the problem was determinate or indeterminate, and such a case would be of little use that I can see.

I do not know whether my letters have borne on your speculations, or whether they have only bothered you, who are already occupied enough. I am not likely however to bother you any more, at least on this subject, for I seem to have stuck fast, and intend to go on with my paper on series.[4]

Yours very truly
G. G. Stokes

1 Kelvin has written at the top of the letter, 'April 7, 47'.

2 Stokes has added a footnote here reading, 'rather if u be the temperature; for u being the temp^{re}, $\varpi(u)$ would not satisfy (3); but $\varpi(u) = \text{cons}^t$ would equally be the eq^n to an isothermal line'.

3 Stokes (1).

4 Stokes (21).

12 Kelvin to Stokes, 7 April 1847

Stokes Collection, K19

Partially printed in H. I. Sharlin, 'William Thomson's dynamical theory: an insight into a scientist's thinking', Annals of Science, **32** (1975), 137.

College, Glasgow,
April 7, 1847

My dear Stokes,

Many thanks for your letters, which have given me plenty of matter for contemplation, in subjects with which I have long been interested. There is a great deal which I would like to say about them, but I do not know where to begin, especially as I am about to start for Ireland in a few hours, to take advantage of half a week's holidays w^h we have. I am not exactly prepared to give you the reference for Dupin's demonstration (unless it is in his *Developements* (?) de Géometrie)[1] but you will find a demonstration by Ellis[2] of what I suppose you want in Gregory's Examples,[3] and you will also find one in the [Cambridge] Mathematical Journal, which I gave.[4]

I have been for a long time thinking on subjects such as those you write about, and helping myself to understand them by illustrations from the theories of heat, electricity, magnetism, and especially galvanism; sometimes also water. I can strongly recommend heat for clearing the head on *all* such considerations, but I suppose you prefer cold water.

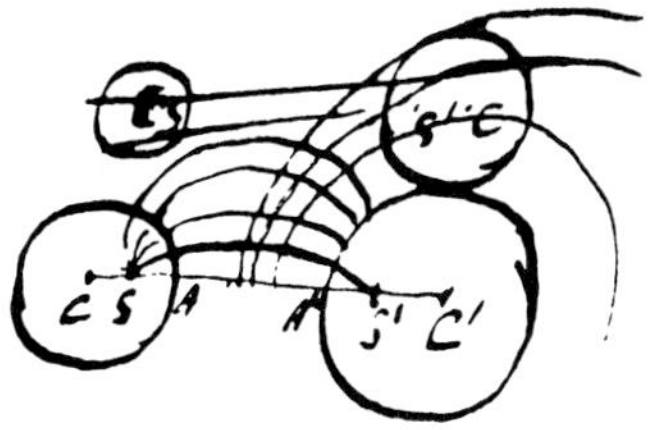

You will be quite delighted if you read a few pages or even a few words of a paper of Gauss's, (Taylor's Scientific Memoirs, vol. II. p. 230)[5] by which you will be supplied by a flood of illustrations for water. For instance, for motion in 2 dimensions, in an infinite space, round circular cores; in CC' take two points S, S' such that $CS{\cdot}CS' = CA^2$, and $C'S{\cdot}C'S' = C'A'^2$. The lines of ⟨motion⟩ displacement will be circular arcs w[h], when continued into the cores, pass through S and S', and the lines of motion are circles each of which is the locus of the vertex of a triangle described on SS', the ratio of the sides being given, the two series of circles being of course orthogonal. Any ring surface generated by lines of motion, may be constructed as a rigid envelope, and the fluid will circulate round it in the interior. There is no reason for saying that ϕ_2 is 'discontinuous', ⟨a⟩ but rather we should say it has an infinite number of values at each point. For if we commence at a point, and call ϕ_2 zero, we may go round the ring continually in the direction of motion and find a continually increasing value of ϕ_2. There is nothing absurd in the infinite series of diff[t] values we thus find for ϕ_2, as we pass again and again through the same point; ⟨but⟩ but if we obtained a corresponding expres[sio]n for p (in a problem of hydrostatics ⟨()⟩ such as the one I set at Peterhouse) or for v (in a problem of heat) then when, in going round the ring, we reach the surface of equilibrium or isoth[l] surf. from w[h] we started, we must place an impermeable barrier, to keep up the diff[ce] betw. the values of p, or betw. those of v.

The condition (w[h] you express by means of 'cores') that the motion may be entirely within, without crossing, a closed surface, ⟨either⟩ in hydrodyn[s], & that ⟨udx⟩ udx + [vdy + wdz] may be a complete diff[l], may be expressed by saying that there must be two distinct ways of passing from any point A to any point A' (& therefore of course an infinite number of ways) such as the dotted line, & the line AKA'; ⟨So⟩ [?] and that we cannot pass from one path, to another, by a series of curves, without drawing curves through the rigid surface. The equation

$$\frac{du}{dx}+\frac{dv}{dy}+\frac{dw}{dz}=0$$

may be satisfied however without ⟨be⟩ [?] any such restriction, as for instance in the case of a fluid revolving with ⟨constant⟩ given angular velocity, in the interior of a sphere.

What I was most anxious about was ϕ_1, as I wished to know whether it is possible to determine it through the interior of any closed surface, *S*, when the flux is given arbitrarily a[c]ross the surface (with the sole restriction that the total flux = 0). It is clearly possible to determine *u*, *v*, *w*, for water, in an infinite number of ways, to satisfy these conditions, as the flux may be arbitrary, across the surface but I wished to know whether it is possible to determine them so that udx + [$vdy + wdz$] may be a complete diff[l] (there would still of course be, in cases such as rings, the arbitrary part ϕ_2). This I could not at once answer from the theory of heat, as it is not (or was not, to me) clear that the flux may be given arbitrarily, for heat. I think I see my way now however, and if I am right, I shall have an immense number of new problems in heat, electricity, and magnetism, to think on, and I shall try to make a proof, suitable for cases of attraction.

In the [Cambridge] mathematical Journal (near the end of Vol. III, and the beginning of Vol. IV.) there are two papers ('note on Isoth[l] Surfaces', and 'On the equations of motion of heat referred to curvil[r] coord[s]')[6] which are similar to what you have written about motion in two dimensions, and motion symm[l] with ref[e] to an axis, in your 2nd and 3[d] letters.

You may easily construct a *tubular case* for ϕ_2, in 3 dimensions, by considering a closed curve of double curvature, *G*. The solid angle, *V*, this subtends at a point *P* is a potential. A line of ⟨motion (or⟩ force (or motion in heat &[c]) will be a closed curve passing round the 1st mentioned. The problem corresponding to this in heat is ⟨the⟩ constructed thus. Let *A* be any surface ⟨by⟩ bounded by *G* as edge, and let the opposite sides of *A*, in the interior of an infinite solid, be kept at the temperatures 2π ⟨1⟩ [?] and -2π . *V* will be the temperature at *P*.

I must reserve further communication till I see you, but still I shall be glad to hear about ϕ_1, although I can wait till I see you. I am very glad you are working about the Trigonometrical series.

Your's very truly,
WILLIAM THOMSON

1 P. C. F. Dupin, *Développements de géométrie, avec des applications à la stabilité des vaisseaux . . . pour faire suite à la Géométrie descriptive . . . de M. Monge* (Paris, 1813).

2 Robert Leslie Ellis (1817–59) was senior wrangler in 1840 and in 1843 became editor of the *Cambridge Mathematical Journal*, giving way to Kelvin in 1846 when the journal became the *Cambridge and Dublin Mathematical Journal.*

3 D. F. Gregory, *Examples of the Processes of the Differential and Integral Calculus* (Cambridge, 1841). Duncan Farquharson Gregory (1813–44) was fifth wrangler in 1837 – the year George Green was fourth – and became the first editor of the *Cambridge Mathematical Journal* the same year. He remained editor until ill health forced him to resign in 1843. Ellis replaced him and wrote a short memoir of him for the journal. (R. L. Ellis, 'Memoir of the Late D. F. Gregory, M. A., Fellow of Trinity College, Cambridge', *Camb. Math. J.*, **4** (1844), 145–52.)

4 Kelvin (9).

5 C. F. Gauss, 'General Theory of Terrestrial Magnetism', Taylor's *Scientific Memoirs*, **2** (1841), 184–251. Kelvin has written between the lines near this point, 'I am very anxious to find whether Gauss has found the opportunity he waited for, of writing more'.

6 Kelvin (6) and (8).

13 Stokes to Kelvin, 10 April 1847

Kelvin Collection, S327

Pembroke College[1]
Cambridge
April 10th 1847.

My dear Thomson,

I have just received your interesting letter, and read it over carefully. I think it best to answer it while I am fresh from it.

I had read your demonstration[2] of Dupin's theorem before I wrote, but it was not what I asked about. Proving that *when* 3 conjugate orthogonal systems of surfaces exist, the surfaces cut one another in their lines of curvature is not quite the same as proving the possibility of the existence of 3 such systems in all cases when one of the three is given arbitrarily. In other words conceive a system of surfaces taken arbitrarily: let P_oP be an orthogonal trajectory to the system, which we may always conceive drawn. Let P_oQ_o be one of the lines of curvature passing through P_o. Let a point P' move along P_oP from P_o to P, and let $P'Q'$ be drawn always a line of curvature of the surface through which P' passes. Is it necessarily true that whatever be the arbitrary nature of the given system of surfaces, the surface generated by $P'Q'$ will everywhere be ⊥ to the given system? It will of course be ultimately ⊥r about the line ⟨PQ⟩ P_oP.

I have not yet read the paper of Gauss's you refer to.[3] I have verified analytically the case of motion you mentioned. The value of ϕ_2 is the difference of two equal values belonging to motion of the kind considered by Euler, the centres of motion being in the points S, S'.

I think that both heat and cold water are very well adapted to give clear notions on these subjects. Sometimes I think one has the advantage and sometimes the other.

As to what you say of the discontinuity of ϕ_2, that depends on the way in which we regard it. If we regard ϕ_2 as merely the result of an integration it is more convenient to regard it as continuous, and having an infinite no. of values for the same point of space. If we regard it as a function of the position of the point of space considered, so to speak, we must look on it as discontinuous. I regarded it in the latter point of view from having been in the habit of so regarding ϕ_1, which in fluid motion has a physical meaning.

I had read your paper[4] (note on Isothermal surfaces) before I wrote. I merely gave the same thing for fluid motion to put it in a different light.

Your mode of determining[5] the circulating motion round the closed curve *G*, which forms the nucleus of an annular core, is extremely beautiful. What an intimate relation there is between the mathematical considerations which are applicable to heat, fluid motion, and attraction!

As ϕ_1 was an old friend of mine, whereas I had hardly considered ϕ_2 at all, I thought that ϕ_2 which was new to me was what you were engaged about.

As to ϕ_1, in the first place it is always determinate, save as to an arbitrary constant added to it, when the motion at the boundary of a closed space *S* is given. For conceive the space filled with incompressible fluid at rest, and let the surface of *S* be moved impulsively in any manner consistent with the condition that ⟨$udx + vdy + wdz$ shall be an exact diff^l^ $d\phi_1$⟩ the total flux shall be 0. Then $udx + vdy + wdz$ will be an exact diff^l^ $d\phi_1$,[6] and $-\zeta\phi_1$ + const. will be the impulsive pressure, which is evidently determinate, save as to an arbitrary constant added to it.

If you wish to see examples of the determination of ⟨*S*⟩ ϕ_1 when the motion at the surface of *S* is given you will find some in my paper On some cases of fluid motion.[7] You will find the solution in the following cases

(1) when *S* is the space comprised between two concentric cylinders, motion in two dimensions. (Article 8). Of course the case in which *S* is a cylinder without a core, or in which *S* is infinite space outside a cylinder is included.

(2) when *S* is the space included between two concentric spheres, including the cases as before (art. 9)

(3) when *S* is infinite space on one side of an infinite plane (art. 10)

(4) where *S* is a rectangular ▱ piped [i.e., parallelepiped] art. 15.

Almost exactly the same analysis will suffice when we transport the origin to a point within the ▱ piped and then make any 1, 2, 3, 4, or 5 of the 6 faces move off to infinity. It will only be necessary to replace some of the infinite series by definite integrals when some of the faces move off to infinity.

The methods however to be given in the paper I am writing[8] will I expect supersede the methods I have just referred to, by enabling us to find ϕ ⟨with⟩ using the same mode of expansion throughout, instead of (in the case of the ▱ piped) 3 different modes of expansion corresponding to the parts of ϕ due to the motion of the 3 pairs of opposite faces respectively.

The case answering to that cont[d] in art. 10 when the motion is in two, instead of three dimensions may be solved in a similar manner by using the particular integral $\phi = C \log r$ instead of $\phi = \frac{C}{r}$.

I do not mean to say that you will not find many of these things elsewhere, but you will have them in a compact form and done shortly in my paper.

I have also considered the motion when *S* is a cyl[r] closed by planes ⊥[r] to its axis, but I have not published the result.

Of course we may have an infinite variety of motions of endless complexity within a closed space *S* if we take the general case in which $udx + vdy + wdz$ is not an exact differential.

I hope to see you soon. Cookson[9] told me you were coming here towards the end of April.

Yours very truly
G. G. STOKES.

P. S. I have thought of the surfaces since and I think I see my way.

1 Kelvin has written at the top of the letter, 'Ap. 12, 1847'.
2 Kelvin (9).
3 See the previous letter.
4 Kelvin (6).
5 Stokes has added a footnote here reading, 'I ought rather to say finding the most general circulating motion about a given nucleus of an annular core, and thereby showing that the motion is not determinate when the nucleus is given. There can be little doubt I think after this that the motion is possible and not determinate when the core is of finite thickness'. See the next letter.
6 Stokes has written above the line here, 'This follows at once from the =[ns] of impulsive motion'.
7 Stokes (13).
8 Stokes (21).
9 Henry Wilkinson Cookson (1810–76) was tutor at Peterhouse while Kelvin was an undergraduate and was master from 1847 until his death. There are several letters in the Kelvin Collection from Cookson to Kelvin's father regarding Kelvin and several more from Cookson to Kelvin himself.

14 STOKES to KELVIN, 12 April 1847

Kelvin Collection, S328

Pembroke College,[1]
Cambridge
April 12, 1847

DEAR THOMSON,

You must have stared when you read my note marked*,[2] I mean the note so marked at the end of my letter of Saturday. After I had sealed ⟨your⟩ my letter just before post time I recollected that I had said you had *determined* the motion round the curve G, and remembering the arbitrary form of the surface of which G is an edge and confounding it with a surface of displacement, with which it has nothing whatsoever to do, I opened my letter and wrote the note*. I thought of my error after I went to bed. Had yesterday been any other day of the week I should have written to you then for fear you should take the trouble to write a long letter to convince me of my error.

I am not yet clear about the surfaces. We have to be sure 4 arbitrary principal radii of curvature when one of the systems is given, but I do not see that that is enough.

Yours very truly
G. G. STOKES

1 Kelvin has written at the top of the letter, 'Ap. 14, 47'.
2 See the previous letter, note 5.

15 KELVIN to STOKES, 14 April 1847
Stokes Collection, K20

College, Glasgow,
April 14, 1847

MY DEAR STOKES,

I was considerably alarmed on receiving your letter to find that I had proved the case of motion you mentioned to be indeterminate, but I was unconscious.

When I received your letter today, it occurred to me that I had most probably committed the same error (it is one w^h I have fallen into and out of again, a great many times) in one of my expressions in my last letter to you. When I spoke about a surface, bounded by an edge, G, and kept with its 2 ⟨surfaces⟩ sides at temperatures 0 and 4π, I do not remember having queried that this surface S is not arbitrary, but is *the* surface, having G for edge, which possesses the property that at each point of it the edge subtends an angle of 2π.[1] It would be an interesting geometrical problem to determine the surface. I have often wondered whether it has any properties; for instance whether its area is a minimum, or what is a minimum with reference to it? Whatever properties it may have, however, each surface of displacement of which this of course is one, will have some corresponding property.

I am afraid I have been misunderstanding about what you mean by ϕ_1. I shall have to get this point cleared when we meet. It did not seem, to me, that the problems in your paper,[2] would have any *direct* (although certainly they are connected in a very interesting way) ⟨with⟩ bearing upon my affair about ϕ_1. I shall have to keep all I have to say, however, till we meet.

I have this morning been showing my class the magnetic curves (i.e. the lines of motion) round a circular galvanic wire.

Your's sincerely
WILLIAM THOMSON

1 Larmor has added a footnote here reading, 'Of course it is now clear that any barrier would make the difference on the two sides equal to 4π; the surface referred to here is that one of the equipotentials for which the value of the solid angle is 2π'.
2 Stokes (13).

16 STOKES to KELVIN, 16 April 1847
Kelvin Collection, S329

Pembroke College[1]
Cambridge
April 16, 1847.

MY DEAR THOMSON,

Your last letter has rather put me to sea again as to your theorem. I *did* suppose you said an arbitrary surface; and on referring to your letter I see you say 'any surface ⟨passing through⟩ bounded by *G*'. The theorem was new to me, but I considered it in this way.

Let m be a mass at the point P, if we are considering attraction, or the coefficient of $\frac{1}{r}$ in the value of ϕ if we are considering fluid motion, the fluid being incompressible, and the motion to or from a centre. Suppose that we take the case of fluid motion. Let $-\frac{m}{r'}$ be the value of ϕ corresponding to motion from or ⟨two⟩ to another centre of motion situated in the p^t P', where $PQ = r$, $P'Q = r'$, Q being any point in the fluid. Then $\frac{m}{r} - \frac{m}{r'}$ will be a potential. Let now m increase beyond all limit while at the same time PP' decreases beyond all limit. Then we pass to the value of ϕ, $\frac{c}{r^2}\cos\theta$, which is still a potential, where θ is increased from PP'. Let the new centre of motion corresponding to $\phi = \frac{c}{r^2}\cos\theta$ be called a centre of the 2nd order.

Conceive an infinite no. of centres of the 2nd order, corresponding to equal values of ϕ, to be equally distributed over any surface, the poles of the centres being all normal to the surface, and turned to the same face of it. The compound ϕ will still be a potential. The part of ϕ due to any element of the surface will $\propto$ area of element $\times$ sine of $\angle$ of emanation $\times$ inverse sq. of distance.

Hence just as in the case of light ϕ will vary as the solid $\angle$ subtended by the curve G. Of course the part of ϕ emanating from the back side of the surface will be negative.

I do not see what the particular surface you mention has to do with the matter more than any other surface of displacement. Of course the arbitrary surface I mentioned may, but need not be a surface of displacement.

To show the thing more clearly let S, S' be sections of the curve G, SaS' the surface you mention. The value of ϕ will not be altered by adding two surfaces of the same form SbS′ one

and one

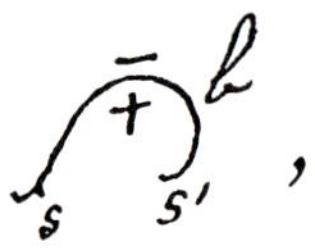

the intensity for each being the same as before. Now the two

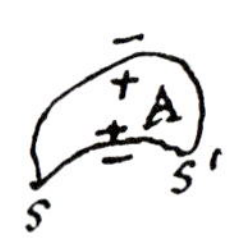

will give a potential $4\pi C$ for any point A within the surfaces, and o for any p^{t} outside. Hence the difference of the two cases

is that ϕ is altered by $4\pi C'$ within the space A, which will make no difference in the motion of the fluid, the only difference being that we must suppose ϕ to alter discontinuously by $4\pi C'$ in passing across the surface SbS' instead of in passing across SaS'. This is of no importance, since the discontinuity of ϕ is merely artificial, introduced to enable us to regard ϕ as a unique function of the coord^{s} of the point considered. It appears then that the surface through G *is* arbitrary, but the motion not.

The value of ϕ external to any closed surface in the case considered is zero, for Q sees the element a on the negative side and b on the positive, and these two elements neutralize each other for the point Q.

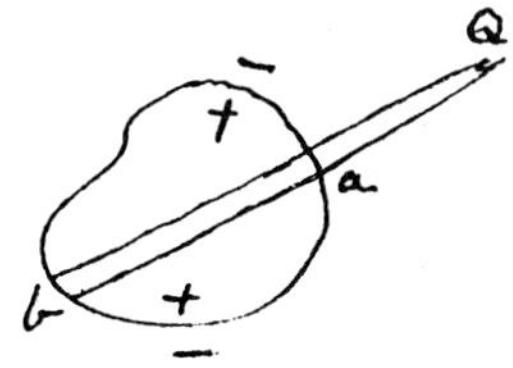

I was thinking of entering into these details in my last, but thought it not worth while, fully believing that you meant the surface to be arbitrary. I was thinking also of illustrating the subject by the illumination produced by a transpartent [*sic*] self-luminous surface which is bright at one side and dark at the other, but thought it not a very happy illustration, inasmuch as it requires us to consider darkness as opposed to light in the same way as +ve and −ve electricities are opposed. We might get rid of this by supposing all space already partially illuminated; but still the illustration is not a very happy one.

I thought of the function resulting from the coalescing of n points such as P as an illustration of a function of r, θ and ϕ which satisfies the eq^{n} $\frac{d^2f}{dr^2}+\frac{r\,df}{\gamma dr}+\&c=0$, and as regards θ and ϕ is composed of Laplace's coeff^{ts} of the 0th, $1^{st}\ldots n^{th}$ orders, but I have not gone on with this, being engaged with other things.

I hope soon to see you when we can talk over these matters freely. I hope you will not give yourself the trouble to write again, as we are so soon to meet.

Fischer has not yet got his German testimonials.[2]

I have written in haste, hoping to be in time for the post. I hope however you will have no difficulty in reading this.

Yours very truly
G. G. STOKES

1 Kelvin has written at the top of the letter, 'Ap. 19, 1847'.
2 This no doubt refers to Fischer's application for the St Andrews professorship, first mentioned in letter 6.

17 KELVIN to STOKES, 20 October 1847

Stokes Collection, K21
Partially printed in H. I. Sharlin, 'William Thomson's dynamical theory: an insight into a scientist's thinking', Annals of Science, **32** (1975), 136.

Fortbreda, Belfast,
Oct. ⟨19⟩ 20, 1847

MY DEAR STOKES,

As I did not find you at Cambridge during my last visit, there is a slight accumulation of hydrodynamical considerations that I wanted to speak to you about, for which the post affords the only outlet. But before I begin, I may tell you of something that I saw in the last No. of the Edin. Phil. Journal, since I arrived in Scotland (where I am to return in a few days). Among a few extracts detailing interesting physical observations on the ocean, from Sir James Ross's Voyages to the Southern Seas, the following observed details with reference to waves are given.

height from (lowest to highest) = 22 f[eet]
veloc. = 89 miles per hour,
length (from wave to wave) = 1910 f[eet]
Ross Vol. I. p. 31.[1]

I perceived a fine instance of elasticity in an incompressible liquid, in a very simple observation made at Paris, on a cup of thick 'chocolat au lait'. When I made the liquid revolve in the cup, by stirring it, and then took out the spoon, ⟨the velocity of rotation diminished very rapidly⟩ the twisting motion (in eddies, and in the general variation of angular vel. on acc[t] of the action of the spoon overcoming the inertia of the liquid, and the fric[n] at the sides) in becoming effaced, always gave rise to several oscillations so that before the liquid began to move as a rigid body, it performed oscillations like an elastic (incompressible) solid.

I have written a few pages on the equation of continuity for the forthcoming N° of the Journal, and I have actually been bold enough to entitle it '⟨M⟩ [?] Notes on Hydrodynamics, N° 1.'[2] Hence I must get some more notes, perhaps from you. I have been thinking on getting something added about $udx + vdy + wdz$, and also perhaps giving an investigation of the equation for the bounding surface, and removing the doubts which Poisson has thrown upon this part of the subject, as seems to me unreasonably.[3] I sh[d] be glad however to hear your opinion. The following occurred to me when I was trying to get an analytical proof of the possibility of distributing matter over a surface, S, so that the force, along the normal, on ⟨a⟩ p[ts] infinitely near the surface may have an arbitrary expression. (If the points considered be within the surface, there will be the restriction that the mean value of the given arb[y] fu[n] for the whole surface may be = 0).

Let u, v, w be the three fu[ns] satisfying the eq[n]

$$\frac{du}{dx}+\frac{dv}{dy}+\frac{dw}{dz}=0$$

for all p[ts] within S, and the condition

$$lu + mv + nw = H$$

at the surface, (l, m, n) being the normal, and H being arb[y] with the sole rest[n] $\int\int H ds = 0$. There exists a solution which renders

$$\int\int\int(u^2 + v^2 + w^2)\, dx\, dy\, dz$$

less than any other. For u, v, w may obviously be taken so as to give this as great a value as we please, but they cannot be taken so as to make it vanish. Hence

$$\delta\left\{\int\int\int(u^2+v^2+w^2)dx\, dy\, dz\ \langle + \rangle\right.$$

$$\left.-\int\int\int 2\lambda\left(\frac{du}{dx}+\frac{dv}{dy}+\frac{dw}{dz}\right)dx\, dy\, dz\right\}=0$$

$$\therefore\ u-\frac{d\lambda}{dx}=0,\ v-\frac{d\lambda}{dy}=0,\ w-\frac{d\lambda}{dz}=0.$$

Hence there is a sol^n for w^h $udx + [vdy + wdz]$ is a comp. diff^l. This sol^n is unique: for it may be easily shown that, if u, v, w be a solution making $udx + [vdy + wdz]$ a compl. dif^l and if u_1, v_1, w_1 be any different solution then

$$\int\int\int(u_1^2\langle dx\rangle+v_1^2+w_1^2)\, dx\, dy\, dz \text{ is} > \langle\text{than}\rangle$$
$$\int\int\int(u^2+[v^2+w^2])\, dx\, dy\, dz.$$

Hence ⟨$u_1dx + [v_1dy + w_1dz]$ cannot be⟩ no different solution from $(u\ v\ w)$ can ⟨ref⟩ [?] make the exp^n a complete diff^l. The most important appl^n of this is to find the dist^n of magnetism through a given mass (bounded by S) so as to make the exter^r potential at the surface have a given exp^n; or rather, this distribution not being unique, to find how it may be done, under various conditions. The cond^n to be satisfied is

$$-\int\int\int\frac{1}{\Delta}\left(\frac{du}{dx}+\frac{dv}{dy}+\frac{dw}{dz}\right)dx\, dy\, dz$$

$$+\int\int\frac{lu+mv+nw}{\Delta}dS=V$$

where $$\Delta^2=(x-\xi)^2+(y-\eta)^2+[(z-\zeta)^2]$$

and V is the arb^y pot^l, at a p^t $(\xi\,\eta\,\zeta)$ of the surface. The most general solution, as may be demonstrated, is found by taking $\frac{du}{dx}+\left[\frac{dv}{dy}+\frac{dw}{dz}\right]$, arb^y throughout S. Then $lu + mv + nw$ is det^d by the above $=^n$. We have then to find $u\ v\ w$ from the cond^{ns} $\frac{du}{dx}+\left[\frac{dv}{dy}+\frac{dw}{dz}\right]$ given through S and $lu + mv + nw$, at the surf. When we have found a solution, to get the most general, for the entire problem, we must add the arb^y solut^n of

$$\frac{du}{dx}+\frac{dv}{dy}+\frac{dw}{dz}=0$$

$$\text{and } lu+mv+nw=0.$$

You will easily see that this analysis, when applied to hydrodynamics, will give a proof, by means of the principle of least action, that $udx + [vdy + wdz]$ is a complete diff^{l} in the case of a mass of fluid, bounded by a surface, (changing of course in form) S, and set in motion by unequal pressure at the surface, thus if t be the time, from rest, we must have

$$\int_0^T dt \int\int\int (u^2+v^2+w^2)\, dx\, dy\, dz$$

a minimum, and the result is easily found, as above. I sh^{d} be glad to hear from you whether this is one of the proofs w^{h} have been given.

If it be required to find the function ϕ such that

$$d\phi = udx + vdy + wdz,$$

$$\frac{du}{dx}+\frac{dv}{dy}+\frac{dw}{dz}$$

$$\text{and } lu+mv+nw=H,$$

the problem is the same as that of finding the impulsive pressure through S, when, by an impulsive pressure at the surface, an initial motion is communicated to a liquid, originally at rest within S so that the normal initial velocity at the surface may be arbitrary. I remember seeing something about 'impulsive pressure' in one of your papers[4] (when I knew nothing of the subject) w^{h} I must look to when I return to Glasgow.

I have had a very pleasant tour in Switzerland, after I saw you last at Cambridge, and next May, when we meet we shall hav⟨ing⟩e something to say about the Mer de Glace and the Faulhorn where I remained (not snowed up) for 2 days, as well as about hydrodynamics. I ⟨gave⟩ left a copy of your paper on solids and fluids,[5] for Liouville. I did not see him, as he has been away from Paris all summer: he returns about the beginning of November.

Your's very sincerely
WILLIAM THOMSON

1 'Scientific Intelligence', *The Edinburgh New Philosophical Journal*, **43** (1847), 403. James Clark Ross, *A Voyage of Discovery and Research in the Southern and Antarctic Regions during the Years 1839–1843*, 2 vols. (London, 1847).
2 Kelvin (29.1).
3 See Kelvin (29.2) which discusses the subject and comments on Poisson's work.
4 See article 3 of Stokes (11).
5 Stokes (11).

18 STOKES to KELVIN, [late October 1847][1]
Kelvin Collection, S330

Pembroke Coll.[2]
Cambridge

DEAR THOMSON,

In all probability you will have heard before this of the death of your Master who died, after a few days illness, of erysipelas in the head.[3] I write so soon in case by any possibility you should not have heard of it, as if you are coming here it might save you a journey to Glasgow to hear of it. But you *must* have heard by this time.

As I only received your letter at Hall time today I have not yet had time to digest it thoroughly.

I find theory gives only 98.0 feet per second for the vel[y] of the waves whereas observ[n] gives 130.5. I have supposed λ the length measured from crest to crest.

The proof you give that $udx + vdy + wdz$ is an exact diff[l] when the motion begins from rest is perfectly new to me, and appears quite satisfactory. Of course you will publish it. To prevent misconception it would be well to state *explicitly* that it only holds on the hypothesis of the absence of tangential action, in w[h] case alone the principle of least action can be applied. You have not it seems considered the case of elastic fluids. I have very little doubt that your proof is quite new.

I have several investigations on hand which are just completed. One relates to the spectrum-bands I showed you and to an analogous exper[t] of Prof. Powell's w[h] he wrote to me about; also to certain bands he describes seen with two prisms producing a partial achromatism.[4] Another relates to Newton's rings. When the $\angle$ of inc[ce] is made to exceed the critical $\angle$ the rings disappear, but the *central spot remains.* I find this to be a necessary consequence of Fresnel's expressions for the intensities, which now become imaginary. My theory makes the permanence of the spot due to the disturbance in the 2nd medium, w[h] is only sensible at distances from the surface comparable with λ. It leaves the question open as to whether the vibrations of plane pol[d] light are in or $\perp$ to the plane of pol[n].[5] I have also succeeded in solving the problem of the ball-pendulum not supposing the pressure equal in all directions, but using the eq[ns] of my paper[6] on the friction of fluids.[7] As I have a chance of seeing you soon I may here conclude. I heard you were likely to come up.

Yours very truly
G. G. STOKES

1 The date of the letter is determined by information in the next two notes plus the fact that it is a response to the previous letter.

2 Kelvin has written at the top of the letter, 'Sat. Oct. 30, 1847 on returning from Camb. forwarded from Fortbreda'.
3 William Hodgson (d. 1847) was master of Peterhouse from 1838 until his death on 16 October 1847.
4 See Stokes (23) and (35) and Baden Powell, 'On a new case of the interference of light', *Phil. Trans.* (1848), 214–26. Powell (1796–1860) graduated with a first class degree in mathematics from Oxford in 1817 and became Savilian professor of geometry at Oxford in 1827. Stokes and Powell corresponded extensively during their collaboration, and Stokes preserved both sides of the correspondence in a letter book. (Stokes Collection, LB2.)
5 See Stokes (26) and (30).
6 Stokes (11).
7 See Stokes (47), which was read to the Cambridge Philosophical Society on 9 December 1850, especially *MPP*, III, 7–8, where he discussed the progress of his work during the 1840s on pendulums and the friction of fluids.

19 Stokes to Kelvin, 25 November 1847.
Kelvin Collection, S331[1]

Pembroke College[2]
Cambridge
Nov 25th 1847.

Dear Thomson,

In considering diffraction last night I thought of something which may be useful to you.

In Electricity you have sometimes to calculate the num[l] value of a series such as

$$u_0 - u_1 + u_2 - u_3 + \dots$$

which is very slowly convergent.

Suppose a few terms, up to u_{x-1}, to have been calculated, req[d] to complete the series by adding or subtracting (as the case may be), R where

$$R = u_x - u_{x+1} + u_{x+2} - \dots$$

Let R be expressed in a series proceeding according to differences of u_x. This may be effected as follows. Let D denote the operation of passing from u_{x+c} to u_{x+1+c} then

$$R = u_x - Du_x = D^2u_x \dots$$

$$= \frac{1}{1+D}u_x$$

$$= \frac{1}{2+\Delta}u_x = \frac{1}{2}u_x - \frac{1}{2^2}\Delta u_x + \frac{1}{2^3}\Delta^2 u_x \dots \quad (1)$$

I do not consider this a rigorous proof because I have not considered the convergency of the series (1) & indeed that is not the only point to consider, for the series $x-\frac{x^3}{\lfloor 3}+\ldots$ though the development of $\epsilon^{-\frac{1}{x^2}}+\sin x$, and convergent, has for some $\sin x$, & the same, or rather something analogous, might be conceived to be true of (1) even if it should be convergent. I believe that in many cases (1) will be at first rapidly convergent and ultimately divergent. Yet in such cases it generally happens that the sum of the conv^{t} terms alone give^3 the approx^{te} value of the quantity sought.

I have tried the formula on the series $\frac{1}{1}-\frac{1}{2}+\frac{1}{3}-\frac{1}{4}\ldots$ I send you the work just as I wrote it that you may see how easy it is.[4] I took 4 terms directly[5] and calculated R from the formula. I ⟨kept⟩ worked with 5 dec. places and got ·69313. The tables give Nap[ier] log. 2 = ·6931472.

You sometimes have to calculate a series

$$u_0-u_1+u_2-u_{4[n?]}\ldots$$

which is ultimately a geom^{l} series whose r[ati]o r can be found directly. If r is small there is no difficulty, but if r is nearly equal to 1 there may be some trouble. I believe in such cases the following formula will be better than (1).

Having calculated $u_0-u_1\ldots\pm u_{x-1}$ directly divide

$u_x\ ,\ u_{x+1}\ ,\ u_{x+2}$ &c.
by $1\ ,\quad r\ ,\quad r^2\ ..$
and let the quotients be
$v_x\quad v_{x+1}\quad v_{x+2}\ \ldots$
so that

$$R=v_x-rv_{x+1}+r^2v_{x+2}\ldots$$

$$=(1-rD+r^2D^2\ldots)u_x[sic]=\frac{1}{1+rD}v_x=\frac{1}{1+r+r\Delta}\langle u_x\rangle \mathrm{v}_x{}^{6}$$

$$=\frac{1}{1+r}\left\{v_x-\frac{r}{1+r}\Delta v_x+\left(\frac{r}{1+r}\right)^2\Delta^2v_x+\ldots\right\}$$

Yours very truly
G.G. STOKES

P.S. We have exactly

$$R=(1-D+D^2\ldots)u_x=\frac{1}{2+\Delta}u_x$$

$$=\left(\frac{1}{2}-\frac{\Delta}{2^2}\ldots\pm\frac{\Delta^{n-1}}{2^n}\right)u_x\mp\frac{\Delta^n}{2^n(2+\Delta)}u_x$$

that is to say the operation $2 + \Delta$ or $1 + D$ performed on both sides will lead to an exact result. If then $E_{x,n}$ be the error committed in stopping after x terms of the original series and n terms of the difference-series we have exactly

$$(1+D)E_{x,n} = \mp\frac{\Delta^n u_x}{2^n}$$

$$\text{or } E_{x,n} + E_{x+1,n} = \mp\frac{\Delta^n u_x}{2^n}$$

In general, or at least in many cases,[7] the approximation will be so much the nearer, for a given n° of terms in the difference-series, as the n° of terms taken in the original series is the greater. In all such cases

$$E_{x+1,y} < E_{x,\langle y\rangle n} \quad \text{(write } n \text{ for } y)^8$$

$$\therefore 2E_{x,y} > E_{x,y} + E_{x+1,y} > 2E_{x+1,y}$$

$$\therefore E_{x,y} > \mp\frac{\Delta^n u_x}{2^{n+1}} \text{ (or <, acc}^{\text{d}} \text{ to signs)}$$

$$E_{x+1,y} < \mp\frac{1}{2^{n+1}}\Delta^n u_x \text{ (or >)}$$

$$\text{or } E_{x,y} < \mp\frac{1}{2^{n+1}}\Delta^n u_{x-1} \text{ (or >)}$$

or $E_{x,y}$ is equal to the product of $\mp\frac{1}{2^{n+1}}$ into a quantity lying between $\Delta^n u_{x-1}$ and $\Delta^n u_x$.

If $E_{x,y}$ should be negative it will only be necessary to change >< into <> and the result will be the same. Hence if we are only sure that $E_{x,n}$ ⟨in⟩ decreases ⟨with x⟩ as x increases (at least for the values $x - 1$, x, $x + 1$, though not necessarily at[9] first,) the differences calculated will enable us to assign a limit to our error.

Since $\frac{E_{x+1,y}}{E_{x,y}} < 1$, $E_{x,y}$ lies between $\mp\frac{\Delta^n u_x}{2^n}$ and $\mp\frac{\Delta^n u_x}{2^{n+1}}$ This may be useful when $\Delta^n u_x$ has been calculated ⟨and⟩ but not $\Delta^n u_{x-1}$, for it may enable us to show, by the value of $\Delta^n u_x$ alone, that the error is insignificant.

Alas for the Hydrodynamics for the [Cambridge and Dublin] math^l journal! What with College audit, Senate-house exam^n coming on – paper on series coming on on Monday week[10] – correspondence with Prof^r Powell pupils &c. I have never set [*sic*] down to write it. Perhaps you may have time to write n° 2; but I suppose you are more occupied than I am.[11]

From some trials which I have since made I fear the method will not do for

Electricity except in certain cases, or at least if it will do that it will not be near so advantageous as in the example I first worked with. Its success depends on u_x being what would be got by interpolating between ... $u_{x-3}, u_{x-1}, u_{x+1}, u_{x+3}$ ... whereas in Electricity you generally have $u_0 u_2$ ... belonging to one system and u_1 u_3 ... to another. Such as it is I send it. I think there are cases it[12] which it may be useful. Thus it may be useful for an electrical point midway between two || planes, and for certain calculations referring to equal spheres equally charged. In the cases in which it will do I believe that though $E_{x,\,n}$, $E_{x+2,n}$ ... decrease, $E_{x,n}$, $E_{x+2,n}$ ... belong to one system and $E_{x+1,n}$, $E_{x+3,n}$ to another, and $E_{x,n}$, $E_{x+1,n}$ may be even nearly equal in magnitude and opposite in sign.

1 S332 in the Kelvin Collection is Stokes's copy of S331. Differences between the copy and the original are pointed out in notes 3, 4, 6, 7, 9, 11, and 12.

2 Kelvin has written at the top of the letter, 'Nov. 27, 1847 (read in train to Edin.)'. Stokes has written at the top of the letter, 'This is in a great measure a mare's nest'.

3 In S332 Stokes has written *gives*.

4 Stokes has enclosed a slip of paper with this work, which in S332 he has included in the text of the letter. The enclosure contains the following figures:

1 ...	1·00000					
2 ...	·50000					
3 ...	·33333					
4 ...	·25000					
5 ...	·20000					
		...3333				
6 ...	·16667		...952			
		...2381		...357		
7 ...	·14286		...595		...159	
		...1786		...198		...80
8 ...	·12500		...397		... 79	
		...1389		...119		
9 ...	·11111		...278			
		...1111				
10 ...	·10000					

·10000	calc^n of *R*.
833	
119	
22	
5	
1	
·10980	+^ve terms of series
1·00000	
·33333	
1·44313	neg d[itt]o
·75	
·69313	

5 Stokes has added a footnote here reading, 'I stopped thus early in order to put the formula to a severer test'.

6 Stokes has written v_x over u_x. In S332 Stokes has made the same correction four times – in the three instances in this line and in the first instance in the next line.

7 Stokes has written 'or at least in many cases' above the line. In S332 he has omitted these words.
8 Stokes apparently substituted n for y and wrote (*write n for y*) after the letter had been completed. Hence, n should be substituted for y in the rest of the letter.
9 In S332 Stokes has written *from the* instead of *at*.
10 Stokes (21), which he read to the Cambridge Philosophical Society on 6 December 1847.
11 Kelvin wrote the second note: Kelvin (29.2), which is dated 10 January 1848. In S332 Stokes has omitted this paragraph and written instead, 'Excuse for not writing no. II Hydrodynamics'.
12 In S332 Stokes has written *in* instead of *it*.

20 STOKES to KELVIN, 28 November 1847
Kelvin Collection, S333

Pembroke College[1]
Cambridge
Nov 28, 1847

DEAR THOMSON,

Did you not say that the potential due to an electrical point between two || planes could be expressed in finite terms for a point (P) in the axis? I have been trying to get it out and cannot. It seems to me that in order that V should be expressible in finite terms it would be necessary that the distance from P should change sign according to the direction in which it is measured, whereas the potential requires that the numerical value of the distance should be taken. If I am right the expression I gave you for V involving cot $\frac{z-z'}{2}$ is wrong.

I ⟨f⟩ have been trying since to get a formula for finding R for a series of the form

$$f(a)-f(a+b)+f(a+b+c)-f(a+2b+c)+f(a+2b+2c)-\ \ldots$$

but the one which I first got after I wrote to you does not satisfy me, and it did not succeed well on an example.

I should be glad of a line to say if I am not right as to the potential.

Yours very truly
G. G. STOKES

1 Kelvin has written at the top of the letter, 'Nov. 29, 1847'.

21 KELVIN to STOKES, 29 November 1847
Stokes Collection, K22

Glasgow College,
Nov. 29, 1847

MY DEAR STOKES

I am much obliged to you for your letters about calculations which may occur in electrical problems. I am afraid I have misled you about the expressibility of the potential on a point in the axis, produced by two parallel plains [*sic*] under the influence of an elec^{l} p^{t}. It is expressible by means of a definite integral, thus.

We have (for the pot^{l} prod^{d} by the electrif^{d} planes)

$$V = \frac{1}{2a-x} - \frac{1}{2c+x} + \frac{1}{2c+2a-x} - \&c$$

$$+ \frac{1}{2b+x} - \frac{1}{2c-x} + \frac{1}{2c+2b+x} - \&c$$

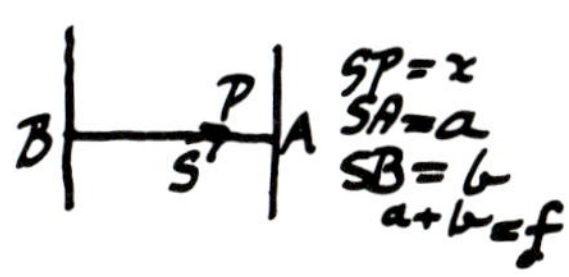

Hence $$\because \int_0^1 t^{m+1}dt = \frac{1}{m},$$

$$2cV = \int_0^1 dt \left\{ t^{\frac{2a-x}{2c}+1} - t^{\frac{2c+x}{2c}+1} + t^{\frac{2a-x}{2c}+2} - \&c \right.$$

$$\left. + t^{\frac{2b+x}{2c}+1} - t^{\frac{2c-x}{2c}+1} + t^{\frac{2b+x}{2c}+2} - \&c \right\}$$

$$= \int_0^1 dt \left\{ \frac{t^{\frac{2a-x}{2c}+1}(\ +\ ) - t^{\frac{2c+x}{2c}+1}}{1-t} \right\}$$

$$+ \int_0^1 dt \left\{ \frac{t^{\frac{2b+x}{2c}+1} - t^{\frac{2c-x}{2c}+1}}{1-t} \right\}$$

If I ever said that the result could be expressed in finite terms, this was all I meant. When however *a*, *b*, & *x* are commensurable (or generally when the fractional indices of *t* are expressible with comme^{ble} numerators & denominators, w^{h} however is probably only possible when *a*, *b* & *x* are commens^{ble} then *V* is actually expressible in finite terms. The only connection between the +ve & −ve terms, in the two parts of the express^{n} for *V*, which enables us to express them by a def. int^{l} is that by considering them together, we get a def. int. of w^{h} the value is finite; while if we had considered the +ve terms & the −ve terms separately ⟨from the⟩ [?] we sh^{d} have had 2 infinite def. int^{ls}, of w^{h} the difference is finite.

I shall try to write a second 'note on hydrodynamics' taking for subject the equation of the surface.[1] I have however at present rather many irons in the fire.

Your's very sincerely
WILLIAM THOMSON

I hope you will have something for No 3 of the same series, ready for the next Number after Feb.[2] Also I wish you would put in something about some of your papers. There is a most absurd paper (fortunately very short) of Bell's 'On the Modulus of Elast^{y} of a rod as det^{d} by its musical note'[3] w^{h} I have committed myself, to print. I must put in some sort of note as a *protest*, along with it. It is at present in the Macmillan's hands. They were to send it to me (as I wished to see it again before publishing it) but I fear they have neglected my commission. Will you get it from them if they have not sent it to me, & after glancing at it, post it for me, yourself?

1 See Kelvin (29.2).
2 See Stokes (16).
3 Andrew Bell, 'On the determination of the modulus of elasticity of a rod of any material, by means of its musical note', *Camb. Dub. Math. J.*, 3 (1848), 63–7. As editor, Kelvin added a note on p. 67. In the Kelvin Collection, there are six letters from Bell, who lived in Edinburgh, to Kelvin, five of them dating between November 1846 and October 1847.

22 KELVIN to STOKES, 1 February 1848
Stokes Collection, K23

Glasgow College,
Feb. 1, 1848

MY DEAR STOKES,

I congratulate you on the completion of your arduous work in the Examination. I am sure you must have had enormous labour, & must be well satisfied now, to have it over. You will I hope find time to write me a line telling how you are satisfied with the working of the new system, & (as far as you think proper) letting me know how the marking was arranged between the different parts of the examination. If the new system is to be permanent it will be of enormous importance to have the system in this respect carried out in the best possible way.[1]

I liked the papers very much, as far as I have been able to judge. I liked your problem paper especially; & I was very glad to see a little of my pet subject, in the last question as well as the hydrodynamics of N°s 11 & 12. N° 20 seems to be an exceedingly nice, & *do-able* problem.[2] I have this moment received a letter from Fischer,[3] who expresses a similar opinion regarding your problem-paper, & remarking its *do-ableness*.

I hope you will write No III. of the 'Notes on Hydrodynamics'.[4] You might continue the plan we sketched out when I saw you last; commencing perhaps with a preface to the effect that the *cinematical* equations, which depend on the characteristics peculiar to fluid bodies have been investigated in the preceding articles, & that you now proceed to apply the ordinary ⟨mechanical⟩ dynamical principles, connecting ⟨the⟩ [?] actual motions, with the forces by w^{h} they are produced or sustained.

I hope now that you have more leisure, you will write soon.

Your's very truly
(in great haste)
WILLIAM THOMSON

P.S. In the July & June N^{os} of Liouville, & in the previous letter, referred to in one of the articles published there, you will have ample reference for the 'electrical images,' if you are writing anything on *your* ⟨equation⟩ [?] images.[5]

1 This paragraph refers to the Cambridge Senate House examination, which in 1848 finished on 21 January. 'The new system' refers to changes introduced in 1848. Since 1839, the examination had consisted of twelve morning and afternoon sessions on six successive days, excluding Sunday. In 1848, the examination was divided into two parts – a three-day period from 6 to 8 January and one of five days from 17 to 21 January. All candidates for an honours bachelor of arts degree, as distinct from an ordinary bachelor of arts degree, took the three-day examination which covered elementary mathematics – that is, the calculus was excluded. Those performing well enough to qualify for honours then took the five-day examination of higher mathematical subjects. On the basis of all eight days' examinations, the honours students were grouped into three classes – wranglers, senior optimes, and junior optimes – and listed by order of merit within each class. This mathematical examination for honours candidates was called the 'mathematical tripos', a term officially, but not actually, synonomous with 'Senate House examination'. From the 1820s, when the term mathematical tripos was introduced, until 1858, the examination for an ordinary degree was also part of the Senate House examination. Although the undisguised pretence was that both sets of students took the same examination, 'the Senate House Examination had in fact divided into two entirely separate examinations – the Mathematical Tripos and the examination for the ordinary degree'. (D. A. Winstanley, *Early Victorian Cambridge* (Cambridge, 1955), 158. See also W. W. Rouse Ball, *Cambridge Papers* (London, 1918), esp., 294–300.) In any case, Kelvin and Stokes appear to have been following common practice in using 'Senate House examination' to mean 'mathematical tripos'.

2 The Senate House examination for 1848 is printed in *The Cambridge University Calendar for the Year 1848* (Cambridge 1848). Although the *Calendar* did not indicate which examiner was responsible for which questions, it is clear that Stokes's problem paper referred to here

was set on the morning of January 19th and that the specific questions mentioned were as follows:

11. A closed vessel is filled with water containing in it a piece of cork which is free to move; if the vessel be suddenly moved forwards by a blow, shew that the cork will shoot forwards relatively to the water.
12. A closed vessel is filled with water which is at rest, and the vessel is then moved in any manner; apply the principle of the conservation of areas to prove that if the vessel have any motion of rotation no finite portion of the water can remain at rest relatively to the vessel.
20. A plane moves so as always to enclose between itself and a given surface S a constant volume; prove that the envelope of the system of such planes is the same as the locus of the centres of gravity of the portions of the planes comprised within S.
22. [The final question.] Prove geometrically, or otherwise, that if g be the attraction which a particle m exerts on a point in a closed surface S, θ the angle between the direction of g and the normal, $d\omega$ an element of S,

$$\iint g\cos\theta d\omega = 4\pi m, \text{ or } = 0,$$

according as m is within or without S, the attraction of m at the distance r being $\frac{m}{r^2}$.

Extend this result to the case of a finite mass cut by S, and thence prove, by taking for S an elementary parallelepiped, that if V be the potential of any mass for an internal particle

$$\frac{d^2V}{dx^2}+\frac{d^2V}{dy^2}+\frac{d^2V}{dz^2}=-4\pi\varrho.$$

These questions are also printed in Stokes's *MPP*, v, 307–9.

3 This letter has not survived in the Kelvin Collection.

4 See Stokes (16).

5 Kelvin (30), and J. Liouville, 'Note au sujet de l'article précédent', in Liouville's *Journal de mathématiques pures et appliquées,* **12** (1847), 265–90. The 'previous letter' is Kelvin (17). Kelvin (17) and (30) and Liouville's article are published together in Kelvin's *PEM*, 144–6, 146–54, and 154–77, respectively.

23 STOKES to KELVIN, 3 February 1848
Kelvin Collection, S334

Pembroke College,[1]
Cambridge
Feb 3^d 184⟨7⟩8

MY DEAR THOMSON,

Here I am free from my labours, and glad I am of it. We had constant examiners' meetings during December, and when the exam^{n} came it was serious work. My own share of papers to look over weighed 6 1/4 stone, and I should think the examiners must each have had about 10 ^{lbs} more. I think the system works well. I had hoped to have found the low men understand their subjects better than I found they did, now that they are not induced to read high. However one must not expect much from them. I think the new system has told with advantage on the higher part of the examination. We are now unfettered in the last 5 days, and can pretty well avoid setting the questions so as to be answered by crammed men.[2]

As to the result I think we got through the shock very comfortably. There was none of the confusion of places that arose the year after the classes were abolished.[3] Men came out on the whole very much as was expected.

As to the marking, the system we went on was to give the first 3 days' subjects very much the same weight as before in the exam^{n}. They were raised perhaps, but raised only very little. In speaking of the marks I exclude the problem papers, which we never consider in marking the book-work. Of course on account of the expansion of the low subjects the disproportion between time and marks was much greater than it used to be under the old system.

Alas for my problem paper! There is one most ridiculous blunder in it, which is prob. 13.[4] I have altered it for the Calendar by substituting ▱^{m} and ellipse for ▱ piped and ellipsoid. The mode of proving the eq^{n}

$$\frac{d^2V}{dx^2}+\frac{d^2V}{dy^2}+\frac{d^2V}{dz^2}=-4\pi\varrho$$

given in the last problem[5] I can hardly think new to you, though I do not think you have given it anywhere, nor do I recollect your speaking of it explicitly. It seems to me far more direct and simple than that given in Pratt,[6] which I suppose is Poisson's.

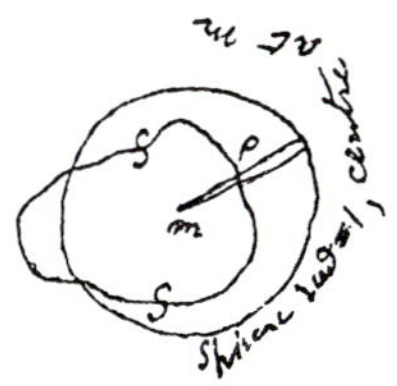

The first part of the problem is very easily proved. For if P be the point in S at which the element $d\omega$ is situated, and if the cone whose base is dw and vertex m, produced if necessary, cut the sphere in the fig. in the element $d\sigma$, we have

$$\cos\theta d\omega = \text{projection of } dw \text{ on plane} \perp \text{ to } mP$$

$$\frac{\cos\theta d\omega}{r^2}=\frac{\text{projection}}{r^2}=d\sigma$$

$$\int\int g\cos\theta d\omega=\int\int\frac{m}{r^2}\cos\theta d\omega=m\int\int d\sigma=4\pi m.$$

If m be without S the elements destroy each other two and two, and the proof is easily extended to the case in which a radius vector cuts S several times.

Hence for a finite mass cut by S

$$\iint g\cos\theta d\omega = 4\pi \times \text{ mass within } S$$

$\iint g\cos\theta d\omega$ is so to speak the flux of potential, and if we take for S an elementary ▱ piped

Flux for the first face $dydz = +Xdydz$ ult[imatel]y

[Flux for] 2nd [face $dydz$] $= -(X + \frac{dX}{dx}dx)\, dydz$

and so for the others. Hence whole flux, or

$$\iint g\cos\theta d\omega = -\left(\frac{dX}{dx} + \frac{dY}{dy} + \frac{dZ}{dz}\right) dxdydz \quad \text{ult}^{y}$$

$$= -\left(\frac{d^2V}{dx^2} + \frac{d^2V}{dy^2} + \frac{d^2V}{dz^2}\right) dxdydz$$

and this $= 4\pi$ included mass

$$= 4\pi\, \varrho\, dxdydz \text{ ult}^{y}$$

so[?] $\frac{d^2V}{dx^2} + \frac{d^2V}{dy^2} + \frac{d^2V}{dz^2} = -4\pi\varrho.$

I am sure you know the geom^{l} proof that $\iint g\cos\theta d\omega = 4\pi m$, or $= 0$, but it is possible you may not have thought of this application: at least I do not remember your mentioning it.

I shall be willing to write N° 3 of Hydrodynamics, the equations of motion.[7] N° 4 might contain the theorem that $udx + vdy + wdz$ is an exact diff^{l} if it is once so, and in particular if the motion begin from rest.[8] Your new proof of the theorem might come in here. Then I should propose the simpler cases of wave motion, which are at the same time the most interesting; then a ball pendulum as a simple example of the working out of a case of resistance to a solid, the images might come after this, and then an example of motion of a fluid within a solid, as the ▱ piped.[9]

If you refer to the Friday m^{g} paper – question 7, you will see another way of getting at the $=^{ns}$ $b = \mp e$, $cf = \pm 1 - e^2$, but this way, besides being less simple than 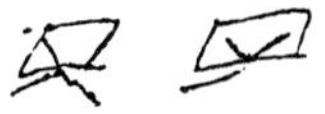 does not determine the sign.[10]

I have several irons in the fire. I have 3 longish papers which I want to write besides what I was writing to you about, and at least one thing besides, not counting the apparent difficulty in the theory of light, which I mean for the [Cambridge and Dublin] math^{l} journal.[11] This relates to the reduction of the 36 constants of an elastic crystallized solid to 21 by a principle employed, almost without explanation, by Green.

Yours very truly
G. G. STOKES

1 Kelvin has written at the top of the letter, 'Feb. 5, 1848'.
2 See the previous letter, note 1. There were two 'examiners' and two 'moderators' for the examination. Stokes was one of the moderators.
3 This occurred in 1839, before which date men were grouped into preliminary classes on the basis of their performances in Latin disputations. Because men in the upper and lower classes were given different questions on the Senate House examination, those in the lower classes had difficulty scoring high enough to pass the examination with honours.
4 As printed in the *Cambridge Calendar for 1848*, p. 355, problem 13 read:
A parallelogram is constructed by drawing tangents at the extremities of two conjugate diameters of an ellipse; prove that the diagonals of the parallelogram form a second system of conjugate diameters, and that the relation between the two systems is reciprocal.
5 See the previous letter, note 2.
6 John Henry Pratt, *The Mathematical Principles of Mechanical Philosophy, and Their Application to the Theory of Universal Gravitation* (Cambridge, 1836). Pratt (d. 1871) was third wrangler in 1833 and 1850 was appointed archdeacon of Calcutta. His *Scripture and Science Not at Variance* (London, 1856) reached its seventh edition in 1872.
7 Stokes (16).
8 Stokes (19).
9 The final two of the six notes on hydrodynamics were Kelvin (40) and Stokes (28).
10 The relevant portions of problem 7 read:
Light is incident obliquely on a thin plate separating two media, or two portions of the same medium.... If the coefficient of vibration be multiplied by b, c for reflection and refraction at the first surface, and by e, f for reflection and refraction at the second, and if the first and third media be the same, shew that the condition that the sum of the intensities of the reflected and transmitted light shall be equal to that of the incident light leads to the relations $b = \mp e$, $cf = \pm(1 - e^2)$. (*Cambridge Calendar for 1848*, 363.)
11 Stokes (26), which involved the problem mentioned in the previous note.

24 KELVIN to STOKES, 5 February 1848

Stokes Collection, K24

Glasgow College,
Feb. 5, 1848

MY DEAR STOKES,

Professor Forbes of Edinburgh, who is at present on a visit with us, has just, *à propos* of your discovery in the theory of elastic solids,[1] told me of an experiment he made a long time ago, & exhibited in his class 10 years ago (without publishing it in any other manner). In OErsted's exper[t] for the compressibility of water, he substituted india rubber for glass, as the bottle ⟨containing⟩ which, along with the water in the external vessel, is compressed; & he found that no water is forced into the bottle of india rubber, nor expelled from it, by the compression.[2] This shows that the compressibility of india rubber differs from that of water by an ⟨quantity⟩ amount inappreciable in an experiment capable of rendering very sensible the ⟨compression of water⟩ difference betw. the compressibility of water & glass.

I wished to ease my mind before going to bed by communicating this to you, and so I shall only add that I like your Hydrodynamical programme very much & that I hope we shall have something of it in each N° [of the *Cambridge and Dublin Mathematical Journal*] for some time, as I am sure it will call attention to a most interesting & much neglected subject; & that Forbes and I have just been talking hydrodynamics to a considerable extent – So wishing you good night & good sleep till chapel time, I remain

Your's sincerely
WILLIAM THOMSON

P.S. Your No 13 is a very pretty problem when corrected.[3] What a pity it will not do for 3 dimensions!

1 See Stokes (11).

2 See Hans Christian Oersted, 'Experiment, die Compression des Wassers zu zeigen', J. S. C. Schweigger's *Journal für Chemie und Physik*, **36** (1822), 332–9, and *Annals of Philosophy*, **5** (1823), 53–6.

3 See the previous letter, note 4.

25 Kelvin to Stokes, 18 February 1848[1]
Stokes Collection, K25

The Editor of the *Cambridge and Dublin Mathematical Journal* thanks Mr. 'Stokes' for his 'very pleasant' communication dated 'Feb. 17.' ⟨Should⟩ his 'uncommonly jolly' paper ⟨be considered suitable for insertion in the Journal, a notice regarding it⟩ will appear in the next Number;[2] ⟨otherwise the Manuscript will be returned as soon as possible.⟩

Glasgow College, 'Feb. 18,' 18'48'[3]

Dear Stokes,

I have made a short insertion, with reference to your illustration; so that the whole passage stands thus, at present – 'The subsidence of the motion in a cup of tea which has been stirred may be mentioned as a familiar instance of friction, or, which is the same, of a deviation from the law of normal pressure; and the absolute regularity of the surface when it comes to rest, ⟨m⟩ whatever may have been the nature of the previous disturbance may be considered as a proof that all tangential force ⟨ceases when⟩ vanishes when the motion ceases.'[4] If you disapprove of this, or if you think it requires farther explanation, or if you can state the same more forcibly, or clearer, will you ⟨ma⟩ send me what correction you think proper, or be ready to make the correction in the proof sheet?

I thought my assertions regarding the true equations of equil^{m} of an elastic solid, & the equation of the surface, as far as you are concerned, to be safe, after you saw the proof sheets.[5] I still hold to my assertion with reference to the latter, as what you tell me in your letter is quite consistent with what I stated. Perhaps I ought to make some note or remark regarding Cauchy's view of the equ^{ns} of an elastic solid. What do you think?

Your's sincerely
William Thomson

1 The first part of this letter is a form letter in which Kelvin has made additions and deletions. His additions are enclosed within inverted commas.
2 Stokes (16).
3 The remainder of the letter is entirely in Kelvin's hand.
4 Stokes (16), in *MPP*, II, 3.
5 See Kelvin (35) and (29.2), both of which appeared in the *Cambridge and Dublin Mathematical Journal* for February 1848.

26 STOKES to KELVIN, 20 April 1848
Kelvin Collection, S335

Pembroke College Cambridge[1]
April 20, 1848

MY DEAR THOMSON,

I have an article almost ready for you on the mode I showed you of obtaining the equations $b = -e$, $cf = 1 - e^2$ (A) in Light.[2] I have entered into some considerations respecting the philosophy of the subject, which perhaps you will not think altogether out of place, though they are not at all necessary for the eq^{ns} (A). I hope to send you the paper in a few days.

As to N° IV of Hydrodynamics I have not written it, but it would not take me long.[3] It was to be on the theorem about $udx + [vdy + wdz]$. I was thinking of first giving Cauchy's proof which you will see the steps of in my paper on the friction of fluids (section II),[4] then my own, which perhaps may be as well given in addition to Cauchy's on account of the length of the expressions which the latter requires. Thus the expression for S contains I think 6 terms each a product of 3 differential coeff^s such as $\frac{dx}{da}\frac{dx}{db}\frac{dx}{dc}\frac{dy}{da}$ &c. Lastly I thought of referring to a new proof to be given by you in the *same number* (?) for the case of incompressible fluids. Do you approve of the plan? and will you have your proof ready?

I see you have been working your Glasgow men with images. I am glad there is a prospect of the images soon making their appearance for our Cambridge men, for it is a very beautiful method, and so simple.

A Professor-Syndicate has recently published its report. Some extensive changes are proposed. It is proposed to have two new triposes for moral sciences (including law, moral philosophy, modern history &c) and for natural sciences (geology, anatomy, comparative d[itt]o, botany, chemistry &c). It is also proposed to compel the Poll men to attend one professor's course of lectures & to pass an exam^n by him.[5]

The Clare Hall fellows have sent in their defence in re Fischer, but I do not know when the Vice-Chancellor[6] is to give his decision as visitor. I suppose you and Fischer will be coming here in May.

Yours very truly,
G. G. STOKES.

1 Kelvin has written at the top of the letter, 'April 22 1848'.
2 Stokes (26).
3 See Stokes (19).
4 Stokes (11).
5 The moral sciences tripos and the natural sciences tripos were introduced in 1851, although it was not until a decade later that a student could earn a degree by taking them. (See D. A.

Winstanley, *Later Victorian Cambridge* (Cambridge, 1947), 184–8.) Poll men were those who took ordinary, as distinct from honours, degrees.

6 Robert Phelps (1808–90) was fifth wrangler in 1833 and vice chancellor in 1844/5 and 1847/8. The issue apparently concerned Fischer's signing religious tests at St Andrews. (See letter 38.)

27 KELVIN to STOKES, 25 April 1848

Stokes Collection, K26

Glasgow College,
April 25, 1848

MY DEAR STOKES

I approve very much of all your plans. You may promise what you refer to, with reference to me, and then my demonstration would follow immediately after your paper. I sh^{d} be glad to have your M.S.S. on $Hydrod^{s}$ as soon as possible, as it would be a guide for me in writing my add^{n}.[1] I intend to be in Cambridge tomorrow week, & perhaps by that time your paper on $Hydrod^{s}$ will be ready. I shall also be glad, if possible, to see a letter I wrote to you last October.

I enclose a second paper on $Elec^{y}$ w^{h} I set today. There was unlimited time, with an hour's interval for dinner, & each candidate gave in (one is still at work, at 8.30) his papers when he had done what he could.

Your's sincerely
WILLIAM THOMSON

P.S. Forbes showed me his experiment on the compressibility of india rubber. I have been lecturing on acoustics for several days (I gave my last lecture today, & am really sorry to be stopped) and I am quite delighted with the subject, & some apparatus I bought for it in Paris last summer. I can count 256 in a second with ease & certainty, & can ⟨find⟩ count the number of vibrations, without missing more than four, performed by a tuning fork in a minute, or in an hour if I please.

1 See Stokes (19), published in November 1848, in which he writes that 'a new proof of the theorem for the case of an incompressible fluid will be given by Professor Thomson in this [*Cambridge and Dublin Mathematical*] *Journal*'. (*MPP*, II, 37.) The addition was not written.

28 KELVIN to STOKES, [c. 9 August 1848][1]
Stokes Collection, K26A

Meadowbank House
Kirknewton
Edinburgh

MY DEAR STOKES

I enclose a short paper which you perhaps [would] read for me in Section A, should the committee agree to receive it. Tomorrow, or the day after I hope to be able to send another paper (on the directive tendency exercised by the earth on irregular masses of soft iron, or of diamagnetic substances) which also [will] be very short and, if possible, tolerably readable.[2]

Will you let me know the fate of the paper I now send (if it is read, whether any discussion arises out of it) and tell me whether I ought to send another copy for the Athenaeum? I think there is not too much 'algebra' for the Athenaeum printers, but if you think there is, I should make an abstract omitting mathematical symbols.

I am sorry not to be present at the meetings at Swansea. I hope you are enjoying it, & I shall be glad to hear from you what you have been doing & seeing.

Your's very truly
WILLIAM THOMSON

P.S. I leave this on Saturday, Aug. 12, after which my address will be Blair Logie, Stirling.

1 The 1848 meeting of the British Association in Swansea began on 9 August 1848.
2 Kelvin (32) and (33). Section A of the British Association was the section for mathematics and physics.

29 STOKES to KELVIN, 21 August 1848
Kelvin Collection, S336

The Hotel, Port Stewart[1]
Col[e]raine Aug 21, 1848

MY DEAR THOMSON,

Partly in consequence of my business as one of the Secretaries of Section A, and partly in consequence of travelling, I have delayed answering your letter. I read both your papers to the section, but they did not elicit any remarks.[2] I doubt if many present were able to follow them. Sir Wm Snow Harris said to me that he had written two letters to you but had not got an answer. He read a paper, or rather gave a lecture, 'On a general law of electrical discharge,' of which you will probably see an abstract in the Athenaeum.[3] As to your abstracts I don't think they will print *any* symbols in the Athenaeum. They seem to object so much to abstruse papers that I doubt if they would print almost anything you could send which would be full enough to let the reader see what you were about. I made what I thought a very printable abstract of some of my communications, and they (I mean rather the editor) thought it too long, as the great mass of his readers could understand nothing about it.

Prof[r] Plücker was at the meeting, and ⟨made⟩ gave an inter[e]sting account of his experiments on the action of magnetism on crystals. The same long ▱ piped, of square section, of a crystal could be made to take either the axial or the equatoreal [*sic*] direction according as it was turned thus

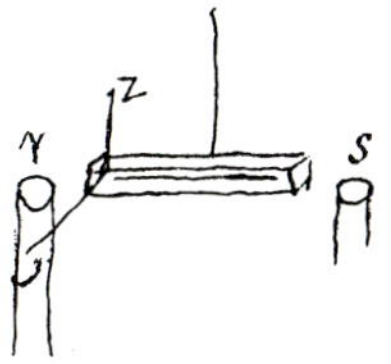

or thus

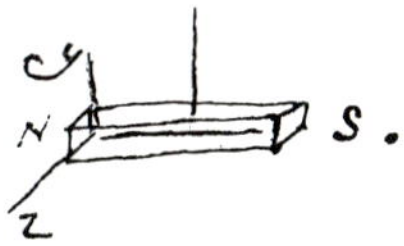

Prof. Plücker said that the optic axes were repelled by the poles. This is true both of calc spar, a neg[ve], and of quartz a pos[ve] crystal. He mentioned one exper[t] which surprised me. As well as I can make out it was this. Brass is magnetic; so that a light brass pan laid on the poles of an electro-magnet is

only pressed more strongly against the iron when the current is put on. But mercury being diamagnetic, if a certain quantity of mercury be put into the dish the whole is repelled {I am not clear what keeps the mercury from flying out leaving the dish behind} but only repelled *to a certain distance* where it remains *suspended in the air in stable equilibrium* like Mahomet's coffin. From this he infers that the diamagnetic force decreases faster than the magnetic as the distance increases. An account of the exper[ts] is coming out in Taylor's scientific memoirs. I believe they, or some of them, have already been published in Poggendor[f]f's Annals.[4]

Yours very truly
G. G. Stokes

P.S. I leave this on Wednesday week, but a letter directed after that to Pembroke College will find me as I have my letters forwarded wherever I may be.

1 Kelvin has written at the top of the letter, 'Aug. 23, 1848'.
2 Kelvin (32) and (33).
3 William Snow Harris, 'On a general law of electrical discharge', *The Athenaeum* (1848), 888. Harris (1791–1867) studied medicine at Edinburgh, but when he married in 1824 he turned to full-time work with electricity. His system of lightning conductors for ships eventually earned him a knighthood in 1847.
4 Julius Plücker, 'On some new relations of the diamagnetic force', *The Athenaeum* (1848), 835–6, and *Brit. Assoc. Rep.* (1848), part 2, p. 2. *The Athenaeum* contains an account of the discussion following Plücker's presentation. See also his 'Ueber die Abstossung der optischen Axen der Crystalle durch die Pole der Magnete', Poggendorff's *Annalen der Physik und Chemie*, **72** (1847), 315–43, and 'Ueber das Verhältniss zwischen Magnetismus und Diamagnetismus', *ibid.*, 343–51. The two papers were published in English in Taylor's *Scientific Memoirs*, 5 (1852), 353–75 and 376–82. Plücker (1801–68) was professor of mathematics from 1836 to 1847 and of physics from 1847 to 1868 at Bonn. There are several letters from him to Stokes in the Stokes Collection and in the Royal Society of London, many concerning Plücker's 'On the magnetic induction of crystals', *Phil. Trans.* (1858), 543–87, which was refereed by both Stokes and Kelvin (Royal Society of London, RR.3.224 and RR.3.222, respectively).

30 KELVIN to STOKES, 23 August 1848
Stokes Collection, K27

Blair Logie,[1]
August 23, 1848

MY DEAR STOKES

Many thanks for your attention to my papers. I could make an abstract of each in which the object might be intelligibly stated, without any proof, or without the introduction of any symbols, but I suppose it is now too late for the Athenaeum. My reason for wishing to have an abstract published there is principally because the Athenaeum is seen on the Continent, and its scientific abstracts are copied into *l'Institut*.[2] I am much obliged to you for your account of Plücker's experiments. I do not assent to his conclusion that the diamagnetic force decreases faster with the distance than the magnetic force. The experiment about the mercury and the brass pan is most extraordinary – scarcely believable I should have thought had the authority been good. I should like to see it. A small ball of magnetic substance (like soft iron) cannot I am almost sure have a pos^{n} of stable equil^{m}, untouched by a solid, or unsuspended by a cord. But I believe I can (at least I think I remember having done so to my own satisfaction) prove that a small diamagnetic ball may be actually held repelled – like Mahomet's coffin, in stable equil^{m}. Are you trying to re-collect the scattered supporters of Irish Nationality and make another effort for independence? Or do you fraternize with the Saxon, the enemy of your country? Have you been seeing the giant's causeway? Have you been bathing in the Atlantic with a good N.W. wind?

Your's very truly
WILLIAM THOMSON

1 Kelvin has written at the top of the letter, 'I received Sir W. Snow Harris's letter containg ⟨some⟩ accounts of some experiments he has made; taking a great deal of trouble very kindly on account of some suggestions I made. I answered his letter last week'. There are two letters from Harris in the Kelvin Collection (H37 and H38) dating from July 1848.

2 *L'Institut; Journal des Académies et Sociétés Scientifiques de la France et de l'Étranger.*

31 STOKES to KELVIN, 28 August 1848
Kelvin Collection, S337

The Hotel, Port Stewart, Col[e]raine[1]
Aug 28th 1848.

MY DEAR THOMSON,

The report of the proceedings of the British Assocn given in the Athenaeum runs through several numbers, and as only one no comes out in a week, and your papers were read on the last day, I dare say there is still time for your abstract. You could write to say that you would prefer the insertion of the abstracts you first sent, but if they would not suit, on account of the introduction of symbols, the abstracts you send might be substituted. I dare say they may print at large the abstract of your 2nd communication, as it contained no symbols, but they are sure to cut out the symbols in the first, and there is danger that in so doing they may cut out all the pith of it.[2] Such was the fate of an abstract of mine on the aberration of light at the Cambridge meeting; the supposition I made – that the ether close to the Earth was at rest relatively to the Earth – was inserted, and the conclusion at which I arrived, namely that the law of aberration would result from this hypothesis provided the motion of the ether was such that $udx + vdy + wdz$ was an exact differential, was wholly omitted; the shell was kept, the kernel thrown away.[3] Of course I do not speak of the Assocn report, where the abstracts ⟨were⟩ are printed at full. If you send the abstracts I speak of, it would be well to write on them 'Not to be forwarded to Profr Phillips'[4] for fear they should get into the Assocn report instead of what you first sent.

I am going on Wednesday to the Giant's Causeway, where I shall remain a few days. Bush Mills is the post town. If you have occasion to write in the course of next month you had best direct to Cambridge as I shall be moving about, but my letters will be forwarded.

Yours very truly
G. G. STOKES.

1 Kelvin has written at the top of the letter, 'August 30 1848'.
2 Kelvin (32) and (33).
3 Stokes (5).
4 John Phillips (1800–74) was curator of the Yorkshire Philosophical Society's museum and was involved in organizing the first meeting of the British Association for the Advancement of Science in York in 1831. He was the association's assistant secretary from its founding until 1859. In 1853 he went to Oxford as deputy reader in geology and later became professor of geology.

32 STOKES to KELVIN, 12 December 1848

Kelvin Collection, S338

Printed in Joseph Larmor, 'An early formulation by Stokes of the theories of the rotatory polarizations of light', *Proc. Camb. Phil. Soc.*, **22** (1924), 76–81. The letter is on pages 78–9.

Pembroke College, Cambridge[1]
Dec 12th 1848.

DEAR THOMSON,

I have hitherto put off writing my paper on Waves,[2] and I fear I shall have no time for it now on account of the Senate-House Examination. However you have one N° of the 'Notes on Hydrodynamics' for the next N° of the journal.[3]

I read my paper 'on the formation of the central spot of Newton's Rings beyond the critical angle' at the Philos^{l} last night.[4] If they print it, as I suppose they will, I hope pretty soon to be able to send you a copy.

An article I saw in the Annales de Chimie for May (?) last[5] set me to work at the circular polarization of magnetized glass. Setting to work inductively, I have arrived at the following equations for the motion of the luminiferous ether – I was going to refer to them, but I have mislaid the paper. I am however nearly sure they were $\{\xi, \eta, \zeta$ being the displacements, and ∇ meaning $\frac{d^2}{dx^2}+\frac{d^2}{dy^2}+\frac{d^2}{dz^2}\}$

$$\frac{d^2\xi}{dt^2}=v^2\nabla\,\xi\langle +A\frac{d}{dt}\langle\rangle + A\frac{d^2}{dtdz}\left(\frac{d\eta}{dt}-\frac{d\zeta}{dy}\right)$$

$$\frac{d^2\eta}{dt^2}=v^2\nabla\,\eta+A\frac{d^2}{dtdz}\left(\frac{d\zeta}{dx}-\frac{d\xi}{dz}\right)$$

$$\frac{d^2\zeta}{dt^2}=v^2\nabla\,\zeta+A\frac{d^2}{dtdz}\left(\frac{d\xi}{dy}-\frac{d\eta}{dx}\right)$$

z being in the direction of the line of magnetic force. In these eq^{ns} ξ, η, ζ are not arbitrary, but connected by the condition of incompressibility

$$\frac{d\xi}{dx}+\frac{d\eta}{dy}+\frac{d\zeta}{dt}=0.$$

The additional terms multiplied by A correspond to a possible system of pressures and tensions, the expressions for which I cannot exactly write down from memory.

If we suppose that the optical effects of two[6] magnetic forces are superposed, it follows that the rotation of the plane of pol^{n} is proportional to the difference of the potentials at the points where the ray enters and quits the glass.

I have a notion as to a mechanical arrangement which would introduce the $\frac{d}{dt}$ in the A terms; in fact which might be conceived to introduce the terms themselves, but I have got nothing very definite on this score. It may be remarked that

$$\frac{d}{dt}\left(\frac{d\eta}{dz}-\frac{d\zeta}{dy}\right)=\langle -2^{d}\rangle-2\omega''' \text{ \&c.}$$

ω''' being an $\angle^{r}$ vel^{y}.

The[7] same method applied to syrop of sugar &c. gives

$$\frac{d^2\xi}{dt^2}=v^2\nabla\xi+A\frac{d}{dx}\left(\frac{d\eta}{dz}-\frac{d\zeta}{dy}\right)$$

$$\frac{d^2\eta}{dt^2}=v^2\nabla\eta+A\frac{d}{dy}\left(\frac{d\zeta}{dx}-\frac{d\xi}{dz}\right)$$

$$\frac{d^2\zeta}{dt^2}=v^2\nabla\zeta+A\frac{d}{dz}\left(\frac{d\xi}{dy}-\frac{d\eta}{dx}\right)$$

The A-terms here again correspond to a very simple system of pressures and tensions; and the whole turns out to be explicable on the supposition that a torsion produces a pressure or tension in the direction of the axis of torsion, as would be the case for instance with a spiral spring, or a medium composed of an immense n^{o} of spiral springs arranged in all directions.

I think you arrived at this conclusion, or something very like it. If so, please let me know, any time before February, that if you have anything to bring forward I may not get before you, or if you don't bring it forward that I may acknowledge it if I write a paper on the subject.

Perhaps you will be able to come to some definite notion as to a mechanical arrangement which would produce the magnetic terms in the first set of equations.

Yours very truly
G. G. Stokes

1 Kelvin has written at the top of the letter, 'Dec 1⟨5⟩4, 1848. Answ[d] Sat. ⟨D⟩ Jan 27, 1849'.
Note that Stokes defines the operator 'del' (∇) in this letter as ∇^2 is usually defined. He also does so in letters 33, 47, 60, 64, and 222, but uses ∇ in the usual way in letter 368.
2 Stokes (28).
3 Kelvin (40).
4 Stokes (30).
5 Pierre Auguste Bertin, 'Sur la polarisation circulaire magnétique', *Annales de Chimie et de Physique*, **23** (1848), 5–32. Bertin (1818–84) was professor of physics at Strasbourg until 1866 and afterwards was at the École Normale.
6 Larmor's version reads *these* instead of *two*.
7 Larmor's version reads *this* instead of *the*.

33 Stokes to Kelvin, 13 December 1848

Kelvin Collection, S339

Printed in Joseph Larmor, 'An early formulation by Stokes of the theories of the rotatory polarizations of light', *Proc. Camb. Phil. Soc.*, **22** (1924), 76–81. The letter is on pages 80–1.

Pembroke College Cambridge[1]
Dec 13th 1848

Dear Thomson,

I have found my papers, and copy the results

Magnetized Glass[2]

$$\frac{d^2\xi}{dt^2} = \langle\omega\rangle v^2\nabla\xi + A\frac{d^2}{dtdz}\left(\frac{d\eta}{dz} - \frac{d\zeta}{dy}\right)$$

$$\frac{d^2\eta}{dt^2} = v^2\nabla\eta + A\frac{d^2}{dtdz}\left(\frac{d\zeta}{dx} - \frac{d\xi}{dz}\right)$$

$$\frac{d^2\zeta}{dt^2} = v^2\nabla\zeta + A\frac{d^2}{dtdz}\left(\frac{d\xi}{dy} - \frac{d\eta}{dx}\right)$$

To the *A*-terms correspond the pressures and tangential forces[3]

$$P_x = P_y = 0$$

$$P_z = 2A\left(\frac{du}{dy} - \frac{dv}{dx}\right) = 4A\omega'''$$

$$t_{xz} = t_{zx} = A\left(\frac{dv}{dz} - \frac{dw}{dy}\right) = 2A\omega'$$

$$t_{yz} = t_{zy} = A\left(\frac{dw}{dx} - \frac{du}{dz}\right) = 2A\omega''$$

$$t_{xy} = t_{yx} = 0\left\{u = \frac{d\xi}{dt},\ v = \frac{d\eta}{dt},\ w = \&c\left[\text{i.e., } \frac{d\zeta}{dt}\right]\right\}$$

Syrop of Sugar &c

$$\frac{d^2\xi}{dt^2} = v^2\nabla\xi + \left(C\frac{d^2}{dt^2} + D\nabla\right)\left(\frac{d\eta}{dz} - \frac{d\zeta}{dy}\right)$$

$$\frac{d^2\eta}{dt^2} = v^2\nabla\eta + \left(C\frac{d^2}{dt^2} + D\nabla\right)\left(\frac{d\zeta}{dx} - \frac{d\xi}{dz}\right)$$

$$\frac{d^2\zeta}{dt^2} = v^2\nabla\zeta + \left(C\frac{d^2}{dt^2} + D\nabla\right)\left(\frac{d\xi}{dy} - \frac{d\eta}{dx}\right)$$

C and D being arbitrary coefficients we[4] may make two hypotheses either that the terms which really exist are

$$C\frac{d^2}{dt^2}\left(\frac{d\eta}{dz} - \frac{d\zeta}{dy}\right) \&c$$

or that they are

$$\langle C\rangle D\left(\frac{d^2}{dx^2} + \frac{d^2}{dy^2} + \frac{d^2}{dz^2}\right)\left(\frac{d\eta}{dz} - \frac{d\zeta}{dy}\right)\&c.$$

The latter seems most easily conceivable mechanically, and these terms would be introduced by the system of ⟨forces⟩ pressures and tang[l] forces[5]

$$P_x = 2A\frac{d\theta_x}{dx}$$

$$P_y = 2A\frac{d\theta_y}{dy}$$

$$P_z = 2A\frac{d\theta_z}{dz}$$

$$t_{yz} = t_{zy} = A\left(\frac{d\theta_z}{dy} + \frac{d\theta_y}{dz}\right)$$

$$t_{zx} = t_{xz} = A\left(\frac{d\theta_x}{dz} + \frac{d\theta_z}{dx}\right)$$

$$t_{xy} = t_{yx} = A\left(\frac{d\theta_y}{dx} + \frac{d\theta_x}{dy}\right)$$

where θ_x θ_y θ_z are the rotations about x, y, z, so that

$$\theta_x = \frac{1}{2}\left(\frac{d\zeta}{dy} - \frac{d\eta}{dz}\right)\&c.$$

These pressures and tensions again arise from 3 principal torsions in rectangular directions (I call $\frac{d\theta_x}{dx}$ a torsion) each producing a pressure or tension in the direction of the axis of torsion.[6]

Yours very truly
G. G. STOKES

1 Kelvin has written at the top of the letter, 'Dec 15, 1848. Answ[d] Sat Jan 27, 1849'.
2 Larmor's version omits the next three equations. For Stokes's use of ∇, see the previous letter, note 1.
3 Larmor includes the following note after ω''': 'The signs of (ω', ω'', ω''') here should be changed'.
4 Larmor begins a new sentence with *we*.
5 Larmor notes that $A = -D$.
6 Larmor adds a note here reading, 'For they are correlative to the stress in an isotropic incompressible elastic solid for which (θ_x, θ_y, θ_z) is the displacement'.

34 KELVIN to STOKES, 4 February 1849

Stokes Collection, K28

2 College, Glasgow
Feb. 4, 1849

MY DEAR STOKES

I feel sure that you will not consider an apology necessary for the delay I have made in answering your letters of the 12th and 13th of last month. I had a great deal to say, and as you gave me some time, I deferred writing until I should have done with a paper on Carnot's Theory of the Motive Power of Heat I was to read at the Royal Soc. of Edinburgh on the 2nd of January.[1] What has happened since you know, and will readily conceive that my mind was very much turned away from matters which at all ordinary times interest me very much.[2] I have been excessively busy of late in recovering from disturbing causes, besides being a good deal occupied in getting out the Number of the

[Cambridge and Dublin Mathematical] Journal w^h is just published, and so I took the longest term you gave me, and had actually commenced writing to you so that you might get my answer at 'the beginning of February' when on Wednesday last I received a letter from Mr Hopkins w^h required immediate attention, and so I was prevented from finishing my letter to you.[3] I cannot delay longer thanking you for your letters w^h interested me exceedingly; and telling you that I claim nothing except what is published in the C. & D. M. J.; and all that has reference to the points you mention is I think in a paper 'on a $Mech^l$ Representation of Electric Magnetic & Galvanic Forces' published this time two years.[4] I have the prospect of another extremely busy week, but I hope very soon to be able to write more at length explaining what I have referred to at the end of the paper & half promised to publish (w^h however I dare say you see through, yourself). When I wrote the paper I had some hope, w^h I still retain, that a satisfactory physical theory of all those agencies, including besides light, is approachable.

Mr Hopkins has probably communicated with you regarding our correspondence. I expect an answer from him very soon.[5] In the mean time believe me

Your's very truly
WILLIAM THOMSON

1 Kelvin (42).
2 Kelvin's father died on 12 January 1849. His death left vacant the Glasgow professorship of mathematics, and the next several letters concern the question of Stokes's application for the vacancy.
3 Hopkins's letter to Kelvin has not survived in the Kelvin Collection or the Kelvin Papers at Glasgow.
4 Kelvin (25).
5 Hopkins wrote to Kelvin on 9 February 1849, discussing the relative merits of Stokes and Hugh Blackburn for the Glasgow position. He wrote that 'if you determine to recognize the *scientific principle* in your election, and to elect a man who is sure hereafter to dignify his position by the highest scientific distinction, Stokes is *unquestionably your man*. If on the contrary you should think right to recognize the primary importance of choosing one who has received his early education in your own institutions, you may well rejoice in such a candidate as Blackburn'. (Kelvin Collection, H126.)

35 STOKES to KELVIN, 12 February 1849
Kelvin Collection, S340

Aughnacloy, C^o Tyrone, Ireland[1]
Feb 12th 184⟨8⟩9.

DEAR THOMSON,

I arrived here safe on Saturday, after a rather rough passage. It was so rainy and cloudy that I saw but little of the scenery of the Clyde, even about

Dumbarton; and as the vessel staid [*sic*] about 2 hours at Greenock it was dark when we passed the mountains on the side of the firth.

Now as to the tests. My brother[2] is decidedly of opinion that the straightforward course is, to decline to take them unless I am prepared to become a thorough presbyterian, which certainly I do not mean to become. I must say that this very much coincides with my own unbiassed opinion: it was all along a very doubtful question with me whether I could sign the test in a lax sense. That being the case I have determined to back out of it in the most polite way I can. If I had not already asked for testimonials there would have been no difficulty.[3] As it is, I suppose my best plan is to write apologies to those whom I have asked to give me testimonials. I would still stand if I thought I had a chance of being elected on condition of not signing the tests, but it seems to me that by standing in that way I should be only making a fool of myself before the electors, & that I would not have the slightest chance.

I would thank you for a line with your advice; but I feel almost certain from what you said that it would be useless for me to stand.

As I shall not return by Glasgow (unless I stand, which I feel almost certain I shall not) I would be obliged to you to forward to me here any letters which may have come for me.

I hope soon to be able to send you the paper on waves I spoke about.[4] I may however have occasion to refer to so⟨o⟩me books which I have not here in which case I must wait till I return to Cambridge.

Yours very truly
G. G. Stokes.

1 Kelvin has written at the top of the letter, 'Feb. 14, 1849'.
2 I do not know which brother this is; all three of Stokes's brothers were Anglican clergymen.
3 The testimonials are now in the Stokes papers at Pembroke College, Cambridge. Airy wrote that Stokes was 'one of the persons on whom the reputation of English Mathematical Science will in great measure depend', while Challis wrote that Stokes's performance on the Smith's prize examination had given Challis 'the impression that he possessed uncommon mathematical ability. The estimation thus formed, has been confirmed...'.
4 Stokes (28).

36 Kelvin to Stokes, [14 February 1849][1]
Stokes Collection, K29

2 College, Glasgow
Wednesday morning

My dear Stokes

Your letter w^{h} I received this morning has put me quite into a state of agitation. I expected possibly to see yourself yesterday, certainly tomorrow, and I thought as you seemed to be convinced about the tests, you would not have felt any difficulty, after consulting with your friends, about coming forward as a candidate. I feel perfectly confident that you would not have occasion to regret having taken the tests sh^{d} they be enforced, & that public opinion, especially the opinion of your colleagues w^{h} is of most importance, would perfectly sanction your conduct, in conforming fully to the Episcopal Church; and that the *amount* of conformity to the Established Church which a conscientious observance by one in your position of the obligations imposed by the tests, would really be in no way inconvenient, or repugnant to your feelings. In fact I believe the amount of conformity w^{h} *you* living in Scotland, would of yourself without any formal obligation, give to the Established Church, would be fully as much as either you or any one of the professors here would think a sufficient fulfilment of ⟨your⟩ the intention expressed in the Formula. Do really take the matter still under consideration and consult your brother and remember that it will be a very serious blow to the interests of this University if an honest member of the Church of England should never be able to be a candidate for any *situation or office* connected with it, however valuable an acquisition he might be; on account of an act of Parliament framed at a period of great political & ecclesiastical excitement; & allowed to continue unmodified in these settled times, merely because the modifications that those who have the interests of the University most at heart would be inclined to have made, are such that only those parts of the Act which at present are practically inoperative, would be abolished.

Since you left, I have spoken to several of the professors with reference to the tests as affecting Episcopalians, & every one of them agreed that they need be no obstacle whatever. In fact I find now, that in my own case even those who thought I had taken the tests in the ordinary way *after* admission, did not think it anything at all to be found fault with that I have been in the habit of regularly ⟨attending⟩ conforming to the Episcopal Church, & not appearing more than once or twice or three times in the course of a session at an Established Church.

Blackburn[2] is here at present, but I have always spoken to him in a perfectly cautious way, as far as regards his competition should you be a candidate. I told him this morning that I had got a letter from you, & that I still did not know whether you would be a candidate or not.

Do reconsider the matter then; & appear here yourself by the 1st opportunity besides answering this if you can by return of post. N. B. Blackburn is of course looked upon as an Episcopalian, yet no objection is *mentioned* even on this account.

I enclose a set of letters which have come for you. Read them & let them ⟨be⟩ convince you that you ought not to draw back. I of course speak but little to electors on the subject at present, but I always ⟨spea⟩ leave the field open for you and I do not think that a *majority* will be in too great a hurry to decide on any body.

Your's sincerely
WILLIAM THOMSON

1 This letter was a response to Stokes's of the 12th, and the 14th was a Wednesday.
2 Hugh Blackburn (1823–1909) was fifth wrangler in 1845 – the year Kelvin was second – and did replace Kelvin's father as professor of mathematics. He and Kelvin were friends as undergraduates, and Blackburn accompanied Kelvin on his trip to Paris in 1845. He retired from the professorship in 1879.

37 STOKES to KELVIN, 16 February 1849
Kelvin Collection, S341

Aughnacloy, Ireland[1]
Feb 16th 1849

DEAR THOMSON,

I was I confess a good deal staggered in my purpose by your powerfully written letter, nevertheless I think I must hold the same opinion still. I have waited so long to consult with some of my relations at odds & ends of times, as there are visitors in the house, and it is now so near dinner that I must give you my reasons tomorrow as the post will be off before we have done. Meanwhile if you had intended to vote for me please keep your vote open till you hear next in case I should change my opinion, but I do not think it likely that I shall. I feel extremely grateful for the interest you have taken in my behalf.

Yours very truly
G. G. STOKES

1 Kelvin has written at the top of the letter, 'Sund Feb 18, 1849'.

38 STOKES to KELVIN, 17 February 1849
Kelvin Collection, S342

Aughnacloy, Ireland[1]
Saturday Feb 17th 1849

DEAR THOMSON,

I promised you yesterday to send you today my reasons for not standing.

There is no doubt that the test means, & was intended in the first instance to mean, that the subscriber is a bonâ fide member of the established Church of Scotland. As I neither am, nor wish to become such, the only question is, how far one would be justified in signing in a lax, rather negative than positive, sense. My brother says that a test imposed by an act may very well become obsolete, but that it is revived by being imposed. I should go further, and say that if the whole imposing body regarded it as a mere form, the subscriber might so regard it too; though even in this case there might be some doubt; & in fact cases are easily conceivable in which the subscriber ought not to regard the subscription as a mere form. But if the imposing body be divided, it then becomes as it seems to me a very different matter. To my own mind, the lawfulness of signing in a lax sense seems ⟨to me⟩ such a nice & doubtful point, and my own interest so clearly lies one way, that I feel that I could not help being somewhat ashamed of what I had done, supposing I were to sign; and if I were taxed with it I should be more disposed to skulk away like a dog with his tail between his legs than to defend my conduct, unless driven into a corner & compeled [*sic*] to fight. You will surely allow this feeling to be a good reason for refusing to sign.

There seems to me to be but one absolutely straight forward course besides declining altogether to stand; and that is to state before the election my objection to take the test, on the ground of being and intending to remain an episcopalian, at the same time professing my non-hostility to the established church of Scotland. If, standing in this way, I should have no chance, the proper answer seems to me to be, not 'Shuffle a little', but, 'Don't stand: if they cared to have you they would take you in this way, especially as they have the precedent of Edinburgh.'[2] To stand in this way I take for granted would be useless.

I have already written to several of those from whom I got testimonials apologizing for having given them the trouble; as, inasmuch as I did not intend to take the test, I should probably not stand.

I showed your letter to my brother. It certainly seemed to have the effect of making him doubtful, from having been decided; still, he evidently still leaned to the opinion that imposing the test revived it.

One reason for not regarding the test as a dead letter is, that it is a privilege to which the Kirk is entitled by law (no privilege I take it to the universities) to have for the professors who instruct the Scotch youth none but members of the Scotch Kirk.

Again, it seems to me to be a little unfair to the Scotch to require them to conform to the Established Church of England in order to get the good things of our universities, and at the same time to seize on the good things of the Scotch Universities without conforming to the Scotch Church. Of course there would not be the practical inconvenience of non-conformity in Scotland that there would be in England.

I do not say that the two last reasons have much weight; still they help to turn the scale.

To turn now from the question of lawfulness to the possible effect of signing, I am not at all sure that I should not get into hot water at Cambridge as Fischer has done.[3] I do not think it likely; at the same time I think it not impossible. Again, I feel that many of my high-church friends at Cambridge and elsewhere would look on me as an apostate; others would think I had done a dirty job, though most, perhaps, of my friends would think me justified. It would not be pleasant to be regarded by any in the way I mentioned, although I should not think so much of this if I were well assured in my own mind.

I should be glad to hear from you whether you think there would be any use in standing in the way I mentioned. I should think most probably not.

Yours very truly
G. G. STOKES

1 Kelvin has written at the top of the letter, 'Feb. 19, 1849'.
2 This probably refers to the appointment in 1838 to the Edinburgh chair of mathematics of Philip Kelland (1808–79), who was senior wrangler in 1834 and who was ordained deacon in the Church of England in 1837 and priest in 1838.
3 See letter 26, note 6.

39 KELVIN to STOKES, 20 February 1849
Stokes Collection, K30

2 College, Glasgow,
Feb. 20, 1849

MY DEAR STOKES

I cannot but regret that your determination is finally against what seems to me to be the course that would have been so much better for us, and for Scotland in general than that which you feel to be most satisfactory to your own honourable feelings.

Now I fear it is too late to do any good by writing a long letter, as your mind is made up. There are insurmountable obstacles at present in the way of our following the precedent of Edinburgh, our constitution is so very different with reference to the electing, and admitting bodies, as I shall explain to you again. There can be no systematic or reliable-upon sinking of the tests with us until the law is altered; & with every well-wisher of the Scotch Universities and good high education for Scotland I hope & trust our lay professorships may by act of parliament be very soon exempted. No case can prove the noxiousness of the law as it is than the present one.

I enclose a letter[1] (from Adams I presume) which, from not knowing how long you might remain at Aughnacloy, or how soon you might be in Glasgow,

I kept. Also you receive[d] the Smith's Prize papers w^h you wished to have returned.

I have just looked again at your letter, especially the last clause. I fear I cannot answer it more explicitly than I have done, and in the present state of the law, and of feeling among a few of our number regarding the tests, that a candidate could not possibly be elected on the public understanding that he would not take the tests. There might not be any considerable opposition to the avoiding of the subscription after the election had been concluded, but alas this must be a matter of uncertainty.

Write soon & tell me your movements & whether you will pass through Glasgow again. I wish you would come, and if you would stay for a few days in the house we should be very glad to receive you.

Believe me,

Your's very sincerely
WILLIAM THOMSON

1 J. C. Adams to Stokes, 15 February 1849, Stokes Collection, A151. Adams indicated that he had sent his testimonial for Stokes, since 'it is not quite certain that the ugly test will be insisted on .... If an anti-test league were got up among the Candidates from Cambridge I think our Northern friends would be induced to give it up, at least as applied to Mathematics'.

40 STOKES to KELVIN, 22 February 1849
Kelvin Collection, S343[1]

Aughnacloy Ireland[2]
Feb 22nd 1849.

DEAR THOMSON,

The reason why I did not return to Glasgow is of course that not being prepared to sign the test I considered it useless to stand. Your last letter confirmed me in my supposition of the uselessness of standing in the way I spoke of; indeed, I am almost sure I heard of the same from your own mouth when I was in Glasgow. My intention is now to go to Dublin, most likely on Monday next, where I shall probably remain two or three days, and then return to Cambridge. I think I could still sign the test – provided I could get *documentary* evidence, in the shape of an entry in the minutes of which I could get an authenticated copy if I wished, to the effect that the electing body considered the negative sort of conformity about which I spoke to you a sufficient fulfilment of the conditions imposed by the test in the case of an episcopalian subscriber about to fill a non-ecclesiastical chair. But the difficulty is that such a resolution would naturally be proposed at the time of *admission*; whereas, having no right to expect that the test would not be required, I could not honourably stand without being prepared to sign the test, even unqualified, if called upon so to do. Indeed, if a majority would agree to admit the qualification I have spoken of, they would probably not stick at the motion which Mr Mac⟨h⟩onochie[3] intends to propose if confident of a majority.

If you have anything to say in answer to this, please direct to me, Hiberinian [i.e., Hibernian] Hotel – Dawson Street – Dublin; but indeed I do not see what you can have to answer that would be of any importance.

Yours very truly
G. G. STOKES

1 S344 in the Kelvin Collection is Stokes's copy of S343.
2 Kelvin has written at the top of the letter, 'Feb 24, 1849'.
3 Allan Alexander Welwood Maconochie (d. 1885) was professor of jurisprudence in Glasgow University from 1842 to 1855.

41 KELVIN to STOKES, 24 February 1849
Stokes Collection, K31

2 College, Glasgow
Feb. 24, 1849

MY DEAR STOKES

I have nothing I fear that will be satisfactory to answer to your last letter, but still I do not like that you should arrive at the Hibernian Hotel and not hear anything from me, and so I write you a line. I do not think that that [*sic*] any entry on the minutes such as you mention or suggest could very well be made, as it would be obviously an assumption on the part of the admitting body to lay down any special interpretation of the tests, when their business is merely to see that they are subscribed, & never afterwards to meddle with the matter. Indeed nobody on earth has anything to do with the observance of the tests but the person himself who has subscribed them, & no court, or body corporate or otherwise has any legal authority in this respect. At the same time I quite agree with you that it is very desirable that the individual members of Faculty should not in conversation, or in their own feelings, cavil at the interpretation acted upon by a colleague, but I think ⟨that⟩ all those who might vote for either you or Blackburn would be quite prepared to find you conforming to the Episcopal Church openly, even after having taken the tests.

I still think the chance would be very good of getting the tests entirely set aside, without any legal measures being had recourse to to enforce them, but feel quite convinced that it will be impossible to get any arrangement of the kind made except by a *coup de main* at the admission. It would be so perfectly easy for any one individual (I believe without expense) to prevent the Faculty from ⟨advisedly⟩ deliberately acting in an illegal manner, that no negociations [*sic*] could go on without *perfect unanimity*, to elect a candidate and admit him without prescribing the tests. On the other hand, when the election is made, and the affair settled, no one would care, comparatively speaking, about the tests, whether they are taken or not. As for *unanimity*, that can scarcely exist in such an election as this, until all be certain how the election is to go; and then it is usual to do the thing ha⟨d⟩ndsomely, by a unanimous vote.

I wish I had something more agre[e]able to write to you, but I really cannot see any more light than when I first wrote to Hopkins. I am very sorry you did not succeed in reconciling yourself to the idea, but I sympathize very much with your feelings.

I remain,

Your's very sincerely
WILLIAM THOMSON

P.S. I write this very late when I am very sleepy, but I daresay you will make it out. I shall look over what I have said tomorrow & add or correct if I can or if necessary.

42 KELVIN to STOKES, 21 March 1849
Stokes Collection, K32

2 College, Glasgow
March 21, 1849

MY DEAR STOKES

I am anxious to get the matter for the next number of the Journal entirely arranged as soon as possible, and I shall be very glad to receive your paper on Waves (Note VI.)[1] as if you have it ready, as soon as you please to send it. Will you have anything else soon for the Journal? Anything for the next number should be sent before Monday the 2nd of April, as I expect by that time to have the programme nearly complete.

Since you have determined not to come forward, Blackburn's chance is nearly a certainty. His opponents are Sandeman,[2] Prof. Gray of Marischal Coll. Aberdeen,[3] Prof. David Thomson of King's Coll, Aberdeen[4] (both Nat. Philosophers) and several others of less note. There can be no doubt I think who is the best of that list.

The election was fixed by a resolution we came to yesterday, to be on Friday the 13th April.

I ought to have thanked you before for your paper on the bands in the spectrum.[5] I am not perfectly clear yet about the explanation, but that is simply because I have not had time to study it sufficiently yet. I can see the 'polarity' perfectly with the apparatus you gave me, and a very ordinary prism.

Your's (in haste)
WILLIAM THOMSON

P.S. We have just heard Macaulay's inaugural address with w^{h} we have been delighted.[6]

1 Stokes (28).
2 Archibald Sandeman (1822–93) was third wrangler in 1846 and became professor of mathematics and natural philosophy at Owens College, Manchester, in 1850.
3 David Gray (d. 1856) was professor of natural philosophy at Marischal College, Aberdeen, and after his death was replaced by Maxwell.
4 David Thomson (1817–80) graduated from Cambridge in 1839 and was professor of natural philosophy at King's College, Aberdeen, from 1845 to 1860, continuing in the University of Aberdeen from 1860 to his death. It was in the consolidation of King's and Marischal Colleges into the University in 1860 that Maxwell lost his position at Marischal while Thomson retained his. Thomson was Faraday's cousin and from 1840 to 1845 had replaced Kelvin's ailing predecessor in the Glasgow professorship of natural philosophy – William Meikleham (d. 1846).
5 Stokes (23).
6 The historian Thomas Babington Macaulay (1800–59) was rector of Glasgow University from 1848 to 1850.

43 STOKES to KELVIN, 22 March 1849
Kelvin Collection, S345

Pembroke College Cambridge[1]
March 22, 1849

DEAR THOMSON,

I send you my long promised paper on waves.[2] It has run to a greater length than I expected, but as it was meant for the *men* not for the *philosophers* I thought it best to be pretty full. It would split nicely into two notes, the first extending to the end of long waves, or it could come as one note breaking off anywhere with 'to be continued.' If however you think it prolix curtail it as you like.

I suppose you will soon be thinking of electing your professor. Blackburn I presume will walk over the course. I have long ceased to think about it except as a thing I once thought of.

We shall soon I expect publish the first report of our mathematical board.[3] I must send you one when it comes out. Nothing done as yet about Fischer. But here comes the porter for the letters.

Yours very truly
G. G. STOKES.

1 Kelvin has written at the top of the letter, 'Mar 24'.
2 Stokes (28).
3 The Board of Mathematical Studies was established in 1848 as part of the reform of the mathematical tripos. Its first report was dated 19 May 1849 and can be found with the minutes of the Board in the Cambridge University Archives. As a moderator for the tripos of 1848, Stokes was a member of the Board.

44 Kelvin to Stokes, 26 March 1849[1]

Stokes Collection, K33

The Editor of the *Cambridge and Dublin Mathematical Journal* thanks Mr. 'Stokes' for his communication dated 'Mar 22.'[2] Should his paper be considered suitable for insertion in the Journal, a notice regarding it will appear in the next Number; otherwise the Manuscript will be returned as soon as possible.

Glasgow College, 'Mar. 26' 18'49'[3]

Dear Stokes

⟨I mu⟩ Our letters must have crossed the consequence of w[h] is that mine was answered before return of post. Either the whole or a part of your paper will of course appear in May.[4] I have just today had an arrival of apparatus from Paris, principally electric, & very charming.

Your's truly
William Thomson

1 The first part of this letter is a form letter in which Kelvin's additions are enclosed within inverted commas.
2 Stokes (28).
3 The rest of the letter is in Kelvin's hand.
4 Actually, it appeared in November. See the next letter.

45 Stokes to Kelvin, 29 March 1849

Kelvin Collection, S346

Pembroke College, Cambridge[1]
March 29, 1849.

Dear Thomson,

We have had a good deal of discussion at the meetings of our mathematical board[2] respecting the Figure of the Earth & Laplace's coefficients. With respect to the first, some thought it a subject which did not at all repay for the trouble it took, but some were very strongly opposed to the rejection of Clairaut's Theorem. As to Laplace's Coefficients, no one I think wished to retain them if it were not for the figure of the Earth. Some were for making the men read the figure of the Earth without Laplace's Coefficients, and some thought they ought not to be prevented, if they read the subject at all, from reading it in the best way. To come to the gist of this note, a happy thought struck me, on following out which I succeeded in proving Clairaut's Theorem without any hypothesis as to the original fluidity of the Earth, or as to the constitution of its interior, provided we assume, as a matter of observation, that the surface is spheroidal, and ⊥ to the direction of gravity. That being the case, I for my part think the Figure of the Earth is not worth reading, as a subject for examination here, inasmuch as the only result of so much mathematics, which can be compared with exper^t, and which is not independent of the hypothesis of original fluidity, is the spheroidal figure; and it does not require much stretch of imagination to conceive that a whirling mass of fluid will assume a spheroidal form when the centrifugal force is small compared with the attraction. I do not speak of the perpendicularity of gravity to the surface as a *result* of the mathematics, because it is one of the conditions we start with; and *so much* would be known by a man who merely knew his Hydrostatics. I am writing a paper on Clairaut's Theorem for the Philosoph^l,[3] in which I introduce Laplace's Coeff^ts, but I propose to write another demonstration without Laplace's Coeff^ts for the math^l journal,[4] for the sake of the men. I should propose, if it be not inconvenient to you, & if I can have my paper ready in time, that it should take the place of my paper on Waves which can wait till next time. I mean supposing 'Clairaut's Theorem' reaches you before 'Waves' is put in type.[5] I do not know whether you may not have anticipated me; perhaps you have known what I tell you about Clairaut's Theorem this long time, as Attractions are a favourite subject with you.

Yours very truly
G. G. STOKES

P.S. I rec^d your note the day after I sent my 'Waves'. I got your other note this morning.

1 Kelvin has written at the top of the letter, 'Mar 31'.

2 The Board of Mathematical Studies. See letter 43, note 3. The Board's second annual report, dated 3 June 1850, stated that Laplace's coefficients 'may be omitted altogether' from the mathematical tripos. Concerning the figure of the earth, it stated:

> Questions in the Theory of Attraction, and of the figure of the Earth considered as homogeneous, may be proposed with advantage.
> The Theory of the Figure of the Earth considered as heterogeneous may be omitted.

By this time, Stokes was a member of the Board as Lucasian professor. (See letter 51.) The report is with the minutes of the Board in the Cambridge University Archives.

3 Stokes (31).

4 Stokes (27).

5 Stokes (27) on Clairaut's theorem was divided between the May and November numbers of the journal, and Stokes (28) on waves appeared in November.

46 Kelvin to Stokes, [31 March 1849][1]
Stokes Collection, K33A

2 College, Glasgow
Saturday evening

My dear Stokes

By all means send your paper on Clairault's theorem[2] for the next number of the journal, if you can have it ready by Wed. Ap. 11, so that I may receive it on Frid. or on Saturday morning. I am not sure how much space there will be, but I shall keep as much as possible for it, or for it & waves. You will not I sh^d think have to occasion any delay about the proof sheets, w^h probably will not require much correction, & I suppose you will be on the spot. I wish the publication to be literally on the 1st of May.

I am very much interested in what you tell me about the consultations of the mathematical board. I agree perfectly with you all that Laplace's coefficients cannot do much good even to the best among the undergraduates. I am sure I wish you would take Trigonometrical series instead. Any one can understand them, & what a beautiful subject it is & how useful not only in high physical science, but in every practical branch Meteorology, Terrestrial magnetism, & astronomy! I think it would be a very great pity however that the Figure of the Earth should be abolished. I would much rather see the Planetary Theory set aside (the Lunar Theory of course being retained) although I think that ⟨it⟩ both subjects ought decidedly to be retained, for the highest class of men. Why not read the Figure of the Earth in Airy's Treatise in the Enc[yclopaedia] Met[ropolitana]?[3] O'Brien[4] is certainly very neat but then there is the fatal objection ⟨g⟩ we all agree in. Airy's Treatise contains actual information on the subject, besides the Theory given in a very complete way (except of course for the assumption that the surfaces of equal dens^y are spheroidal) and in a form w^h renders it perfectly intelligible to any industrious reader. It all contains the most interesting ⟨p⟩ [?] investigation regarding the form of a revolving homogeneous fluid w^h is omitted by O'Brien (not given even by Pratt, if I remember right).[5] I discovered yesterday that Airy's treatise may be had separately from Griffin and Co. proprietors of the Enc. Met., for 4/, as I had occasion to buy several of the articles to make *select* volumes, for prizes. Airy's Tract[6] on the figure of the Earth is *I* think unreadable, & far more difficult than his Treatise in the Enc. Met. so I found at least, a long time ago when I was engaged in writing an Essay[7] on the subject, for a prize prescribed by this University. At that time I could make nothing of Pratt's treatment of the subject, nor of any other involving Laplace's coeff^ts; I was disgusted with Airy's Tract; but I was much interested & pleased with his Treatise.

Your's very truly
in great haste
William Thomson

1 March 31st was the first Saturday after March 29th, the date of the previous letter.
2 Stokes (27).
3 George Biddell Airy, 'Figure of the earth', in Edward Smedley, Hugh James Rose, and Henry John Rose, eds., *Encyclopaedia Metropolitana*, 20 vols. (London, 1845), v, 165–240.
4 Matthew O'Brien, *Mathematical Tracts* (Cambridge, 1840). O'Brien (1814–55) was third wrangler in 1838 and professor of natural philosophy and astronomy at King's College, London, from 1844 to 1854.
5 John Henry Pratt, *The Mathematical Principles of Mechanical Philosophy, and Their Application to the Theory of Universal Gravitation* (Cambridge, 1836).
6 George Biddell Airy, *Mathematical Tracts on the Lunar and Planetary Theories, the Figure of the Earth, Precession and Nutation, the Calculus of Variations, and the Undulatory Theory of Optics*, 2nd edn (Cambridge and London, 1831).
7 The notebook containing Kelvin's essay on the figure of the earth is in the Kelvin Collection, NB11.

47 STOKES to KELVIN, 11 April 1849

Kelvin Collection, S347

Pembroke Coll. Cambridge[1]
Wednesday Ap 11, 1849.

DEAR THOMSON,

It is just post time. I send my Clairaut's theorem &c. &c.[2] I have gone a little into potentials. I have given the demonstration you know of the eq^{n} $\nabla V = -4\pi\rho$.[3] I don't know whether we are at cross purposes, but if you have written it for this n^{o} of the Journal you can strike it out. I fear I have at first been too prolix. I was writing for 'men' (not savans) but I felt as if I was writing for mathematical babes. If you have a convenient reference you can avoid the foot note at p. 10 (see back of page) or at least the analysis in it. I fear my dissertation on potentials is somewhat of a rigmarole.

Yours very truly
G. G. STOKES

1 Kelvin has written at the top of the letter, 'Ap 13'.
2 Stokes (27).
3 For Stokes's use of ∇, see letter 32, note 1.

48 KELVIN to STOKES, 21 April 1849

Stokes Collection, K34

2 College, Glasgow,[1]
April 21, 1849

MY DEAR STOKES

I this evening ⟨eve⟩ [?] send off your paper on Attr^{ns} &c[2] to the printers. Will you (if you can possibly spare the time) get the 3^{d} Vol of Taylor's

Scientific Memoirs for the sake of reference;[3] & call on the printers and get your M.S.S. from them in such a way as not to interfere with their working upon it (for instance after they give up work for the evening they would send it to you & then call for it at the College porter's next morning at 6 o'clock)? I would like you to see some notes I have added in pen (I have been very stupid not to put them on the blank sides of your leaves where they would have been much more distinct) and some suggestions in pencil with ref^{ce} to w^{h} you will perhaps make some remarks. There is no need for me explaining to *you* my object in wanting references &c. As far as I am concerned myself I claim nothing but relationship. All the theorems in question w^{h} I had found out myself & published had been all given before by Green; & in those I valued most I had been triply anticipated; but I think it will add to the interest of your paper, besides bringing mine[4] into notice, to give references from one to another.

The printers have been behaving very badly & are I fear terribly behind so I hope you will both exhort them, & not keep them waiting for your proof. The number *must* appear on the 1st of May.

We elected Blackburn yesterday week, & we prescribed 'de Mathematices ad Naturam Explorandam Usu' as a subject on which he is to write a Latin Essay to prove himself competent. We are to hear it tomorrow[5] & if we are satisfied with his performance we are to admit him, provided he shall take the tests when the Presbytery shall meet.

I expect to arrive in Cambridge on Friday week.

Your's in much haste to get to bed.

WILLIAM THOMSON

1 Stokes has written the following on the envelope:

Poggendor[f]f's Annalen der Physik und Chemie. Band 49. page 241.

16th meeting. report 59. Hopkins. Ordnance Survey. Shanklin Down.

The first reference is to Johann Gottfried Galle, 'Ueber Höfe und Nebensonnen', Poggendorff's *Annalen der Physik und Chemie*, **49** (1840), 241–91. Galle (1812–1910) was an assistant at the Berlin Observatory from 1835 to 1851 and was professor and director of the observatory at Breslau from 1851 to 1897. The second reference is to William Hopkins, 'On certain deviations of the plumb-line from its mean direction, as observed in the neighbourhood of Shanklin Down, in the Isle of Wight, during the progress of the Ordnance Survey', *Brit. Assoc. Rep.* (1846), part 2, p. 59.

2 Stokes (27).

3 In his (27), Stokes cited C. F. Gauss, 'General propositions relating to attractive and repulsive forces acting in the inverse ratio of the square of the distance', Taylor's *Scientific Memoirs*, 3 (1843), 153–96. (*MPP*, II, 105.)

4 Kelvin (2), cited in Stokes (27).

5 Kelvin has written between the lines here, 'on Monday (unless after mid-night the day after tomorrow is tomorrow'.

49 STOKES to KELVIN, 26 April 1849
Kelvin Collection, S348

Pembroke College, Cambridge[1]
April 26th 1849.

DEAR THOMSON,

When I asked whether you had anticipated me I spoke merely with reference to the idea of applying the theory of potentials to the determination of the variation of gravity at the surface of the Earth. I never dreamt of claiming originality on the ground of the propositions in attractions, which were given merely for the sake of the 'men', who could hardly be expected to be acquainted with them. Some of the propositions may be proved in a manner a trifle different from what others have done but that is nothing. The only thing in the 'Propositions' for which I take any credit is the proof of the equation

$$\frac{d^2V}{dx^2}+\frac{d^2V}{dy^2}+\frac{d^2V}{dz^2}=-4\pi\varrho$$

and with respect to that the credit is rather due to you & I hope I have done you justice. I have added a note to the M.S. explaining why I gave the propositions at all.[2]

I kept the M.S. one evening in the way you wished, sending it off in the morning, having made the alterations before I went to bed.

I have asked for two proofs of my paper, in order to send you one if there be time. I know you get a proof of the whole but I thought it might save time to send you one as soon as my own was struck off.

Yours very truly
G. G. STOKES.

1 Kelvin has written at the top of the letter, 'Ap 29, 1849'.
2 Stokes (27). The note is in *MPP*, II, 105.

50 KELVIN to STOKES, [29 April 1849][1]
Stokes Collection, K34A

2 College, Sunday evening[2]

MY DEAR STOKES

On account of my movements to terminate on Friday evening in Cambridge, I have directed the printers to strike off the 7th sheet at your *imprimatur*, which you will be able to give I shd think within a few hours of receiving the revise. As far as my corrections are concerned the only thing to be attended to is that my changes (which are rather roughly marked in the proof) in the note on p. 197 are not bungled by the printers.[3] I have cancelled the second note for w^h there is no occasion after your note p. 194.[4] In the very hurried glance through your M.S.S. w^h was all I have been able to give, it struck me that you had done me more than justice in your note near the end,[5] and I intended to suggest some alteration in the proof sheet, but I see the printers have not got so far and we can talk over the matter leisurely. I am glad they have got in Clairault's theorem however as that gives some completeness the want of w^h might have been felt on acc^t of the long interval till the next number appears.

I have directed the printers also to send you a proof of the cover. You will oblige me much (and the British mathematical public too by so allowing the Number to appear at the promised time or within 24 h. of it) by glancing at the Table of contents and at the notices just to see that there is nothing absurd.

Ought it not to be Clairault? (French books will be correct, or sufficiently correct on that p^t).

I see no heading on the proof I have received but no doubt you have set that right.

I have such a day of it tomorrow; & on Tuesday after our 'Distribution of Prizes' I leave town, that I cannot delay till tomorrow writing all this *business*.

Your's very sincerely
WILLIAM THOMSON

1 April 29th was the Sunday before Kelvin's publication deadline of May 1st. (See letter 48.)
2 Kelvin has written at the top of the letter, 'If you have any thing to tell me before I see you, address Meadowbank House, Kirknewton, Edinburgh'.
3 There is no note on page 197 of Stokes (27) in the *Cambridge and Dublin Mathematical Journal*, and Kelvin was probably referring to his note which appears on page 198 (or in Stokes's *MPP*, II, 108).
4 Stokes's note is in *MPP*, II, 104.
5 Kelvin is mentioned in two notes 'near the end' of Stokes (27) – on pages 213 and 219 in the original and in Stokes's *MPP*, II, 124 and 130.

51 STOKES to KELVIN, 23 October 1849
Kelvin Collection, S349

Pembroke College Cambridge[1]
Oct 23^d 1849.

DEAR THOMSON,

Budd[2] has returned, & I have handed him over the paper you left me for his examination. I have not yet looked into it myself.

I have convinced myself of the following theorem, to communicate which is the principal object of my writing.

Let S be a closed surface containing within it attracting matter anyhow distributed: we may always, without affecting the attraction outside *S*, distribute the matter in *open surfaces*, lying in general within *S*; and this mode of distribution (which I propose to call *fontal*) is unique.

It is to be understood *first* that singular lines and points are included as limiting cases of surfaces, *secondly* that a point of zero-density on a closed surface is to be regarded as an infinitely small hole, and such a surface is to be regarded as open.

In the case of a conducting sphere in communication with the ground, and influenced by an electrical point, the image of the point is the fontal distribution of the electricity induced on the surface.

Yours very truly
G. G. STOKES,
Lucasian Professor of Mathematics
in the University of Cambridge.

P.S. I was elected and admitted this morning. There was no other candidate.

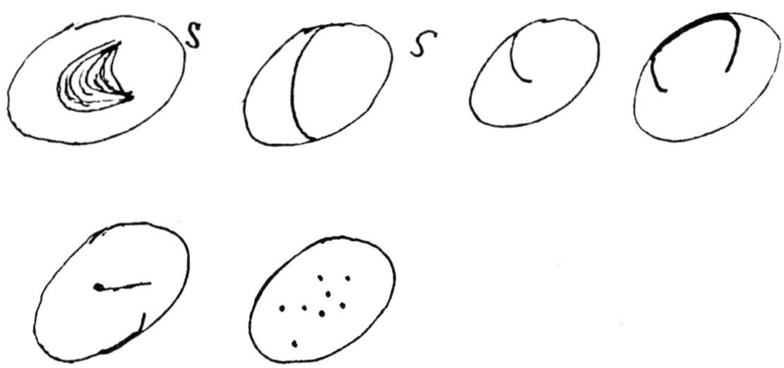

Figures of fontal distributions

1 Kelvin has written at the top of the letter, 'Oct 25, 1849'.
2 Charles Octavius Budd (c. 1821–90) was third wrangler in 1844 and became a fellow of Stokes's college, Pembroke, in 1848.

52 STOKES to KELVIN, 25 October 1849
Kelvin Collection, S350

Pembroke College, Cambridge[1]
Oct 25th 1849

DEAR THOMSON,

I sent off my theorem before it was sufficiently matured. I wanted to announce to you my election, & at the same time to communicate the theorem, which I believed to be true, though there was one step about which I had some misgivings, not having obtained a mathematical or physical demonstration ⟨demonstration⟩ of it, although from the way I looked on it I certainly thought it was true. It *is* true that the mass may be distributed in open surfaces, (including singular lines and points,) or in closed surfaces where the density vanishes at one or more isolated points, or along one or more lines. It *is* true also that a distribution in open surfaces (at least when a certain condition of max^{m} is ⟨st⟩ [?] satisfied, which I suppose to be satisfied,) is unique. The doubtful point is, whether a closed surface with a zero-density at a point or along a line in it is to be regarded as an open surface, to which the character of uniqueness applies, or whether such a distribution, in order to fulfil the requisite condition of max^{m}, must be exchanged for another in really open surfaces. I now see that among the infinite n^{o} of equivalent distributions, a distribution in open surfaces, among which is admitted a closed surface with a zero-density at one point, is not necessarily unique. I have a glimmering of a demonstration that such a surface does not satisfy the requisite condition of max^{m}.

The quantity which is a max^{m} is the mean potential through[ou]t the space bounded by *S*. This quantity is $+^{ve}$ and finite, & ∴ admits of a max^{m} and min^{m}. The min^{m} gives a distribution confined to *S*, and is unique; but it is the max^{m} I at present most care about.

I hope it may not be an ill omen that the letter in which I signed myself Lucasian Professor should contain a false theorem. I would not have communicated such an immature speculation to one with whom I was less intimate, nor even to you, only that I wanted to announce my election, & thought I might as well do the two things at once.

Yours very truly
G. G. STOKES.

1 Kelvin has written at the top of the letter, 'Oct 2⟨6⟩ 7, 1849'.

53 KELVIN to STOKES, 25 October 1849
Stokes Collection, K35

2 College, Glasgow
Oct 25, 1849

MY DEAR PROFESSOR

I have been daily expecting to hear of the election of a Lucasian Professor, and whenever the Times has been in my hands I have looked for such a proceeding in the University Intelligence, and now I am glad to be able to congratulate you on the result. I am very glad for the sake of mathematics as well as for your own that you have got it as you will now have every inducement to go on as you have been doing, and we shall feel much surer of you than we could when you had only your fellowship to connect you with Cambridge. No wonder you have just discovered a theorem! Are you going to deliver lectures on physical mathematics, or on what branch, & when?

I have often tried without success ⟨ive⟩ to arrive at some positive conclusion on the subject of your discovery, as you may see by Notes on Some Points in the Theory of Heat in the fourth (possibly the third) volume of the Math. Journal.[1]

I thought I had arrived at a contrary conclusion to your's by considering arbitrary distributions of matter on a closed surface of any kind, for instance a spherical surface, but now I *see the hole* by w^{h} I suppose I must get out of my mistake. Instead of a spherical surface with a given arbitrary distribution of matter over it, we may substitute a point at the centre of the sphere, and a distribution on the spherical surface with a vanishing density at *one point*. This ⟨last⟩ arrangement will be determinate provided negative densities be excluded from your charged fontal surfaces. I cannot however see that your mode of distribution in general is unique. For, to keep to the case of a spherical surface, we may take any excentric point within it, place a certain quantity of matter at that point, and destroy its action on points without the sphere by distributing an equal quantity of the opposite kind of matter over the spherical surface acc^{g} [to] the inverse cube of dist[ance]. This quantity may be so chosen that the dens^{y} at one point of the sp^{l} surf. may be zero, and at all others positive. Thus we have a complex 'fontal distribution' consisting of a p^{t} and a sp^{l} surface with 'an infinitely small hole' in it; and yet such a fontal distribution, to produce a given potential at all points ext^{l} to the sp^{l} surface, is not unique. All this is applicable to any closed surface, even discontinuous, as a cube; (except of course that the law will not be that of the inverse cube of the dist^{ce} as above.) I shall be very glad to hear from you again on this subject, and if I have not taken up correctly what you told me, to be set right.

The conclusion with reference to sp^{l} surfaces that I rested in formerly was this.

Let $A_0Q_0 + A_1Q_1 + A_2Q_2 + \ldots$ represent ⟨at⟩ the ⟨el⟩ dens^{y} at any p^{t} ⟨*D*⟩ [?] *E* of a sph^{l} surface, Q_0, Q_1, &c being coeff^{ts} in the expansion of $(1 - 2xp + x^2)^{-1/2}$, where p is the cosine of the angle betw. a rad[iu]s thr. *E* and any axis, different for the different terms of the series. Then if A_0, A_1, A_2, &c ultimately converge less rapidly than any convergent geom^{l} prog[ressio]n (i.e. if $\frac{A_n}{A_{n+1}} = 1$ when $n = \infty$) it will be impossible to produce the same pot^{l} for p^{ts} without the given surf., by any dist[ributio]n whatever on any surface lying wholly within the given surface. On the other hand if $\frac{A_n}{A_{n+1}} = e$,[2] when $n = \infty$) e being < 1; the same pot^{l} on all p^{ts} without the given surf. may be produced by a dist^{n} on a conc. sph^{l} surf. of rad. *ea*, *a* being the rad. of the given surf.; or by a dist^{n} on any closed surface drawn between the two. If you look at the paper I referred to above you will see that I guessed wrong about the complex surface consisting of two sph^{l} surf^{s}.[3] I have always intended, since I knew this to put a note in the Journal to correct my false conjecture, but now I think I may wait till I get to the problem of the two spheres in the series of papers I am publishing on elec^{y}.

You must excuse the confusedness of this, w^{h} I sh^{d} re-write if I had more time.

Your's sincerely,
WILLIAM THOMSON

To The Lucasian Professor

P.S. I have just received a letter from the Savilian Professor, of Oxford,[4] in w^{h} incredulity as to Prof. Challis's 'new equation', so we have now 3 professors against one.[5]

1 Kelvin (10).
2 Above the line here, Kelvin has deleted two or three letters which are now illegible.
3 Kelvin added a correction to the version of his (10) which appeared in *MPP*, 1, 39–47. The correction is dated 26 June 1881 and is on pages 45–7.
4 There are no letters in the Kelvin Collection or the Kelvin Papers at Glasgow from either Baden Powell, Savilian professor of geometry, or William Fishburn Donkin (1814–69), Savilian professor of astronomy from 1842 until his death. However, Kelvin probably meant Donkin. Donkin's publications were more mathematical than Powell's, and he published articles in Kelvin's *Cambridge and Dublin Mathematical Journal* in 1850 and 1851, suggesting a connection between the two men.
5 See James Challis, 'On a new fundamental equation in hydrodynamics', *Camb. Phil. Soc. Trans.*, **8** (1849), 31–43 [read in 1843]; 'On a new equation in hydrodynamics', *Phil. Mag.*, **36** (1850), 295–302; and 'On the principles of hydrodynamics', *Phil. Mag.*, **1** (1851), 26–38, 231–41, 477–8, and **4** (1852), 438–50. For Stokes's response, see Stokes (45), entitled 'On the alleged necessity for a new general equation in hydrodynamics'.

54 KELVIN to STOKES, 27 October 1849
Stokes Collection, K36

2 College, Glasgow
Oct 27, 1849

MY DEAR STOKES

I am much interested in the subject of your two letters, and I hope as soon as you can, you will give me a sketch of your investigations.

I am very curious to know what you think, or what you guess, might be the nature of the fontal distribution to produce a constant potential at the surface of a cube. I used always to think the most *primitive* distribution ⟨was⟩ must be a dist^{n} on the surface itself in this case. Certainly ⟨a cubic⟩ the surface of a cube is what we might call the *primitive* of ⟨a⟩ the series of isothermal surfaces external to it; but you have suggested ideas which never occurred to me before. Allowing some matter to remain distributed ⟨of⟩ over the surface of the cube, ⟨and there is clearly⟩ we may have some in the interior, and so get a *closer* or more compact distribution, to produce the given potential. I am greatly puzzled in the treatment of your problem to make the mean potential within S a maximum or minimum, & I shall be very glad to hear [f]rom you anything about it. I suppose, besides the cond^{n} that the potential has a given value at each point of the surface, you lay it down that there must be no negative matter. Without this latter condition we might, ⟨produce⟩ by dist^{ns} within S ⟨giving⟩ producing the given superficial potential, make the mean interior potential have as great a positive, or as great a negative value as we please.

Green's & Gauss's ⟨maximum⟩ *minimum* considerations have suggested to me to try what function it is that is a minimum when all the matter to produce a constant potential, or, again, a given variable potential, at the surface S is actually distributed on S, instead of through the interior. I have often thought, and I still am inclined to think that, when the potential is to be constant at S, and the matter to be condensed from an infinitely diffuse state, the least possible amount of work will be required, if we bring the matter only to the surface S; or,in fewer words, that the *mechanical value* of the superficial distribution is less than that of any other which produces a given cons^{t} pot^{l} at S. When the given pot^{l} is not cons^{t} it must be some other function, which it will be easy to discover by analogy from Gauss's $\int\int(V - 2\Omega)\zeta d\sigma$. I think it very likely that, in the passage I have marked with a marginal bar, the interlined word 'cons^{t}' may be omitted,[1] provided we consider the whole space within S to be subject to the force of external matter producing a pot^{l} ⟨Ω⟩ equal to the given variable potential, at the surface. From your letter which I received today, it would appear that the superficial distribution has also the property of making the mean interior potential a maximum. Is this certainly true? If it is it is most interesting; but at present I am greatly puzzled.

I am at present in the middle of a paper on Magnetism the first part of w^{h} was communicated last June, and the place where I stopped, in my writing is exactly where considerations like these occur.[2] I have had all the matter ready, and a good deal of it roughly put on paper for a long time, and so I hope soon to get it reduced to publishable form. I am the more anxious on this account to know more of what you have been telling me about, and I hope you will write soon. I shall of course avoid anything suggested by what you have written to me, unless you have published something in time to allow me a proper reference, but in the mean time I should like to find out how much connection there is between our speculations on this subject. I think you know nearly all I have upon it, but if you care I shall write out for you as briefly as possible the principal points of the investigations I intend to publish in this part of my paper.

I had a letter from Cayley today in w^{h} he said he had heard that Stokes of Caius [i.e., W. H. Stokes] had been elected Lucasian Professor![3] Will you put him right?

Your's sincerely
WILLIAM THOMSON

1 The passage which Kelvin has marked with a bar is: *fewer words, that .... When the given pot^{l} is not.* The word *cons^{t}* in the passage is written above the line.

2 Kelvin (53), which was read to the Royal Society in two parts on 21 June 1849 and 20 June 1850.

3 The letter from Arthur Cayley to Kelvin has not survived in the Kelvin Collection or the Kelvin Papers at Glasgow, but the subsequent letter from Stokes to Cayley, dated 29 October 1849, has survived (Stokes Collection, C200) and is printed in Stokes's *Memoir*, 1, 381–2. Cayley (1821–95) was senior wrangler in 1842, was called to the bar in London in 1849, and in 1863 became the first Sadlerian professor of pure mathematics at Cambridge. A substantial correspondence survives between him and both Stokes and Kelvin. The already mentioned letter from Stokes to Cayley nicely indicates the difference in interests between the mathematician Cayley and the two natural philosophers. Stokes wrote that 'Thomson and I are at present writing to each other about potentials. I think that potentials may throw light on the interpretation of $f(x + \sqrt{-1}\, y)$. How horrible you would think it to prove, even in one's own mind, a proposition in pure mathematics by means of physics'.

55 STOKES to KELVIN, 29 October 1849

Kelvin Collection, S351[1]

Pembroke College, Cambridge[2]
Oct 29th 1849.

DEAR THOMSON,

I fear my 'discovery,' as you call it, is an abortion, brought into premature existence by my haste to communicate it to you. Only that I wanted to write to announce my election, I would have waited till it was more mature, or else put it off *sine die*. I will now communicate the ideas by which I have been led towards some such theorem, and so lay it by for the present. My first letter deserves no better fate than the fire.

In the first place I will observe that I have all along supposed the mass to be positive. You have considered potentials more with reference to electricity, I have been considering them of late more with reference to gravitation. In fact, my train of ideas was suggested to me by something which had occurred to me while answering a letter of Airy's relating to Clairaut's Theorem.[3]

Consider the various equivalent distributions of positive matter within a closed surface S. Consider also the integral I defined by the $=^{n}$

$$I = \iiint \text{potential} \times \text{diff}^{l} \text{ of volume}$$

taken within S, or the mean potential to which it is proportional. It is evident that I must be $+^{ve}$, and likewise finite, since that would be the case even if the whole mass were collected in a point. Hence I must admit of a $\max^{m}$ and $\min^{m}$.

Let S' be a sph^{l} surface having for centre an internal point P, and lying wholly within S, or at most only touching it. A mass m placed at P and an equal mass distributed uniformly on S' would be equivalent distributions. The part of I due to the mass m and to the space between S and S' will be the same in the two cases. For the part due to m & the space within S', we shall have if a be the radius of S'.

in the first case

$$4\pi\int_0^a \frac{m}{r}\cdot r^2dr = 2\pi ma^2$$

In the 2nd

$$4\pi\int_0^a \frac{m}{a}\cdot r^2dr = \frac{4}{3}\pi ma^2,$$

which is less than the other. More generally, let S' be any closed surface surrounding P and interior to S, or at least nowhere exterior to it, $-\varrho$ the electrical density on S' due to $+m$ at P. ϱ will be always $+^{ve}$, or at least never $-^{ve}$, being equal to zero only in the case of a reentrant angle, cusp, or conical point. Then m at P and ϱ over S' will be equivalent distributions, and the difference of the potentials, which is that due to $+m$ at P & $-\varrho$ over S', ⟨increa⟩ decreases from $+\infty$ at P to O at S'; and therefore the part of I due to m at P is greater than if m were distributed at S' according to the density ϱ.

{If negative mass were admissible m might have any value from $-\infty$ to $+\infty$, and consequently the same would be true of I, & there would be no $\max^{m}$ and $\min^{m}$}

A. When I is a $\min^{m}$ the matter is confined to S.

For if there were any matter m at an internal point P, I might be diminished by distributing m as before over any closed surface surrounding P and not lying outside S.

B. This distribution is unique.

For if there were two such D, D', $D-D'$ would give a zero-potential without S and we should have possible an electrical distribution on the wall of a cavity in a conducting solid, without any internal electricity to induce such a distribution.

C. The surface-density is everywhere $+^{ve}$, except at what I shall call for brevity *salient points*, where the density is equal to zero, unless the mass were originally, in whole or in part, superficially distributed in such a manner as to give a ⟨zero⟩ finite density at such points.

By salient points I mean the points which belong to the edge of a salient angle or cusp, formed by two curved surfaces, or the vertex of a salient conical point or break.

D. When I is a $\max^{m}$ there is no matter distributed over a space.

For if there were, it would be possible to draw a sphere in this space, in w^{h} the density is everywhere not less than a finite quantity $\varrho_\circ$, and the mass corresponding to the volume of this sphere and the density $\varrho_\circ$ could be collected at the centre without either affecting the $attr^{n}$ outside S or introducing $-^{ve}$ matter within S. But this change would increase I.

Hence when I is a $\max^{m}$ the distribution can only be in *surfaces*, including lines and points.

E. These surfaces cannot be closed, with a den^{y} everywhere $+^{ve}$.

For if possible Let [*sic*] S' be such a surface, let P be an internal point, $-\varrho$ the electricity induced on S' by $+m$ at P. Then ϱ will be positive, and will $\propto m$, and therefore by taking m small enough we may remove the mass m corresponding to den^{y} ϱ from S' and place it at P without introducing negative matter. But this change will increase I.

F. Hitch in E. If S' have a re-entrant point or points, ϱ will there be infinite, and the demonstration will fail.

G. This distribution (that for w^{h} I is a $\max^{m}$) is unique (? ? ?)[4]

I think it likely that the distribution for which I is a $\max^{m}$ is unique but I think there must be some character of such a distribution besides its being confined to open surfaces. For let the tail ends A, A'; B, B' of two ⟨sm⟩ symmetrically

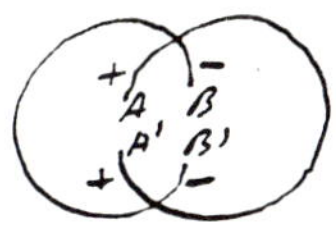

situated sph^{l} surfaces be non-conducting & let them be charged with $electr^{y}$ symmetrically as in the figure. Then if the rest of the segments be conducting, neg^{ve} electricity will be induced on the left surface & $+^{ve}$ on the right and the external potential will be zero so that two distributions such as

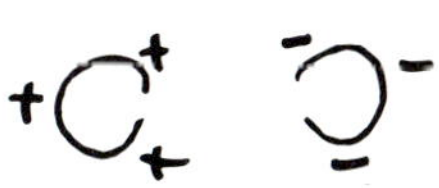

are equivalent.

Pray excuse the scrawl of the writing and the immaturity of the matter and believe me

Yours very truly
G. G. STOKES.

P.S. It was reasoning quite similar to what you have employed that led me to see that I was in error as to the equivalence of a point of zero density and a pole.

1 S352 in the Kelvin Collection is Stokes's copy of S351. See note 4.

2 Kelvin has written at the top of the letter, 'Oct 31, 1849'.

3 See Stokes to Airy, 9 October 1849, in Stokes's *Memoir*, II, 163, and in the Stokes Collection, A261. It is an answer to Airy's letters of 5 and 8 October 1849 (Stokes Collection, A259 and A260).

4 The next portion of the letter is missing. It apparently consisted of one folded sheet with four pages. The above portion of the letter consists of two such sheets whose eight pages are number 1 to 8. The following portion consists of two pages numbered 13 and 14. The missing portion can be partially reconstructed from S352 which is Stokes's rough copy of S351. The following runs from Stokes's point *G* to the end of the copy:

G This distribution is unique (? ? ?)
If poss[ible] ... D and D' $D - D'$ cannot leave isolated masses

nor tails

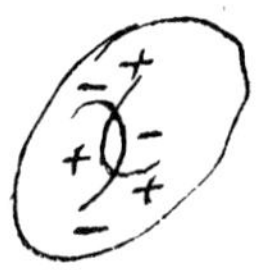

nor come so

for then $\frac{D+D'}{2}$ would give I the same max^{m} value contrary to E.
On account of the hitch F it is not proved that $\frac{D+D'}{2}$ cannot be so

The points left requiring exam[n] are
1st case of a closed surface with a den[y] =0 at one point
2 . . . with a re-entrant point

Closed surface with a zero-den[y] at Q. Try patching the zero-density point & placing a mass *m* at *P*.

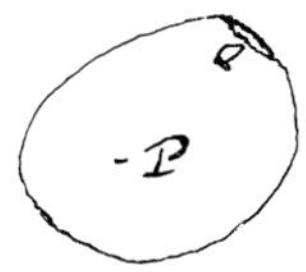

d[itt]o for re-entrant.

56 Stokes to Kelvin, 29 October 1849
Kelvin Collection, S353

Dear Thomson,

In speaking of the fontal distribution in the case of the cube I believe I wrote 8 planes when I should have written 6. You will no doubt see that it is an error.

Yours very truly
G. G. Stokes

Pembroke Coll.
Oct 29th 1849

The whole of each plane joining two opposite edges (or rather of that part of it which lies within the cube) is not to be taken, but only two △[s], namely those which have the centre for vertex & the edges for bases.[1]

P.S. N[o] 2. I received your second letter at Hall-time. I have already as you see explained my views, imperfect as they are.

As to the cube, I think a very probable fontal distribution to be a distribution in 8 planes, passing each through two opposite edges, & of course through the centre, and thus forming the triangular faces of 6 pyramids on square bases, having the centre of the cube for a common vertex, the superficial density (or mass referred to a unit of surface) becoming infinite for the edges, and the edge-density again becoming infinite at the corners, where the edges stop.

As to what you say about the *mechanical value* of a surface distribution, does it not follow at once from a general mechanical principle, combined with the consideration of electricity? If the work done (in expanding, suppose, from a condensed state) could be a max^{m} for an interior distribution, such a distribution of electricity within a conducting solid would be in stable equilibrium, whereas it cannot be in eq^{m} at all, stable or unstable. Your conclusion with respect to the case in which the external potential is variable follows, if I mistake not, at once from the same consideration; but here the work to be considered is that due to the mutual forces of the matter to be distributed + the work of the external forces on this matter.

The latter is expressed by $\int\int\int V\rho dxdydz$ + const V being the potential arising from the given potential at the surface, and $\rho dxdydz$ an element of the mass to be distributed.

If I said the mean potential was a maximum when the matter was distributed at the surface it was a slip of the pen: I meant a minimum.

I will write tonight to Cayley to set him right.[2]

Pray make free use of my imperfect investigation if it be of any use to you.

1 The remainder of the letter is written on one sheet with four pages, numbered beginning with 15. That is, it is a continuation of the previous letter. (See note 4 of the previous letter.) The first part of this letter is written on a small sheet of paper pasted to the larger sheet.

2 See letter 54, note 3.

57 Kelvin to Stokes, 19 November 1849

Stokes Collection, K37

2 College, Glasgow
Nov. 19, 1849

My dear Stokes

I am quite ashamed of having been so dilatory in thanking you for your long letter of Oct 29. I cannot now delay any longer, as I have besides to thank you for [your] paper[1] on the Railway equation, w^{h} I received on Saturday night. I am quite delighted with your plates, and I hope they will also appear in connection with Willis's Report.[2]

In consequence of some conversation I had with one of my brothers [James Thomson], who is an engineer, on the subject of Professor Willis's and your researches, it occurred to me that a bridge, if without inertia, might be made of such a form, slightly convex, that when a train passes over it at any velocity whatever, the trajectory will be a straight line. The form of the bridge would be merely the equil^{m} curve inverted. Hence if it were not for the inertia of the bridge, it might be so made that there would be no more tendency to break at high velocities than at low. The inertia of the bridge will however interfere, probably in practice to such an extent that this idea will be of no use. If you see

Prof. Willis will you ⟨say⟩ tell him of this and ask whether he thinks anything could be made of it? At present all I can see is that the road way of a bridge might be so curved that a train of a certain weight passing over it at a certain velocity, ⟨c⟩ would not by its own inertia help to break the bridge. The solution of the problem would not I think be very difficult, but I have not yet had time to attempt it.

I read your long letter of the 29th, with much interest, and I think it is very likely that ⟨it⟩ your *discovery* may lead to extremely curious results, but I do not yet see any application with reference to magnetism & electromagnetism with w^h I am a good deal concerned at present. Do you think it w^d be possible actually to investigate the fontal distribution in the case of the cube. There will of course be left a distribution of matter on the square faces (I suppose with a vanishing density along the edges) w^h, along with the interior distribution (on the six planes, according to your conjecture), will produce a constant potential at the surface of the cube.

I send, by the same post copies of two papers[3] by me & my brother.

Your's ever
WILLIAM THOMSON

P.S. I am very glad you are to lecture on Hydrostatics & Optics!

1 Stokes (32).

2 As Stokes explained at the beginning of his (32), the paper resulted from a request made by Robert Willis with regard to 'some experiments which have been performed by a Royal Commission, of which Professor Willis is a member, appointed on the 27th of August, 1847, "for the purpose of inquiring into the conditions to be observed by engineers in the application of iron in structures exposed to violent concussions and vibration" '. Willis (1800–75) was ninth wrangler in 1826 and held the Jacksonian professorship of applied mechanics at Cambridge from 1837 until his death. In 1841 he published his *Principles of Mechanism* (London and Cambridge, 1841), and his contribution to the Royal Commission resulted in an appendix to the Commission's report entitled, 'On the effects produced by causing weights to travel over elastic bars'. His appendix was reprinted in Peter Barlow, *Treatise on the Strength of Timber, Cast Iron, Malleable Iron, and Other Materials*, a new edition revised by I. F. Heather (London, 1851). Stokes apparently attended Willis's lectures as an undergraduate. (See what appear to be his notes on the lectures in the Stokes Collection, PA25 and PA26.)

3 James Thomson, 'Theoretical considerations on the effect of pressure in lowering the freezing point of water', *Trans. Roy. Soc. Edinb.*, **16** (1849), 575–80, and Kelvin (42), which immediately preceded his brother's article in the Edinburgh *Transactions*.

58 STOKES to KELVIN, 15 January 1850
Kelvin Collection, S354

Pembroke College, Cambridge[1]
Jany 15th 1850

DEAR THOMSON,

An idea has occurred to me with reference to magnetism which leads to a very simple demonstration of a general theorem which Gauss has proved in a particular case by means of Laplace's Coeff^ts^. The theorem is this. Let *S* be the surface of any magnetic body, as irregular as you please, or any surface enclosing it. Conceive *S* covered with a single system of lines of any form, and suppose the component of the force along the tangent at each point of each of these lines known. If the lines should be ring-shaped it is necessary moreover to know the component along an arbitrary line going from one ring-centre to another. Then we have data for determining the potential everywhere outside *S*.

I dare say this is obvious to you. If not I will send you the demonstration; but probably you are already acquainted with the theorem.

Yours very truly
G. G. STOKES

1 Kelvin has written at the top of the letter, 'Jan. 17, 1850'.

59 KELVIN to STOKES, [19 January 1850][1]
Stokes Collection, K37A

2 College, Glasgow
Saturday evening

MY DEAR STOKES

I have been intending for several days to write to you to tell you that I have experimentally verified my brother's deduction from Carnot's Theory with reference to the freezing point of water; and now I have your letter of the 15th to answer besides.

I think I see how the theorem you enunciate may be demonstrated, by taking the set of lines along w^h^ the comp^t^ of the force is given as one set of coordinates, and any other set each of which is a trajectory of all the others, as the second set of coordinates, to fix the position of points on the surface. If λ & μ denote the parameters of a p^{t} P, & L and M the comp^{s} of the force at P in the tang. pl., when resolved into three comp^{s}, respectively along the tang^{s} to the two curves intersecting in P, and the normal, and if V be the potential at P, we sh^{d} have

$$dV = L \cdot \phi(\lambda, \mu)\, d\lambda + M \cdot \psi(\lambda, \mu)\, d\mu$$

if we consider only points in the surface,

$$\phi(\lambda, \mu) \quad \& \quad \psi(\lambda, \mu)$$

are fu^{ns} which will be determinable when the two systems of curves are given. If then the 1st set of curves be given & the 2nd set chosen arbitrarily, we shall have

$$\frac{dV}{d\lambda} = L \cdot \phi(\lambda, \mu)$$

$$\& \quad V = \int L \cdot \phi(\lambda, \mu)\, d\lambda$$

where everything is known in the second member, & the arb^{y} const. of integr^{n} a fu^{n} of μ to be determined I suppose by the given value of M along one of the coordinates μ.

Some time ago I got upon some very interesting work relative to magnetic distributions which has made me want very much to decide upon names for the 2 kinds of distributions for w^{h} $udx + vdy + wdz$ is a complete differential, and

$$\frac{du}{dx} + \frac{dv}{dy} + \frac{dw}{dz} = 0.$$

A long time ago I began to call the former 'lamellar' to myself & in my own notes; and I had just begun, about a month ago to call the latter 'filamentary' but I have not been at all satisfied with this name, and I am now nearly resolved to call it 'solenoidal'. Do the names convey any corresponding ideas to your mind? & do you recommend me to stick to them? I get on dreadfully slowly in writing out my magnetism even when I get to work in earnest; and I scarcely ever get putting pen to paper on the subject. I hope soon however to get a chapter on electromagnets & on solenoidal & lamellar distributions ready, and that with what I sent before to the R. S. might be published as a first Part.[2]

I got an ether thermometer made for the experiment on the freezing point of water (by Mansell, my assistant)[3] which gives about 2 in., w^{h} is divided into 71 parts on the tube, for 1° Fah. This inst^{t} I had sealed herm[eticall]y in a glass tube & surrounded with clean ice & water in a compression vessel (Oersted's app[aratu]s; which I have from Pixii)[4] & put in a lead ring to keep a small space thr. w^h readings were expected, clear from ice. 8.1 atmospheres (added) made the column of ether sink $7\frac{1}{2}$ divisions, & 16.8 atmosp^s (added to the at^c pressure) made it sink $16\frac{1}{2}$ divisions!

observed pressures	observed depressions in temp^e	depressions in temp^e acc^g to theory on the hypoth. that the observed pressures were correct	diff^e
8.1	.106	.109	−.003
16.8	.232	.227	+.005

I attribute the extreme closeness of the agreement partly to chance as the measurement parts of the experiment were rude to a degree. I made the experiment on Thursday week last. On Monday a notice of it will be read at the R.S.E.

Tell me when you write next how your preparations for lectures are advancing, & give me Camb. news in general.

Your's ever
WILLIAM THOMSON

P.S. Can you tell me about the velocity of sound or give me references? What is the difficulty?[5]

1 January 19th was the first Saturday after the 15th.
2 Kelvin (53).
3 Kelvin (47) is the published version of these experiments, and in it Kelvin identifies 'Mr Robert Mansell of Glasgow'. (See *MPP*, I, 166.) The paper was read to the Royal Society of Edinburgh on 21 January 1850.
4 Hippolyte Pixii was an instrument maker in Paris, known especially for his electro-magnetic machine. See his 'Nouvelle machine éléctromagnétic', *Annales de Chimie et de Physique*, 50 (1832), 322–4.
5 See Stokes (33) and (34) which represent another disagreement between him and Challis.

60 STOKES to KELVIN, [*c.* 22 January 1850][1]
Kelvin Collection, S558

the surface we add a total flux ⟨pr⟩ proportional to $-C$; & since the grand total flux $=0$, only one value of C is admissible.

I congratulate you and your brother on the success of the heat experiment, which is a very remarkable one. I will tell Hopkins about it. It goes rather contrary to his notions, or rather conjectures, relative to the possible solidity of the centre of the Earth (depending on pressure) while the surrounding parts are in a state of fusion. I don't know that however; for some substances contract in freezing (like mercury) and in such cases I presume pressure would favour solidity.

I have not begun to put my lectures together yet, nor have I as yet made any progress worth mentioning, except ordering a few things in addition to what I am taking from Challis.

As to the difficulty about the vel^y of sound *I* don't know what it is: you must ask Challis that. I suppose you allude to the controversy which he has been carrying on first with Airy & then with me.[2] If so the papers are to be found in the Phil^l Magazine. I will write again and send you the ⟨news⟩ references.

That V[3] is given externally if either V or $\dfrac{dV}{dn}$ is given at the surface may be proved very readily thus.

If possible let there be two solutions V_1 V_2 and let

$$V_1 - V_2 = V'$$

then outside the surface S

$$\nabla V_1 = 0, \quad \nabla V_2 = 0 \quad \therefore \ \nabla V' = 0$$

Now

$$\iiint \nabla V' \cdot V' \, dx\, dy\, dz = \iiint \left(\frac{d^2 V'}{dx^2} + \frac{d^2 V'}{dy^2} + \frac{d^2 V'}{dz^2} \right) V' \, dx\, dy\, dz$$

$$= -\left\{ \iint \frac{dV'}{dx} V' \, dy\, dz + \iint \frac{dV'}{dy} V' \, dz\, dx + \iint \frac{dV'}{dz} [V'] \, dx\, dy \right\}$$

$$+ \iiint \left\{ \left(\frac{dV'}{dx} \right)^2 + \ldots + \ldots \right\} dx\, dy\, dz$$

$$= -\iint \frac{dV'}{dn} \cdot V' \cdot \langle dx\, dy\, dz \rangle \, dS + \iiint \left\{ \left(\frac{dV'}{dx} \right)^2 + \left(\frac{dV'}{dy} \right)^2 + \left(\frac{dV'}{dz} \right)^2 \right\} dx\, dy\, dz$$

Now on the 1st supposition V' is given at the surface [and on the] 2nd [supposition] $\dfrac{dV'}{dn}$ [is given at the surface.]

$\therefore$ in both cases

$$\iint \frac{dV'}{dn} V' \, dS = 0$$

Also $\nabla V' = 0$

$$\therefore \iiint \left\{ \left(\frac{dV'}{dx}\right)^2 + \left(\frac{dV'}{dy}\right)^2 + \left(\frac{dV'}{dz}\right)^2 \right\} dx\, dy\, dz = 0$$

$$\therefore \frac{dV'}{dx} = \frac{dV'}{dy} = \frac{dV'}{dz} = 0$$

but V, and $\therefore$ V_1, V_2, and their diff^{ce} V' vanishes at an inf^{te} distance. $\therefore$ $V' = 0$.

This is a slight extension of a thing I saw in a paper of Liouville's. I dare say it is not new to you.

Yours very truly
G. G. Stokes

1 The first portion of the letter, which would have contained the date, is missing. However, the letter is a reply to the previous one.
2 Stokes (33) and (34). See also, Challis, 'Theoretical determination of the velocity of sound', *Phil. Mag.*, **32** (1848), 276–84; Airy, 'Remarks on Prof. Challis's "Theoretical determination of the velocity of sound"', *ibid.*, **32** (1848), 339–43; and Challis, 'On the velocity of sound, in reply to Prof. Airy', *ibid.*, **32** (1848), 494–9.
3 Stokes has added a footnote reading, 'I mean this to apply to attr^{n} generally acc^{g} to the inverse sq of the distance, and not to magnetism merely'.
4 For Stokes's use of ∇, see letter 32, note 1.

61 Kelvin to Stokes, 25 January 1850

Stokes Collection, K38

3 Park Place, Edinburgh,
Jan 25, 1850

My dear Stokes

When I wrote to you last I saw my way to the determination of V, with an arbitrary constant, for any point of the surface, with your data, but I omitted to say anything about the arbitrary constant (and I may even have said that V was *completely* determinate, which of course is not the case, by the integration of $Ld\lambda + Md\mu$) from my being in a hurry when I wrote; but still I think the way that occurred to me was founded on the same principles as the demonstration you give. Your demonstration amounts to integrating the expression Fds for the broken curve SNP; w^{h} will, with an arb^{y} cons^{t}, give the value of V at any p^{t} P, provided the value of F be given for every p^{t} of the series of curves, and for every p^{t} of one curve, SK cutting across all of them (when the series all pass thr. one point, the int^{n} may be performed along each curve of the series from that point, without the necessity of considering a curve cutting across them) and this is the simplest view that can be given of the ⟨system⟩ principle involved in the integration of $Ld\lambda + Md\mu$ when it is the diff^{l} of a fu^{n}

of two independent variables, by ⟨fin⟩ [?] integrating the term $Ld\lambda$, and determining the arb^y fu^n of μ by the cond^n $\frac{dV}{d\mu}=M$. (Since $Ld\lambda+Md\mu$ is a compl. diff^l;) if this cond^n is satisfied for a particular value of λ, it will be satisfied universally, and therefore the determination of the arbitrary fu^n of μ may be effected if the value of M be given for every value of μ & one part^r value of λ. The introduction of the coordinate system is of course quite superfluous in the magnetic problem, but it occurred to me, & I put it down in a hurry, from an imperfect recollection of Gauss's demonstration in the case of a sph^l surface γ, ⟨as⟩ for data, the N & S. or the E & W. comp^{ts} of the force.

N.B. *S* is the letter at the ⟨pole⟩ vanishing p^t of the ovals

I am quite accustomed to neglect arbitrary constants in relation to magnetic potentials; always considering them as determinate by the condition $\iint N\cdot d\sigma=0$.

I am afraid my names 'lamellar' & 'solenoidal' do not convey much meaning without explanation, but I find them very convenient, & I am much tempted to use them. A few words will be enough to explain them to you.

If $udx+[vdy+wdz]$ be a comp^l dif^l let it be denoted by $d\phi$: consider the entire series of surf^s of w^h $\phi=C$ is the general equ^n. The whole magnet may be divided into infinitely thin shells by these surfaces. Each shell is normally magnetized with an intensity $\{V(u^2+v^2+w^2)\}$ inversely proportional to the thickness in different parts. A shell of matter, thus magnetized I call a uniform magnetic shell;[1] & therefore if ⟨a⟩ the distribution of magnetism in a magnet satisfy the cond^n $udx+[vdy+wdz]=d\phi$, the magnet may be divided into uniform magnetic shells; whence it may be said to have a lamellar distribution of magnetism.

If $\frac{du}{dx}+\frac{dv}{dy}+\frac{dw}{dz}=0$; take an infinitely small tube $(\sigma\,\omega\,\lambda\,\eta\,\nu)$ bounded by lines of magnetization ($\frac{dx}{u}=\frac{dy}{v}=\frac{dz}{w}$, for a line of magn^n). The matter contained in this is magnetized longitudinally, with an intensity inversely proportional to the section of the tube in different places. Now a bent or straight rod, whether uniformly thick or not, so magnetized, may be called a uniform magnetic solenoid (by bor[r]owing Ampère's term from electro-magnetism). Hence if $\frac{du}{dx}+\left[\frac{dv}{dy}+\frac{dw}{dz}\right]=0$ the magnet may be divided into uniform magnetic solenoids, whence I propose to call such a distribution solenoidal.

The magnetic force produced by a lamellarly magnetized ⟨m⟩ body, may be also produced by galvanic circuits coinciding with the edges of the shells of w^h it is composed; the external effects of a body possessing a solenoidal distribution of magnetism may be produced by an imaginary magnetic matter distributed over the ends of the solenoids; i.e. over the surface of the body.

⟨The ends of the sol⟩

⟨An infinite number of⟩ A lamellar, and consequently an infinite number of lamellar distributions may be found to produce a given superficial external potential; but the edges of the shells, as determined by the superficial values of ϕ, are determinate. A ⟨n infinite num⟩ solenoidal & consequently an infinite number of solenoidal distributions may be found to produce a given external potential; but the ends of the ⟨bars⟩ solenoids will give a determinate superficial distribution of imaginary magnetic matter (the density at any p^t of the surface being equal to $lu+mv+nw$, or the normal $comp^t$ of the magnetization).

There is one & only one solenoidal lamellar distribution, w^h produces given external effects.

When the $dist^n$ is not lamellar; $galv^c$ currents superficially & internally distributed, may be found to produce the same effects of force, both internal & external (if the ⟨internal⟩ force ⟨upon⟩ at any internal point of a magnet be defined as the resultant force at a point in a *split perpendicular* to the lines of magnetization). The components of the galvanism in this distribution, at any point x, y, z, ⟨of⟩ within the magnet will be

$$\frac{dv}{dz}-\frac{dw}{dy}, \quad \frac{dw}{dx}-\frac{du}{dz}, \quad \frac{du}{dy}-\frac{dv}{dx};$$

& the components of the superficial galvanism at any point x, y, z of the surface will be

$$-(vn-wm), \quad -(wl-un), \quad -(um-vl).$$

Again, when the distribution is not solenoidal, magnetic matter, superficially & internally distributed, may be found to produce the same effects of force, both internal & external (if the force at any internal point of a magnet be defined as the resultant force at a point in a split *along* the lines of magnetization). The density of the internal distribution, at any point x, y, z, within the magnet, is

$$-\left(\frac{du}{dx}+\frac{dv}{dy}+\frac{dw}{dz}\right),$$

and the superficial density, at any point P of the surface, is $lu+mv+nw$.

An infinity of distributions, of magnetism neither lamellar nor solenoidal, may be found to produce any given external ⟨or internal⟩ potential.

⟨A single, determinate, distribution of magnetism through all space, may be found to produce any given potential at ever [?]⟩ ⟨through all space.⟩

I am in Edinburgh, on a visit to Professor Forbes for 2 days, w^h being the last Friday & Saturday of January, are holidays at Glasgow. I return to Glasgow this evening (Saturday).

Your's very truly
WILLIAM THOMSON

P.S. The tangential $comp^s$ of the $result^t$ forces at two points, P, and P', infinitely near one another, on two sides of a surface possessing a $superf^l$ $dist^n$ of imaginary magnetic matter, ⟨differ⟩ are the same in magnitude & direction; the normal components differ by $4\pi\zeta$, if ζ be the $superf^l$ density.

The normal components & the components parallel to the $direc^n$ of the current, of the resultant forces at two points P & P', infinitely near one another on two sides of a surface possessing a galvanic $dist^n$ (or a $dist^n$ of galvanic wires) are the same in magnitude & direction; the components along the tangent plane of the surface & $\perp$ to the ⟨lines of⟩ direction of the current, differ by $4\pi\gamma$, if γ be the superficial density or intensity of the galvanism, i.e. if γ be such that $\gamma \cdot b$ is the strength of the galvanism in a band of the surface, bounded on ea. side by lines of galvanism, and of the infinitely small breadth b, in the neighbourhood of P & P'.

1 Kelvin has added a footnote here reading, 'The pot^l at P due to a uniform magnetic shell is $propor^l$ to the solid angle sub-tended by the edge of the shell at P'.

62 STOKES to KELVIN, 5 March 1850
Kelvin Collection, S355

Pembroke College, Cambridge[1]
March 5, 1850

MY DEAR THOMSON,

I have some how or other put off writing to you for a long time.

As to the difficulty (or what $Prof^r$ Challis considers a difficulty) in the theory of sound you will find the controversy in the 33^d and some following volumes of the Philosophical Magazine.[2] I think it not worth while to send a list of the articles, as you have probably no leisure for the thing at present, and as you will probably be here in May or the end of April we can, all well, talk over it then.

It was really carrying coals to Newcastle to send a simple magnetic theorem to you. It seems that what I considered the point of the thing was in your eyes so microscopic that you could not see it without the use of a magnifying glass.

You will see in the last N° of the Phil. Mag. that Prof^r Tardy of Messina has attacked pretty sharply Prof^r Challis and his new equation.[3]

The 3^d part of Moigno's Repertoire d'optique has just come out.[4] He finds a 4th part is still wanting. It seems M. Fizeau has lately determined *directly* the velocity of propagation of light![5] His result agrees very closely indeed with that derived from the eclipses of Jupiter's satellites and from the coeff^t of aberration.

I am at present engaged in getting together materials for my lectures which are to begin on the 29th of April. I hope however to be able to send you before very long a paper (I mean a copy of a printed paper) on the dynamical theory of diffraction.[6]

Yours very truly
G. G. STOKES

1 Kelvin has written at the top of the letter, 'March 7, 1850'.
2 See letter 60, note 2.
3 Placido Tardy, 'Some observations on a new equation in hydrodynamics', *Phil. Mag.*, **36** (1850), 171–8. Tardy (1816–1914) was professor of infinitesimal calculus at the University of Messina from 1841 to 1851 and afterwards at the Marine School at Genoa.
4 L'Abbé Moigno, *Répertoire d'Optique Moderne ou Analyse Complète des Travaux Modernes Relatifs aux Phénomènes de la Lumière*, 4 vols. (Paris, 1847–50). François Napoléon Marie l'Abbé Moigno (1804–84) lived at St Denis in Paris and until 1862 was editor of *Cosmos, revue encyclopédique hebdomadaire des progrés des sciences*, which was first published in 1852.
5 Hippolyte Louis Fizeau, 'Sur une expérience relative à la vitesse de propagation de la lumière', *Comptes Rendus*, **29** (1849), 90–2.
6 Stokes (40).

63 KELVIN to STOKES, 2 July 1850
Stokes Collection, K39

9 Barton Street,
Westminster
July 2, 1850

MY DEAR STOKES

As I have not a copy of your paper on the Equil. & Motion of Elastic Solids,[1] nor any other work of reference for the purpose, by me, I shall be much obliged by your sending me the *equations of equil^m* of ⟨an⟩ a non crystalline elastic solid under the action of any forces, and the formulae for the mutual actions betw. any two contiguous portions of the body. I have been trying but as yet without success, to make out something about the integration of the equations for the case of a solid of any form, with each point of its surface displaced to a given extent & in a given direction from its natural position. I think I see how it can be done when the solid is a rectangular parallelepiped, but not in a very inviting way. It was reading your paper on diffraction[2] on my way from Cambridge that made me take up the subject again.

Do you know that the condition that $\alpha dx + \beta dy + \gamma dz$ may be the diff[l] of a function of two indep[t] variables for all points of a surface is

$$l\left(\frac{d\beta}{dz}-\frac{d\gamma}{dy}\right)+m\left(\frac{d\gamma}{dx}-\frac{d\alpha}{dz}\right)+n\left(\frac{d\alpha}{dy}-\frac{d\beta}{dx}\right)=0?$$

I made this out some weeks ago with ref[ce] to electromagnetism. With ref[ce] to an elastic solid, the cond[n] may be expressed thus – the resultant axis of rotation at any point of the surface must be perp[r] to the normal.

Your's very truly
WILLIAM THOMSON

P.S. The following is also interesting, & is of importance with reference to both physical subjects.

$$\int(\alpha dx+\beta dy+\gamma dz)=\pm\iint\left\{l\left(\frac{d\beta}{dz}-\frac{d\gamma}{dy}\right)+m\left(\frac{d\gamma}{dx}-\frac{d\alpha}{dz}\right)+n\left(\frac{d\alpha}{dy}-\frac{d\beta}{dx}\right)\right\}dS$$

where l, m, n denote the dir[n] cosines of a normal through any el[t] dS of a ⟨limited⟩ surface; & the integ[n] in the sec[d] member is performed over a portion of this surface bounded by a curve round w[h] the int[n] in the 1st member is performed.[3]

1 Stokes (11).
2 Stokes (40).
3 Stokes included the equation in this postscript on the Smith's prize examination for 1854 (the year Maxwell took the examination), and it has become known as Stokes's Theorem. (See Larmor's footnote in Stokes's *MPP*, v, 320–1.)

64 STOKES to KELVIN, 4 July 1850
Kelvin Collection, S356

Pembroke College Cambridge[1]
July 4th 1850

DEAR THOMSON,

The equations of equilibrium are

$$\varrho X+(A+B)\frac{d\delta}{dx}+\nabla\alpha=0$$

$$\varrho Y + (A+B)\frac{d\delta}{dy} + \nabla\beta = 0$$

$$\varrho Z + (A+B)\frac{d\delta}{dz} + \nabla\gamma = 0$$

where

$$\delta = \tfrac{1}{3}\left(\frac{d\alpha}{dx} + \frac{d\beta}{dy} + \frac{d\gamma}{dz}\right)$$

$$\nabla = \frac{d^2}{dx^2} + \frac{d^2}{dy^2} + \frac{d^2}{dz^2}$$

Also if $P_1 P_2 P_3$ are the 3 normal pressures and $T_1 T_2 T_3$ the 3 tangential pressures on planes $\|$ to the coor[dinat]e planes

$$P_1 = -\langle m\rangle A\delta - 2B\left(\frac{d\alpha}{dx} - \delta\right)$$

$$P_2 = -\langle m\rangle A\delta - 2B\left(\frac{d\beta}{dy} - \delta\right)$$

$$P_3 = -A\delta - 2C\left(\frac{d\gamma}{dz} - \delta\right)$$

$$T_1 = -B\left(\frac{d\beta}{dz} + \frac{d\gamma}{dy}\right)$$

$T_2 = \&c. \quad T_3 = \&c.$[2]

$T_{\langle 3\rangle 1}$ is here the tangential component of the pressure on a plane $\perp$ to y resolved $\|$ to z

= the tang[l] component of the pressure on a plane $\perp$ to z resolved $\|$ to y

Also the forces P_1 P_2 P_3 T_1 T_2 T_3 are reckoned positive in a manner analogous to p in Hydrodynamics.

I mean to write to a friend of mine [H. Cox] in London who has got a copy of the paper[3] to ask him to send it to you if he has done with it.

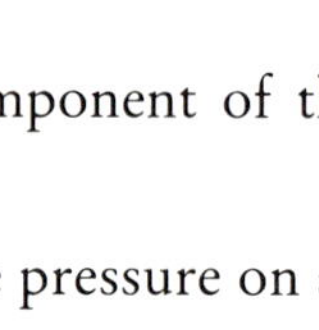

I don't see how images will apply here. It seems to me that if you take F' for the image of F in a plane there will be no tangential forces along the plane but there will be normal pressures. But if you take F'' there will be no normal pressures but there will be tangential forces.

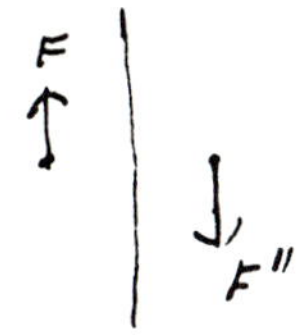

The theorems which you communicated are very elegant and are new to me. I have demonstrated them for myself, the first by the calculus of variations, the second (which includes the first as a particular case) by simple considerations like what I have employed at the beginning of my paper. I suppose from what you say that you arrived at these theorems by working with magnetic ideas.

I forgot when you were here to give you £8 for a commission for a Pembroke man a friend of mine Dr Goode or more properly Mr Goode for he is only a L.M. and M.B.[4] I told him I might be going to Paris and if I did not I thought you would have no objection to execute his commission. It is for an Oberhauser's microscope.[5] Goode has written to Oberhauser, so you will have no trouble about choosing. I believe you don't go to Paris now till after the meetings at Edinburgh so I can give you the money there if I don't go to London.

Yours very truly
G. G. Stokes

1 Kelvin has written at the top of the letter, 'July 5'. For Stokes's use of ∇, see letter 32, note 1.
2 Stokes has written upside down in the corner of the page, 'The pressure on any plane (l, m, n) is $lP_1 + mT_3 + nT_2 \parallel$ to x. &c $\parallel$ to y. &c $\parallel$ to z'.
3 Stokes (11).
4 Henry Goode was an undergraduate friend of Stokes and took his B.A. in 1842 as sixteenth wrangler. He received from Cambridge an M.B. in 1846 and a Medical License in 1849. There are over forty letters from him to Stokes in the Stokes Collection, the earliest dating from 1840.
5 Georg Oberhäuser (1798–1868) was an optician in Paris.

65 Kelvin to Stokes, 5 July 1850

Stokes Collection, K40

9 Barton Street, Westminster
July 5, 1850

Dear Stokes

Many thanks for the formulae, which, especially those turned upside down in the right hand corner of your second page, are just what I want.

Do you chance to know how the optic axes of a crystal of Sulphate of

Barytes (a prism with a rhombic base) lie? Is it correct that they are in a plane perpendicular to the base, through its long diagonal? In a paper by Tyndall and Knoblauch in the Phil Mag. for July,[1] this is stated to be the case; but experiments are described which if correct, show that the axes of greatest and least magnetic induction are along the two diagonals of the base (all that I refer to is in the last paragraph of p ⟨20⟩ [?] 5). This of course makes one imagine that the optic axes must lie in the plane of the base.

I have, ever since Faraday's[2] experiments on diamagnetics, thought that experimenters would be much the better of having some of the obvious conclusions of mathematical theory of induction brought before them, and the subsequent researches on magnetic induction in crystals make me feel this more strongly. The paper I have just seen by T & K has struck me very much in this respect and has made me determine to write a short paper on magnetic induction in crystalline substances for the Phil. Mag.[3] The whole that I shall give is quite obvious, and you of course would have no difficulty in working it out, on the assumption of three principal axes of induction at right angles to one another. I am not (acc^g to my present intention) give [*sic*] a demons^n that there are three such axes & that they are at right angles founded on the principle of the superposition of magnetic inductions but I have such a proof, which I think is quite satisfactory, & w^h I intend to give when I get to induction in my long paper.[4]

If a powder consisting of flat plates or scales (powdered mica e.g.) or rhomboids of calc spar, or pounded crystals of bismuth (w^h I think would consist of flat scales) be mixed up as uniformly as possible in dough, and a ball of the dough squeezed flat, how would the fragments be arranged as regards angular position? Will the planes of the plates or the largest sections of the rhomboids be parallel to the line of compression or will they be perpendicular to it? What would a set of little bars in sim^r circumstances do? (I care more about the plates ⟨)⟩ however) I wish you would tell me. That is quite a problem in your way as you are doing fluids with friction. N.B. never mind inertia! I wish you would tell me. You will see why I want to know, if you look at T and K^s paper. If the plates do not arrange themselves right, the only way left to explain T & K's facts will be to deny them. Faraday knew a long time ago (see his first paper on diamagnetics)[5] that when the magnetic induction is excessively feeble in any substance, the mutual actions of the diff^t magnetized portions are insensible compared with the magnetizing force, but no such refined ideas seem to affect Messrs T & K's reasonings.

Your's very truly
WILLIAM THOMSON

1 John Tyndall and Hermann Knoblauch, 'Second memoir on the magneto-optic properties of crystals, and the relation of magnetism and diamagnetism to molecular arrangement', *Phil.*

Mag., **37** (1850), 1–33. Karl Hermann Knoblauch (1820–95) was professor extraordinary at the University of Marburg from 1849 to 1853. Tyndall (1820–93) received his Ph.D. at Marburg in 1850 and eventually succeeded Faraday as director of the Royal Institution. He was involved in scientific controversies with Kelvin, and, although he corresponded with both Kelvin and Stokes, his correspondence with Stokes was much more extensive than that with Kelvin. The two sides of the Tyndall–Stokes correspondence are preserved in the Royal Institution and the Stokes Collection. Three letters between Kelvin and Tyndall were published in the *Phil. Mag.* in 1855 and 1856 were reprinted in Kelvin's *PEM*, 540–7.

2 Michael Faraday (1791–1867) was, of course, a major figure in early Victorian science, and his ideas established areas of research for both Kelvin and Maxwell, though not Stokes. There are a few letters from Faraday to Kelvin and Stokes, respectively, in the Kelvin and Stokes Collections.

3 See Kelvin (43) and (50).

4 Apparently an intended continuation of Kelvin (53) which had been read to the Royal Society in two parts on 21 June 1849 and 20 June 1850.

5 See L. Pearce Williams, *Michael Faraday: A Biography* (New York, n.d.), 381–94, for Faraday's discovery of diamagnetism and Kelvin's role in it.

66 STOKES to KELVIN, 8 July 1850

Kelvin Collection, S357

Pembroke College Cambridge[1]
July 8th 1850

MY DEAR THOMSON,

Before this reaches you I hope you will have got a copy of my paper.[2] I had a letter from Cox[3] by this afternoon's post in which he said he should send you the copy I lent him which his friend had done with.

As to Sulphate of Barytes the optic axes lie in a plane perpendicular to the two faces which are usually most developed and one of which would be called in common parlance the base. I believe however that the crystallographic base (the primitive form being a right rhombic prism) is the plane of the optic axes.

Has Plücker *made out* that the optic axis of a crystal has an invariable magnetic relation? I saw in the Phil. Mag. ⟨s⟩ a paper by some one, I forget who, who stated the result of *his* experiments to be that in I think six specimens of Iceland spar the optic axis took an equatoreal [*sic*] position as with Plücker and in seven took on the contrary an axial position.[4]

As to the dough, if we suppose it compressed between two smooth parallel planes (and tapping it with the finger must produce much the same effect) and if we neglect irregularities depending on the bulging &c. at the edges the result will simply be that all lines $\parallel$ to the axis of compression will be shortened in a given ratio (say $1:k$ where $k<1$) remaining $\parallel$ to themselves and all lines $\perp$ the axis will be lengthened in the ratio $\sqrt{k}$ to 1. Hence the direction-cosines of a small plane of particles which at first were l, m, n suppose will become proportional to

$$\sqrt{k}\cdot l \qquad \sqrt{k}\cdot m \qquad \frac{n}{k}$$

and if θ be the $\angle$ between the normal in its new position and the axis

$$\cos\theta = \frac{\frac{n}{k}}{\sqrt{kl^2 + km^2 + \frac{n^2}{k^2}}}$$

$$= \frac{\langle n\rangle \cos\theta_\circ}{\sqrt{k^3(1-\cos^2\theta_\circ)+\cos^2\theta_\circ}} \qquad (n=\cos\theta_\circ)$$

$$= \frac{\cos\theta_\circ}{\sqrt{\{1-\sin^2\theta_\circ(1-k^3)\}}}$$

and all the normals will be driven further from the axis (except in the particular case in w^h^ $\langle\cos\rangle$ $\theta = 0$ or $= \frac{\pi}{2}$.

If now we assume that an inf^y^ small and inf^y^ thin lamina will be displaced like a plane of particles, if we have originally a ⟨n⟩ very great n° of such laminae having on the average no particular arrangement as regards orientation, after compression there will be a crowding of the planes on the average towards the plane ⊥ to the axis of compression. It appears in a similar way that bars would on the average arrange themselves more ⊥ than ‖ to the axis of compression. I do not know whether this solution of the problem will be sufficient for your purpose. I will conclude as it is just post time.

Yours very truly
G. G. Stokes

1 Kelvin has written at the top of the letter, 'July 9'.
2 Stokes (11).
3 Homersham Cox (1821–97) took his B.A. at Cambridge in 1844, was admitted at the Inner Temple in 1845, and was called to the bar in 1851. There are several letters from Cox to Stokes in the Stokes Collection, although the one mentioned here is not.
4 These results were reported by Tyndall and Knoblauch in their 'On the deportment of crystalline bodies between the poles of a magnet', *Phil. Mag.*, **36** (1850), 178–83. Their paper referred to in the previous letter, note 1, cites this one on page 3.

67 KELVIN to STOKES, 15 July 1850
Stokes Collection, K41

9 Barton Street, Westminster,
July 15, 1850

DEAR STOKES,

When you see Cox, or write to him, will you say that I feel very much obliged to him for the copy of your paper[1] which he left for me one day when he called here and for the copy of his own paper on the Impact of Elastic Beams,[2] which I have read with interest? As he did not leave his address I have not been able to thank him myself. Before I leave London I should like to let him have his copy of your paper as he may perhaps want it before I return, or I might forget it if I do not send it back to him now.

If I do not hear from you to the contrary, I shall assume that the optic axes of a crystal of sulphate of barytes are in the plane of the base of the primitive form (the right rhombic prism) and I shall (if I write the paper I proposed, for the Phil. Mag.) correct Messrs Tyndall & Knoblauch on this point.[3] I guess from the magnetic ⟨relations of⟩ phenomena relative to the axis of the prism, and the longer & shorter diagonals of its base mentioned by them as observed in their experiments, and their relation to the magnetic phenomena of calc spar, and the facts (?) that sulphate of barytes is positive & that it is diamagnetic, w^h they affirm; that the long diagonal of the base bisects the acute angle between the optic axes, contrary to the assertion ⟨in⟩ at the top of p. 6 (Phil. Mag. July). The facts of which I want your confirmation are the following three –

1. Sulphate of barytes is a +^ve crystal.
2. The longer diagonal of the rhombic base of a crystal in the primitive form bisects the acute angle betw. the optic axes.
3. The shorter diagonal bisects the obtuse angle betw. the optic axes.

I shall be very glad, & much obliged if you will make me certain on all these points, and I shall (most unwarrantably) assume that silence gives consent, and make the assertion on the credit of the Lucasian Professor (without however quoting my authority) that N°s 2 & 3 are true although contrary to the statements of Messrs T. & K.

Would you have any objections to glance at the revises of the 3^d & 4th sheets of the forthcoming N^o of the Journal if Metcalfe and Palmer[4] send them to you, either at Cambridge or where-ever you may be, and, after seeing that the authors' corrections marked on the proofs (w^h w^d be sent along) are properly made, return them to the printers with your *imprimatur* marked on each? Also, if you are at Cambridge, will you open any packets w^h the porter of Peterhouse may give you marked on the outside 'for the mathematical journal,' and acknowledge the receipts to the authors by printed forms w^{h} I should send you if you are willing & if you are likely to be in Cambridge?

I am now almost settled in the plan of leaving England about the 25th of July, passing rapidly through Paris to Bordeaux, & Lus [i.e., Luz?], and making a tour in the Pyrenees; and spending a week in Paris on my return, early in September, to purchase apparatus for illustrating optical lectures to a large class. Do you think you could, or would be inclined to meet me in Paris, or anywhere on my journey? If you are not to be in Paris I shall gladly undertake your commission. I have not been to Darker's[5] yet, but I shall go almost immediately & take your things. You did not say the necessity of taking them to him was immediate so I hope the delay I have made is not of consequence.

Your's ever
W. THOMSON

P.S. I forgot to say in answer to a question of yours that I think Plücker & others have established the fact that a sphere of calc spar, suspended with its crystalline axis horizontal, and free to turn round a vertical axis, would assume a position with that axis perpendicular to the direction of the horizontal comp^{t} of the magnetic force. People have confused phenomena resulting from the attr^{n} or repuls^{n} of parts of the masses operated on towards or from places of stronger force, with those w^{h} result from ⟨different⟩ crystalline structure, immensely in their experiments, and so have arrived at all sorts of apparent anomalies. Faraday is the only experimenter so far as I know who understands how to distinguish them.

I thought as you ⟨had⟩ do, that there would be a crowding of the planes of the plates perpendicular to the lines of compression, but this will not do for explaining Messrs T. & K^{s} experiments w^{h} I believe would require a preponderance of planes $\|^{l}$ to the lines of compression. In reality however I doubt very much their experiments. Their own explanations & expectations regarding the dough are I believe extremely false.

1 Stokes (11).

2 Homersham Cox, 'Impact of elastic beams', *Trans. Camb. Phil. Soc.*, **9** (1856), 73–8. Read in 1849.

3 Neither Kelvin (48), published in the *Philosophical Magazine* in October 1850, nor Kelvin (50), published in the *Philosophical Magazine* in March 1851, mentioned the paper by Tyndall and Knoblauch (see letter 65, note 1). However, Kelvin (50) did lead to an exchange between Kelvin and Tyndall (see Kelvin's *PEM*, 481–2n).

4 Metcalfe and Palmer were Cambridge printers who printed the *Cambridge and Dublin Mathematical Journal*.

5 William H. Darker was a London optician who made scientific instruments for Stokes and Kelvin.

68 STOKES to KELVIN, 17 July 1850
Kelvin Collection, S358

Pembroke College Cambridge[1]
July 17th 1850

MY DEAR THOMSON,

I have satisfied myself by my own observation respecting the three points under discussion.

I have in my possession a crystal of Sulphate of Barytes cut so as to show both optic axes. The reason why I could not tell how the optic axes lay was, that I was not sufficiently acquainted with the crystalline form to make it out from my specimen, in which some of the faces were, or appeared to have been, obliterated by cutting. Since I received your letter this morning, I have chipped off a piece or two, which show the planes of cleavage without injuring the specimen. I was unwilling before to do anything which might injure the specimen, as it was ground by Kingsley[2] who made a present of it to Prof^r Challis, who handed it down to me as belonging to the Professor who undertook the optical lectures.

I have obtained two easy cleavages, giving brilliant planes inclined to each other at an angle very sensibly acute, as may be seen without any measurement though the little planes are not near each other. I have also obtained a third cleavage, though with somewhat less facility, giving a plane perpendicular to the two former. This sufficiently indicates the crystalline form. It is stated in Phillips's Mineralogy[3] that occasionally the crystals admit of division $\|$ to the lesser diagonal of the prism (the prism which is taken for the primitive form) but that this is not a regular cleavage, as a close inspection of the uneven plane so produced will evince. Now the only indication of cleavage which my crystal presented till I chipped it was certain internal planes, indicating a cleavage $\|$ to the plane of the optic axes. This cleavage was $\perp$ to 4 artificial and 2 natural faces, of which the two latter might (without violating the laws of crystallography) have been planes truncating the acute or obtuse edges of the primitive form, on the supposition that this cleavage plane was in fact the base. I accordingly presumed that it was. I now see my error, and that the cleavage (indicated by the planes) which I took for the base is in fact the cleavage $\|$ to the shorter diagonal mentioned by Phillips.

Having now ascertained the primitive form I am able to assert

(1) that the plane of the optic axes is *perpendicular* to the ⟨plane of the optic axes⟩ base

(2) That the line bisecting the *acute* angle of the optic axes is $\|$ to the *shorter* diagonal of the rhombic base. The longer diagonal is of course $\perp$ to the plane of the optic axes, and consequently is the axis of mean (optical) elasticity.

I have also ascertained the class to which the crystal belongs in the following manner.

1. It is requisite in the first place to define the terms positive and negative, which ⟨relate⟩ properly relate to uniaxal crystals, and the use of which must be to a certain extent conventional when they are applied to biaxal crystals. Imagine the optical constants of a biaxal crystal to change continuously in such a manner that the ⟨p⟩ [?] crystal passes into uniaxal by the approach of its optic axes to the line bisecting the ⟨actute⟩ *acute* angle between them, a line which ultimately becomes the single optic axis. Then the original biaxal crystal may be called $+^{ve}$ or $-^{ve}$ according as the resulting uniaxal crystal is $+^{ve}$ or $-^{ve}$.

2. In sulphate of Barytes the ∠ between the optic axes is not nearly equal to 80°, so that even without measurement it is easy to tell which line bisects the acute and which the obtuse angle between the optic axes.

The distribution of the tints in the ring-systems shown by the crystal already mentioned showed at once that the red axes lay closer to the middle line than the blue axes. This being the case, the distribution of the tints showed instantly on which side of the optic axis the middle line lay in a thinner crystal w^{h} exhibited only one optic axis, and which was more convenient in the subsequent observation.

3. Light pol^{d} by reflexion was transmitted through a Fresnel's rhomb in w^{h} the plane of reflexion had an azimuth of about +45°, measured from the plane of primitive pol^{n}. I measure azimuths in the direction of the hands of a watch to an observer placed so as to receive the light into his eye. Viewing the plane pol^{d} light seen on looking aside the rhomb through the crystal, and analyzing it with a Nicol's prism, I turned the prism till the dark brush was formed, and ∴ the field without the crystal would have been dark, and then turned the crystal in azimuth till the brush lay in the plane of primitive pol^{n}, and ∴ the optic axes also lay in that plane. On moving the whole (the crystal and analyzer) ∥ to itself so as to look through the rhomb the ring-system was changed from

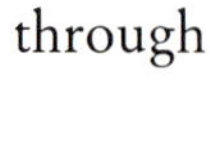

into

and the position of the dark spot (I only regard the side on which it lay) was

BR the optic axis, *S* the spot, *M*, (which must have lain on the side R_1) the

middle line. Now at *S* the crystal must have just neutralized the rhomb, and ∴ the ray pol^{d} in the plane *AA*

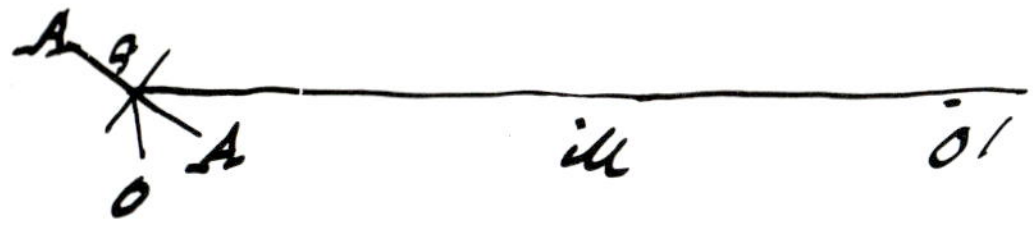

must have been *accelerated* a quarter of an undulation relatively to the other.

Now when *O*, *O′* coalesce in *M*, the ray pol^{d} in the plane *AA* becomes the ordinary ray of the uniaxal crystal thus formed

(*AA* will be inclined at an ∠ of 45° to ⟨the⟩ *OO′* the plane of primitive pol^{n}) and therefore the crystal is of the $+^{ve}$ class.

I shall have occasion to write to Cox in a few days when I shall give him your message. His present address is Wentworth Castle, Barnsley.

The copy of my paper[4] which he had does not belong to him. It is one I lent him. I like to keep one copy of my own papers. If I had had a spare one Prof^{r} Forbes should have had it. Cox has done with the paper so you need not return it to him. If you have done with the paper you can leave it with Darker when you call and ask him to keep it for me till I call on him.

I shall be happy to look at the revises of the 3^{d} and 4th sheets of the forthcoming N° of the Journal if it be not requisite to do so between about the first week and near the end of August when I shall probably be travelling about the Highlands having no fixed address. I am not likely to be in Cambridge after this month till October or thereabouts.

I send you the 1st half of a £5 note. I will send the 2nd half and a P.O. order for £3 when I hear from you. This is for the microscope. Of course if there is anything extra for duty or so forth I will pay you when you return. My friend Mr Goode has written to Oberhauser, so there will be no choosing in the case.

As I expect to be far north in September I think it extremely unlikely that I shall join you in Paris. There is no hurry about Darker.

Would it be convenient to take a few memoirs for me to Paris for Moigno and Cauchy? If so I will send you them by railway. I see M. Franck[5] 69 Rue Richlieu is Moigno's present publisher. I dare say he would have no objection to receive the papers for Moigno.

Yours very truly
G. G. Stokes

1 Kelvin has written at the top of the letter, 'Lond Thurs. Jul. 18'.

2 William Towler Kingsley (1815–1916) was thirty-fifth wrangler in 1838 and was rector of South Kilvington, Yorkshire, from 1859 to 1916. He is mentioned in Stokes (42) as having assisted Stokes in his researches on fluorescence. (*MPP*, III, 288.)

3 William Phillips, *Outlines of Mineralogy and Geology* (London, 1815), 4th edn in 1826. Phillips (1775–1828) took over his father's printing and bookselling business in London, devoting his spare time to scientific studies. He was a founding member of the Geological Society of London in 1807 and became a fellow of the Royal Society in 1827.

4 Stokes (11).

5 The Paris publisher A. Franck published all four volumes of Moigno's *Répertoire d'Optique Moderne ou Analyse Complète des Travaux Modernes Relatifs aux Phénomènes de la Lumière*, 4 vols. (Paris, 1847–50).

69 KELVIN to STOKES, [18 and 19 July 1850][1]

Stokes Collection, K42B

Friday evening[2]
9 Barton Street

DEAR STOKES

I this morning received the first half of a £5 note from you. Address the second to care of the Rev W. W. Herringham,[3] Borley Rectory, Sudbury, Suffolk, and make the P.O. order payable at Sudbury. I shall be there till Wednesday next at least. It is just possible I may have to forego making my tour after all, in which case I should be at the meeting of the Brit. Ass^{n} at Edinburgh, where I should hope to meet you.[4]

Many thanks for your letter and the trouble you have taken. I have not got learning it all yet as I have been very busy & out nearly the whole day. I was at Darker's before breakfast & did not get your letter till I came in. So I shall keep your paper[5] for the present unless you want it in w^{h} case I shall send it by post. Mine is lent to Clerk Maxwell.[6]

I shall be here for a day on my way to Paris & shall be most happy to take your papers. If you address them here it will be all right as they will lie for me. If I do not go to Paris now it is almost certain that I shall go at the end of Septemb.

Your's ever
in much haste
WILLIAM THOMSON

1 Kelvin wrote this letter on Thursday the 18th. On the 19th he added 'Friday evening, 9 Barton Street' and posted it. In letter 72, Stokes correctly explained the relationship between Kelvin's three previous letters (69, 70, and 71).

2 Kelvin has written at the top of the letter, 'by a negative biaxal crystal I understood one in w^{h} the line bisecting the acute angle betw. the axes corresp^{ds} to the one axis of calc spar; by a *positive*, one in w^{h} that line corresponds to the one axis of quartz'.

3 William Walton Herringham (1824–1905), an undergraduate friend of Kelvin's, was curate at Borley from 1848 to 1853 and later was rector in Nottinghamshire and then in Somerset.

There are several letters from him to Kelvin in the Kelvin Collection, dating mainly from 1845 to 1847.

4 The British Association meeting began at Edinburgh on July 21st.

5 Stokes (11).

6 James Clerk Maxwell (1831–79) was second wrangler in 1854 and was professor of natural philosophy at Marischal College, Aberdeen, from 1856 to 1860 and at King's College, London, from 1860 to 1865, before becoming Cambridge's first Cavendish professor of experimental physics in 1871. Much of the correspondence from Maxwell to Stokes and Kelvin in the Stokes and Kelvin Collections was published by Joseph Larmor in Stokes's *Memoir*, II, 1–45, and in Larmor (ed.), *Origins of Clerk Maxwell's Electric Ideas, as Described in Familiar Letters to William Thomson* (Cambridge, 1937). There are also several unpublished letters from Maxwell to Kelvin in the Kelvin Papers in the Glasgow University Library. It will be noted that this letter is much earlier than the earliest surviving Maxwell–Kelvin letter (Maxwell to Kelvin, 20 February 1854, Kelvin Collection, M87, printed in *Origins of Clerk Maxwell's Electric Ideas*, pp. 3–4).

70 KELVIN to STOKES, [19 July 1850][1]

Stokes Collection, K42

Greenhithe, Friday morning
Rail. St^n 7.45

DEAR STOKES

I forgot to leave out a letter[2] for you to be posted, acknowledging the receipt of the 1st half of a £5 note & asking you to send the 2nd half to me care of the Rev. W. W. Herringham, Borley Rectory, Sudbury, Suffolk,and to make the P.O. order payable at Sudbury. Your papers for Paris may be addressed to 9 Barton St Westm^r.

I am here to see 'the Retribution' *swing*, for det^g the dev^n of his compass.

Your's ever
W THOMSON

1 See letter 72.

2 Letter 69.

71 KELVIN to STOKES, [19 July 1850][1]

Stokes Collection, K42A

9 Barton Street
Friday afternoon

DEAR STOKES

On arriving from Greenhithe I found a letter announcing a change of plans on the part of my friend whom I hoped to join in the Pyrenees, in consequence of which any continental excursion I may be able to make will be deferred till after the meeting in Edin. where I now hope to be & to meet you there. I write this to let you know that it will be not necessary to send your papers here, as you can give me them in Edinburgh. Neither will it be necessary to send the money order, & even the bank note can wait to be united till we meet.

Your's ever truly
WILLIAM THOMSON

1 See letter 72.

72 STOKES to KELVIN, 20 July 1850
Kelvin Collection, S359

Pembroke College Cambridge[1]
July 20th 1850

DEAR THOMSON,

I got your three notes this morning, which at first puzzled me, as the middle by date seemed to contain the latest intelligence. However I have made it out now. 'Friday evening' is written with slightly different ink, and you evidently wrote it after your return from Greenhithe on the note you spoke of as having forgotten to post. I have had a letter from my friend Goode,[2] in which he says he thinks he can get an earlier opportunity. However you need not send back the half note till you hear again.

I am in no hurry for my paper.[3] You can give me it in Edinburgh.

As to $+^{ve}$ or $-^{ve}$ biaxal crystal, I merely gave a definition for the sake of precision. When the ∠ between the optic axes is small, or moderately small, as is the case with sulphate of Barytes, there is no doubt about the meaning of the term; but when the ∠ is near 90° it is possible that the same crystal might be called positive or negative according as one or other conventional definition was adopted.

Yours very truly
G. G. STOKES

1 Kelvin has written at the top of the letter, 'Mond Jul. 22 Borley'.
2 The letter is not in the Stokes Collection.
3 Stokes (11).

73 KELVIN to STOKES, 28 November 1850
Stokes Collection, K43

2 College, Glasgow
Nov 28, 1850

DEAR STOKES

Many thanks for the inquiries you have been making on my account regarding the engagements of a moderator.[1] I fear it will be impossible for me ever to be free enough during the winter 6 months to meet them. I think it is *just possible* that I might manage to get a substitute to perform my duties here, and leave of absence, for two months but I do not think it would be possible for me to be at Cambridge again, for the poll examinations, before the 1st of May.

I shall be very glad if you will keep Wilbraham's paper till we settle whether it is to be published or not.[2] If you chance to meet the author at Cambridge you might perhaps communicate some ideas as to the true theory to him. I scarcely think there can be anything in the paper which w[d] justify its being published but if you think differently & recommend it for publication I shall of course be very glad to put it into the Journal. I am excessively busy now, between lecturing on magnetism & on mechanics, & trying to make out something of Faraday's new discoveries, and of the possibility of accounting for the diurnal & other variations of the needle by the magnetization of the atmosphere.[3] I am as yet very much in the dark.

Your's very truly
WILLIAM THOMSON

1 There were two 'moderators' and two 'examiners' each year for the mathematical tripos. Kelvin examined for the tripos only one time – in 1874 when he was 'additional examiner' in charge of questions on heat, electricity, and magnetism, subjects added to the tripos the previous year.

2 Possibly Henry Wilbraham, 'Two arithmetical theorems', *Camb. Dub. Math. J.*, **6** (1851), 32–5. The paper is dated 21 June 1850 and was published in February 1851. Wilbraham published two other articles in the *Cambridge and Dublin Mathematical Journal*, but they both dated from 1851. Wilbraham (1825–83) was seventh wrangler in 1846, was admitted at Lincoln's Inn in 1848, and was called to the bar in 1851.

3 See letter 75, note 6.

74 STOKES to KELVIN, 6 January 1851
Kelvin Collection, S360

Pembroke College Cambridge[1]
Jan[y] 6th 1851

MY DEAR THOMSON,

Do you know de Senarmont's papers on the propagation of heat by conduction in crystals, in the Annales de Chimie for 1847 2nd half 1848 1st half & 2nd half?[2] A short time ago I took it into my head to work at the theory, though I knew a paper had been written on the subject, which I had stumbled on in looking for something else but had not read. However as I am fond of doing things in my own way I thought I would not look at it till I had tried my hand at ⟨it⟩ the subject, though of course it would not be worth while to spend much time over a thing that had been already done. After working a day or two I got over or rather eluded my principal difficulty, & the rest was pretty plain sailing. On referring to the memoir, or rather memoirs (for I found there were two) I found the principal results there already, as might have been expected. The memoirs are by Duhamel, & are to be found in the Polytechnic Journal Cah. 21 p 356 and Cah ⟨2⟩ 32 p. 155.[3]

Duhamel according to the French fashion works on the hypothesis of intra molecular radiation, which for my own part I don't much like making the results depend upon. I had some thoughts of drawing up a short paper on the subject for the Camb. Math. Journal.[4] I should like to know whether the subject is new to you, & if so whether you think you would like a communication on the subject.

Yours very truly
G. G. Stokes.

P.S. Have you seen Prof^r Challis's awful heterodoxy in the present n^o of the Phil. Mag.[5] I am half inclined to take up arms, but I fear the controversy would be endless.

1 Kelvin has written at the top of the letter, 'Jan 8, 1851 answ^d Jan 13'.
2 Henri Hureau de Sénarmont, 'Mémoire sur la conductibilité des substances cristallisées pour la chaleur', *Annales de Chimie et de Physique*, 3rd series, **21** (1847), 457–76; 'Mémoire sur la conductibilité des corps cristallisées pour la chaleur', *ibid.*, **22** (1848), 179–211; and 'Experiences sur les modifications que les agents mécaniques impriment à la conductibilité des corps homogènes pour la chaleur', *ibid.*, **23** (1848), 257–67. Sénarmont (1808–62) was professor of mineralogy at the École des Mines.
3 Jean Marie Constant Duhamel, 'Sur les équations générales de la propagation de la chaleur dans les corps solides dont la conductibilité n'est pas la même dans tous les sens', *Journal de l'École Polytechnique*, **13** (1832), cahier 21, 356–99, and 'Sur la propagation de la chaleur dans les cristaux', *ibid.*, **19** (1848), cahier 32, 155–88. Duhamel (1797–1872) taught at the École Polytechnique from 1830 to 1869, except for one year.
4 Stokes (44), which refers to the five papers listed in notes 2 and 3.
5 James Challis, 'On the Principles of Hydrodynamics', *Phil. Mag.*, **1** (1851), 26–38.

75 Kelvin to Stokes, 13 January 1851
Stokes Collection, K44

2 College, Glasgow
Jan 13, 1851

Dear Stokes

I am quite shocked to find I have allowed so many days to pass without answ^g your letter of the 6th.

I had a general notion of the existence of Senarmont's papers on the propagation of heat in crystals, and I saw Duhamel's lately, in looking thr. some Cahiers of the J. de l'Ec. Polyt. in Paris, but I have read none of either paper.[1] I fancy I have a general notion of the subject too, but I have not tried to work it out at all, although I have been working at mag^{c} induc^{n} in crystals, w^{h} if the superpos^{n} of mag^{c} induc^{ns} be assumed in one & the indep^{t} superpos^{n} of dist^{ns} & corresponding mot^{ns} of heat, in the other theory, ⟨be assumed⟩ must probably follow corresponding laws. I have not however attempted any demons^{n} to show that the 9 constants are reducible to 6 independent, except in the theory of magn^{c} induction. I despatched, for the Philosophical Magazine,[2] some weeks ago, an abstract of what I communicated in part to the Brit. Ass^{n}[3] in Edinburgh on this subject.

I hope you will send a paper of the kind you propose for the C&DMJ,[4] as you know what has been done on the subject besides having worked it out yourself.

I saw but did not read Prof Challis's last.[5] How curiously hard it is to convince of what is as plain as a pike-staff to everyone else who has looked at it, that he is wrong!

How are you pleased with the Senate House papers as yet? I have seen the 1st set, w^{h} appear to be very nice as far as I have looked into them.

What do you think of Faraday's last?[6] Do you agree to his ⟨account⟩ theory of the diurnal &c. perturbations of the needle?

Your's very truly
WILLIAM THOMSON.

P.S. I hope you are going to give some more Hydrodynamics to the Journal. Have you been working at anything new?

P.S.[7] If you see Cox, will you tell him that I have rec^{d} his paper on Taylor's Theorem, w^{h} I sh^{d} think will make a very good 'Math Note.'[8]

W.T.

1 See the previous letter, notes 2 and 3.
2 Kelvin (50).
3 Kelvin (43).
4 Stokes (44).
5 See the previous letter, note 5.
6 Michael Faraday, 'Experimental researches in electricity. – twenty-sixth series', *Phil. Trans.* (1851), 29–84, especially 42–84 on atmospheric magnetism.
7 Kelvin has written the second postscript inside the flap of the envelope.
8 Homersham Cox, 'A demonstration of Taylor's theorem', *Camb. Dub. Math. J.*, **6** (1851), 80–1.

76 KELVIN to STOKES, 3 February 1851
Stokes Collection, K45

2 College, Glasgow
Feb 3, 1851

DEAR STOKES

I had a letter from Cox a few days ago by w^h I was informed that he was to be in Camb. on Sat.[1] If he is still there, will you give him the enclosed? If not, & if you know his address, will you forward it by post, taking out the proof & cancelling the PS in the note accompanying it, if he is farther, in point of time, than London. If this last be the case, the proof is to go direct from you (if you will oblige me by sending it) to the printers, & my note is to go to Cox, as he asked a question w^h I answer in the note. Will you read the note & tell me if you think it is not right as regards the right of the public to help themsel⟨f⟩ves from the C&DMJ or other like $public^{ns}$?

Thanks for your examination paper w^h contained some questions that interested me. I have an unfinished 'Note on Hydrodynamics' written just two years ago on the 'work' relations of the motion of a compressible fluid w^h w^d have enabled any one who might have read it to answer your No 16 if it had been published, w^h it is not.[2] Did you get any answers to that? Who are the Smith's Prize men?[3]

Your's very truly
William Thomson

1 The letter is not in the Kelvin Collection or the Kelvin Papers at Glasgow.

2 Stokes's Smith's prize examination paper for 1851 is printed in his *MPP*, v, 312–14. The Smith's prizes were established by the will of Robert Smith (1689–1768), Plumian professor and master of Trinity College. There were two prizes awarded each year on the basis of an examination for which the Lucasian professor was usually one of the examiners. After 1883, the prizes were awarded for an essay rather than on examination results. All of Stokes's Smith's prize papers are in *MPP*, v, 309–68.

3 They were N. M. Ferrers and George Valentine Yool (c. 1829–1907). Yool, third wrangler and second Smith's prizeman in 1851, placed in the first class of the natural sciences tripos in 1852, entered Lincoln's Inn in 1851, and was called to the bar in 1856.

77 KELVIN to STOKES, 25 February 1851
Stokes Collection, K46

2 College, Glasgow
Feb. 25, 1851

MY DEAR STOKES

I shall be very glad to get publishing your paper[1] in the Journal, as I am very desirous of getting such papers on physical subjects sometimes in place of the endless algebra & combinations w^h so abound.

I have been very much occupied or else I should have written sooner to acknowledge your last communications. Now I write when I have a favour to ask viz. that you will look at the enclosed w^h I received lately from Rankine.[2] I presume you have already seen the Feb. Number (It must surely have been published by this time, although they have not sent it to me yet) w^h contains Rankine's paper.[3] You will probably be interested in the § 'on the results of the hypothesis of atomic centres' and if you read that & look at my note appended at the end of the entire article, you will see what gave rise to the enclosed.

In Fourier p 521 you will find ⟨an⟩ this equation

$$\frac{dv}{dt} = a\frac{d^2v}{dx^2} + b\frac{d^4v}{dx^4} + c\frac{d^6v}{dx^6} + \&$$

which he says 'exprime ce que serait le mouvement de la chaleur dans les corps solides, si la transmission instantanée n'était pas bornée à une distance infiniment petite. ⟨'⟩ On a un⟨e⟩ exemple de ce genre de question⟨s⟩ dans le mouvement de la chaleur lumineuse qui pénètre les milieux diaphanes.'[4] No reference in this part of the work, nor as far as I remember, in any other, to any published investigation on the subject of the propagation of heat in rock salt or the like, and I believe what I have quoted is about all he has on it in the big book. I am very glad you put the investigations ⟨of the⟩ regarding the conduction of heat on their right basis instead of confusing them with molecular radiation hypotheses. Fourier gives both ways for the conduc^n in non-crystalline matter. What I do in magnetism is quite ⟨simi⟩ analogous. I dispense with the hypothesis of mag^c fluids & substitute the principle of the independent superposition of magnetic inductions (w^h is only approximately, or restrictedly, true in reality, although a necessary consequence of Poisson's hypotheses) Something in your paper[5] of 'rotatory' motion of heat w^h I saw near the end struck me as corresponding to the proof I give that the 9 constants of mag^c ind^n are reducible to 3. You will see an acc^t of what I have got on the subj^t in the Phil Mag March.[6]

Your's ever
WILLIAM THOMSON

1 Stokes (44).
2 Probably W. J. M. Rankine, 'Laws of elasticity of solid bodies', *Camb. Dub. Math. J.*, **6** (1851), 178–81 and 185–6, both parts of which are dated February 1851. Rankine to Kelvin, 15 February 1851, Kelvin Collection, R22, enclosed Rankine's paper and explained what he thought Kelvin had misunderstood in it.
3 William John Macquorn Rankine, 'Laws of elasticity of solid bodies', *Camb. Dub. Math. J.*, **6** (1851), 47–80. Rankine (1820–72), both physicist and engineer, was at Glasgow with Kelvin as professor of civil engineering and mechanics from 1855 until his death.

4 Joseph Fourier, *Théorie Analytique de la Chaleur* (Paris, 1822). The equation is on page 521; the quotation on pages 522–3. The quotation reads *extrêmement* instead of *infiniment*.
5 Stokes (44).
6 Kelvin (50).

78 KELVIN to STOKES, 10 April 1851
Stokes Collection, K47

2 College, Glasgow
Ap. 10, 1851

DEAR STOKES

Did you receive a short paper by Rankine on Elasticity about the 25th of Feb. last?[1] If you did & if you think it ought to be published (whether you are fully convinced by it or not) will you give it to Metcalfe & Palmer & tell them that I shall direct them as to its locality?

Do you think, when air is compressed any appreciable portion of the mechanical effect produced by the work spent is statical?

Your's very truly
WILLIAM THOMSON

1 See the previous letter, note 2.

79 KELVIN to STOKES, 21 April 1851
Stokes Collection, K48

Ap 21, 1851

DEAR STOKES

Thanks for your correction on my Mag[c] paper,[1] which I forgot to acknowledge before. Of course your correction is required. The sentence as it stands is inconsistent with itself.

Your's
W. THOMSON

1 Kelvin (50). Kelvin's acknowledgement of Stokes's correction is in *PEM*, 481–2n.

80 KELVIN to STOKES, 21 April 1851
Stokes Collection, K49

2 College, Glasgow
April 21, 1851

MY DEAR STOKES

I have been excessively overwhelmed with other occupations for a long time or I should have sooner thanked you for all the trouble you have taken at my request regarding Rankine's paper.[1] In consequence of your report, especially of that part of it which is contained in your last letter, I have directed the printers to put in type Rankine's paper from the copy you sent them.

Now I have a very serious matter to talk to you about. – I have been finding the management of the Journal more and more trying to me for some years back on account of my numerous duties here and the distance I am from Cambridge, and I am beginning to be anxious to retire. By next October I shall have held the post of Editor for 6 years, and in that time brought out 6 volumes, or $\frac{3}{4}$ of the amount brought ⟨ough⟩ [?] out by my predecessors in the eight preceding years. Do you feel inclined to take a turn of Editorship? I have found the duties not on the whole onerous, even although I have been away from Cambridge so much, and I have always found them very pleasant. My mathematical friends have always been most kind in helping me, by giving me reports on papers, and I have never once found it necessary to wade through a paper on a subject that was at all out of my way, as you, and others have formed a sufficient *council* to enable me always without difficulty to find a willing referee. I am now however beginning to think that I might be allowed to retire on half pay in consideration of long services. I certainly shall not retire however unless I can get the interests of the Journal competently provided for in a successor.

I ought to let you know that I am always getting complaints such as in the enclosed letter from the publishers,[2] and that the existence of the Journal is a perpetual struggle. There is no pecuniary risk incurred by the Editor, as the publishers undertake all that; and there is only the expense of postage actually incurred by him. I find that considerable, but I get Crelle's Journal (& Liouville's too, when I go to Paris & claim it)[3] to set off against that; & ⟨that⟩ it would be next to o if I lived at Cambridge.

If you were undertaking the Editorship,[4] I should ⟨of course⟩ write to all my volunteer councillors, requesting them to continue their good offices, which of course they would agree to do. I myself would be very glad to undertake my share of reporting although unfortunately there are but few papers of the kind of w^{h} I should be most competent to express an opinion.

Regarding the publishers' complaints, it is clear that the Journal *must* be kept up; but the publishers cannot be expected to continue it at a loss. If we could get more papers useful for undergraduates it might sell much better than it does; but I think some of the contributors, ⟨even⟩ [?] & some of the supporters of science to be found in Cambridge & elsewhere, might be found willing to subscribe enough (very little if any w^{d} be required) to make two ends meet.

Will you take all this into consideration, & let me know what you think of it?

Yours very truly
William Thomson

1 See letter 77, note 2.
2 Kelvin was perhaps referring to a letter to him from Macmillan and Co. dated 18 May 1850, which is now in the Stokes Collection, PA1317. The letter reads in part, 'You will see [from the enclosed accounts] that it is not only losing but *increasing* in loss every late number. The loss upon the late numbers seems to be about £6. a number. We fancy that decreasing the number of copies printed to 250, and making another push for advertisements might perhaps make the succeeding numbers pay'.
3 Crelle's *Journal für die Reine und Angewandte Mathematik* and Liouville's *Journal de mathématiques pures et appliquées*. August Leopold Crelle (1780–1855) was a civil engineer who studied mathematics independently, eventually taking a Ph.D. at the University of Heidelberg at the age of thirty-six. He founded his journal in 1826 and edited it through fifty-two volumes.
4 Stokes did not become editor of the *Cambridge and Dublin Mathematical Journal*. The volume for 1853 was co-edited by Kelvin and Norman Macleod Ferrers, and Ferrers edited the 1854 volume by himself. The next year the journal merged into *The Quarterly Journal of Pure and Applied Mathematics*, which through its first fifteen volumes was co-edited by Ferrers and J. J. Sylvester, 'assisted by' Stokes, Cayley, and Charles Hermite. The editorial staff changed after the fifteenth volume in 1878, but Ferrers remained as a co-editor until the twenty-fifth volume in 1891. Ferrers (1829–1903) was senior wrangler in 1851 and master of Caius College from 1880 until his death. Hermite (1822–1901) held from 1848 a series of positions at the École Polytechnique, becoming in 1869 professor of analysis at the École Polytechnique and at the Faculté des Sciences. He held the first professorship until 1876, the second until 1897.

81 Kelvin to Stokes, 9 May 1851

Stokes Collection, K50

9 Barton Street
Westminster
May 9, 1851

My dear Stokes

I sh^d^ be much obliged by your writing me a few lines to say whether you have thought at all on the proposal I made regarding the C&DMJ the last time I wrote to you. I do not want to ask you to make up your mind in a hurry, but as I set out on Monday for a tour in Italy, I sh^d^ like to know before I go, whether you entertain the idea at all.

Perhaps some time when you are in to Macmillans you would say to them that I wrote to you on the subject of their letter,[1] and that, although I have not answered it (because I have not known what to say) I am not neglecting the subject.

What do you think of the pendulum? Do you think there is anything Hooke's-joint-ish about it in passing the 4 cardinal points?

I am exceedingly glad to hear you have applied for £200 for fluid friction[2] w^h^ well deserves it & from what Sabine[3] tells me I think is pretty sure to get it. Hopkins too I am glad to hear is going to get experiments done on the effect of pressure on the melting p^{ts}. Joule[4] & I have applied for £100 for experiments to test the relation between the heat absorbed & the mechanical effect emitted by air expanding at a constant temperature, which we propose to do by making it waste all its work on fluid friction. What do you think of Clausius' third mode of expansion!!! Phil. Mag. Mar 1851.[5] It is a bore to have to write a defense of what is as clear as a pike staff against objections which must be appreciated by all who admit the truth of Newton's first law of motion with *nee* in place of 'nisi quatenus.' However I suppose the insult cannot be shirked, but I shall content myself with a shot in the air.[6]

I saw tonight at the Royal Institution a beautiful contrivance of ⟨a⟩ Wheatstone's to illustrate the pendulum.[7] It is a spiral stretched ⟨like a⟩ for a vibrating cord which answers better than a simple cord because it can be drawn aside further; and requires less tension to bring its elasticity property into play – one end of it is fixed at the centre of a hoop moveable about a vertical axis & the other is adjustable to any point of the circumference. The action is beautiful when the line of the cord is 60° from the axis.

Col. Sabine showed me the 15 names recommended for Fellows by the council w^{h} included your's & mine.[8] This address will do for me till Monday morning.

Your's sincerely
WILLIAM THOMSON

1 See the previous letter, note 2.
2 This application (and the one mentioned later in the paragraph) was for money from the newly established government grant programme administered by the Royal Society. Until 1876, the Society's Government Grant Committee had £1000 per year to disperse. (See Roy M. Macleod, 'The Royal Society and the Government Grant: Notes on the administration of scientific research, 1849–1914', *Minerva*, **14** (1971), 323–58.)
3 Edward Sabine (1788–1883) was an artillery officer who had graduated from the Royal Military Academy at Woolwich. His first scientific expeditions in the late 1810s were supported by the Royal Society of which he eventually become secretary (1827–30), foreign secretary (1845–50), treasurer (1850–61), and president (1861–71).
4 James Prescott Joule (1818–89) worked in his own laboratory near Manchester, and, in their studies of heat and energy, he and Kelvin formed one of the great collaborations in the history of science.
5 R. Clausius, 'On the deportment of vapour during its expansion under different circumstances', *Phil. Mag.*, **1** (1851), 398–405.
6 Kelvin (50A).
7 Charles Wheatstone (1802–75) was professor of experimental physics at King's College, London, from 1834 until his death.
8 Both Stokes and Kelvin became fellows of the Royal Society on 5 June 1851.

82 STOKES to KELVIN, 10 May 1851
Kelvin Collection, S361

Pembroke College Cambridge[1]
May 10th 1851

MY DEAR THOMSON,

Almost the time your letter reached me I was engaged in a set of optical experts with which I was greatly interested, and then my lectures were close at hand so that I had to set to work at preparing for them, so that I have hardly been able to think calmly about the subject. I must say I rather dread the thoughts of it, still you certainly have worked your time and it is fair you should be relieved. Besides the 'dread' another point which makes me doubtful is this. Hopkins is going to resign the secretaryship of the Camb. Phil. Soc. and means to propose me to fill his place.[2] Would these not then be rival interests? the Camb. Phil. Trans. and the C&D.M.J.?

Did the thought ever occur to you of our being joint Editors? Or do you think the thing would be better done if one man had the sole contract?

I can hardly say I am prepared to accept yet I hardly like to refuse, you have worked so long at it. I should not at any rate like to accept without first speaking to Hopkins on the subject.

I have not looked at Clausius's[3] whatever it is I am so busy preparing for my daily lectures.

I wish you a pleasant tour in Italy and hope you will not be overroasted. You are running away from the total eclipse.[4]

Yours very truly
G. G. STOKES

1 Kelvin has written at the top of the letter, 'Aug 13'.
2 In 1851 Hopkins became president of the Cambridge Philosophical Society, and Stokes became one of the three secretaries.
3 See the previous letter, note 5.
4 The total solar eclipse of 28 July 1851 was observable in Sweden and Norway. (A. M. Clerke, *A Popular History of Astronomy during the Nineteenth Century* (London, 1908), 69–70.)

83 STOKES to KELVIN, 14 August 1851
Kelvin Collection, S362

Pembroke Coll. Cambridge[1]
Aug 14 1851

MY DEAR THOMSON,

Here I am and am likely to remain, for the instrument makers do not seem inclined to let me have my pendulums. I must write again & hurry them.

However I have not been idle in the scientific line; for I have been following out a rather remarkable physical discovery 'Tho' I says it that oughtn't' which I made about a week after Easter. I have communicated it to Fischer & will do so to you when you come; but unless the sun shines I cannot show you the experiments.[2]

Do you know the address of Griffin,[3] a chymist who formerly lived in Glasgow? If you happen to have a directory in the house where you are staying will you have the goodness to look?

Yours very truly
G. G. STOKES

1 Kelvin has written at the top of the letter, 'Lond Aug 15'.
2 See Stokes (42), which was read to the Royal Society 27 May 1852.
3 Alfred Thomas Griffin, 38 Houndsditch, is the only chemist named Griffin listed in the *London Post Office Directory*.

84 STOKES to KELVIN, 30 August 1851
Kelvin Collection, S363

Pembroke Coll. Cambridge[1]
August 30, 1851

MY DEAR THOMSON,

I think you said you had *got* Silbermann's map of the chemical lines.[2] If so I would feel obliged to you to compare it with the lines such as I saw them. I send you a copy of my sketch. The sketch is pretty truthful as regards the relative intensities of the lines. It is strange that 6 is omitted in Draper's map.[3] The reason probably is that in his photographs 6 was resolved into lines which individually were too small to notice. I have seen it resolved. I am puzzled about Draper's group N. The distances from H and M and the breadth of the group accord better with the supposition that his group N is made up of the lines 18–21. But granting that what appeared to him as 3 appeared to me as 4, how could he have left out 22 and 23? Yet again he has actually left out 6, which is far more conspicuous than the group 3, 4, 5 i.e. L. Can it be the difference of climate?

If you have *not* got Silbermann's map, and if Forbes be in Edinburgh, I should be obliged to you to forward the ⟨map⟩ sketch to him, and say I should be much obliged if he could assist me in the identification.

My sketch satisfies very fairly the condition that if you view it from a little distance with the eyes partially closed it brings out those parts which appear with a lower power.

I have seen as far as the pair 7, 8 directly with the eye, by using a deep blue glass.

You need not return the sketch. I shall know the lines by the numbers.

Yours very truly
G. G. STOKES.

P.S. If Becquerel's map be published I should thank you to let me know.[4] But I believe it is not.

1 Kelvin has written at the top of the letter, 'Glasg Mond Sep 15 returning fr. Ardgartan'.
2 Johann Theobald Silbermann (1806–65) was a laboratory assistant and then curator of collections at the Conservatoire des Arts et Métiers. In his (42), Stokes said Kelvin had furnished him with a copy of Silbermann's map. (See Stokes's *MPP*, III, 287.)
3 See J. W. Draper, 'On a new system of inactive tithonographic spaces in the solar spectrum analogous to the fixed lines of Fraunhofer', *Phil. Mag.*, **22** (1843), 360–4, cited in Stokes (42) in *MPP*, III, 285. John William Draper (1811–82), though born in England, was professor of chemistry at New York University.
4 There is a map published with E. Becquerel, 'Memoir on the constitution of the solar spectrum, presented to the Academy of Sciences at the meeting of the 13th of June, 1842', Taylor's *Scientific Memoirs*, **3** (1843), 537–57. Alexandre Edmond Becquerel (1820–91) became professor of physics in the Conservatoire des Arts et Métiers in 1852 and in 1878 succeeded his father – Antoine César (1788–1878) – as professor of physics at the Muséum d'Histoire Naturelle. His son – Henri (1852–1908) – discovered radioactivity in 1896.

85 KELVIN to STOKES, 18 September 1851

Stokes Collection, K51

2 College, Glasgow,
Sep 18, 1851

MY DEAR STOKES

Your letter reached me a fortnight after date, on account of an accidental delay, and I am afraid you must have thought me very much to blame for not answering it sooner. I have been trying to compare ⟨your⟩ the sketch you sent me with Silberma[n]n's painting, but I have not succeeded to my own satisfaction in identifying any of the lines. I think the most satisfactory way to you will be for me to send you a sketch taken from Silbermann's picture, in pencil in the same style as yours. For convenience in copying, it will be made on the same scale as the picture, & therefore much longer than yours, but you will get an idea of the scale by two or three of the principal ordinary lines being shown. The complete copying of the lines will only commence with H, unless you w^d like a copy in pencil of the whole, w^h I shall get done if you please.

Do you think the differences between your lines & those of Draper may not depend on the glass used, which may have a great effect in absorbing chemical rays!

Your's very truly
WILLIAM THOMSON

P.S. I forgot to say that I hope to have a sketch posted for you tomorrow or next day.

86 STOKES to KELVIN, 25 September 1851
Kelvin Collection, S364

Craven Hotel Charing +[1]
London Sept 25th 1851

MY DEAR THOMSON,

I am greatly obliged to you for the copy of Silbermann's map. I am sorry I gave you so much trouble, especially as after all I am unable to identify the lines. I have not compared your sketch with my own, still less with the actual object, since I had left Cambridge before your letter reached it. However I have got the lines, as they appeared to me, very well by heart, so that I can compare the lines in your sketch with what I saw nearly as well as if I had the actual object before me, except of course as to detailed measures. As to your conjecture respecting the cause of the discrepancy, no solid or liquid media, so far as is yet known, produce any such fine and apparently capricious bands by their absorption. It would therefore be a violent supposition to make, to attribute the discrepancy to this cause. But the two sketches hardly differ more, perhaps not so much as I have seen *the same object* (the fixed lines in the visible spectrum produced by the absorption of nitrous acid gas) differ according as it was seen in detail (by using a narrow slit & high mag. power) or in comparatively little detail (as seen by widening the slit, using a low mag. power, and throwing the eye-piece a very little out of focus). To me therefore the discrepancy is nothing surprising. But in this case an additional source of discrepancy seems to exist in the circumstance that the original of one drawing was produced by the deposition of mercurial vapour. This process is well adapted to take *black* lines, whether coarse or fine, but does not seem at all so well adapted to give faithfully more *minima* of intensity. The omission or all but omission of such minima might might [*sic*] a great difference in a sketch considered as an artistic representation of an object.

I am able to identify my lines 1 to 7, but there I stick. If I can get sun-light when I return to Cambridge I must try whether the light will allow me to use a purer spectrum.

I got the silk for you when I was in London before, but to avoid loading my pockets I left it at a shop from whence I was to receive a parcel, but they forgot to send it. I send it now, having got it yesterday at the shop. I don't know whether it is the right kind. I got it at Pearsall's; they said it was the finest imported.

I mean to go to Chester on Saturday & start with only a knapsack for Wales on Monday. I saw Prof^{r} Forbes in the exhib^{n} today.[2]

Yours very truly
G. G. STOKES.

1 Kelvin has written at the top of the letter, 'Sep 27, 1851'.
2 No doubt Stokes meant the Great Exhibition of 1851, declared open by Queen Victoria on 1 May 1851.

87 KELVIN to STOKES, 8 November 1851
Stokes Collection, K52

2 College, Glasgow[1]
Nov 8, 1851

DEAR STOKES

I broke your confidence so far as to tell Prof Maconochie your secret regarding light, he being Professor of Laws and having nothing to do with the scientific world in general. He is however a very skilful amateur of photography, and when I told him about your substances which are made visible by chemical light, he told [me] that many years ago he had observed that some of those substances which are rendered 'phosphorescent' by common day light are rendered so when held in the chemical part of the spectrum and perfectly shaded from all visible light. It is so long since he made the experiments that he could not (immediately at least) give me any farther particulars. It struck us both very much that that fact must be closely allied to your discovery. If I get any particulars regarding the experiments Mr Maconochie made I shall inform you of them.

I am going to Edinburgh next Thursday to spend two or three days with Forbes. I shall say nothing of your discovery unless you tell me I may. When is it to be published? Or is it published?[2] I want to tell it to my students sometime during the session.

I have been greatly engrossed with electrodynamics especially in connection with the principle of mechanical effect. I think I have got a good foundation for a theory of the mechanical effect of thermoelectric currents. What would you say to finding the law of density of saturated steam by observations on the difference of potential between E & E' when the junctions of the copper wires C, C' with a bismuth wire B are kept at temperatures S and T respectively?

This could be done if it were true that there is no electromotive force due to the difference of temperature in different parts of the same metals. As however I do not believe this to be true, a little more is required than merely the difference of potential at E & E', but I believe that (theoretically) it might all be supplied by thermoelectric observations alone.

What do you think is the mechanical value of a current of given strength ⟨suless⟩ [?] left without ⟨a⟩ sustaining electromotive force in a closed ⟨linear⟩ conductor of given form? It is the quantity of work that would be *gained* by bringing an infinite number of infinitely thin closed linear conductors together, with a constant current sustained (no matter how provided it be constant) in each, so as to form the given conductor with its actual distribution of currents. I am quite convinced[3] that diamagnetics are only bodies less magnetizable than space (i.e. I suppose the luminiferous medium).

Your's always truly
WILLIAM THOMSON

1 Kelvin has written at the top of the letter, 'If you chance to post a letter for me on Wednesday or later in the week address to 3 Park Place Edinburgh'.
2 Stokes read his (42) to the Royal Society on 27 May 1852.
3 Kelvin has added a footnote reading, 'If any substance was diamagnetizable, and was a little (ever so little) slow in taking the magnetization due to its position; then a ball of the substance supported in a magnetic field by its centre of gravity would of its self turn ever so fast in any direction'.

88 STOKES to KELVIN, 15 November 1851

Kelvin Collection, S365

Pembroke College Cambridge[1]
Nov 15th 1851

MY DEAR THOMSON,

By the time this reaches Scotland I suppose you will have returned to Glasgow, so I write there.

The result of Profr Machonochie's [i.e., Maconochie's] expert is already known, and in print, (see Moigno's repertoire 3d part p. 987.)[2] I don't know whether Machonochie or Becquerel had the precedence in point of time.

I have thought all along that of phenomena already known phosphorescence by 'insolation' bore the closest resemblance to my discovery. *Theoretical* reasons lead me to suppose that the difference is in reality less than it appears to be experimentally. Experimentally the leisurely character of the one phenomenon & the instan[tan]eous character of the other seem to indicate an essential difference. *Theoretical* reasons lead me to believe that the commencement and cessation of the illn (in my experts) really occupies a time which is large compared with the period of a luminous vibration, though whether it can be made sensible in experiment is altogether another question.

I have not yet gone on with my paper[3] since my return from Wales. I hope almost immediately to set about it again.

The difference of potentials at the points *E*, *E′* in your figure, and the law of density of saturated steam certainly seem to me very heterogeneous. I am however too ignorant of electricity to form a firm notion of what might or might not be expected to be done.

As to your question about the work of an unsustained electric current, I am too ignorant of the subject to say anything worth listening to. It certainly seems to me that a sustained and an unsustained current are as it were quantities of a different nature, and that you cannot pass from the one to the other without adopting some special physical theory relating to the nature of a current.* Does not the latter involve an idea analogous to inertia which is wholly wanting in the former? (*nor even then without introducing some constant of unknown numerical value, which would have to be determined, if it could be determined at all, by some special experiment.) Excuse these worthless remarks and believe me

Yours very truly
G. G. Stokes

1 Kelvin has written at the top of the letter, 'Nov 18, 1851'.

2 L'Abbé Moigno, *Répertoire d'Optique Moderne ou Analyse Complète des Travaux Modernes Relatifs aux Phénomènes de la Lumière*, 4 vols. (Paris, 1847–50).

3 Stokes (42).

89 Kelvin to Stokes, 13 January 1852

Stokes Collection, K53

2 College, Glasgow
Jan 13, 1852

My dear Stokes

Do you know any experimental data from which the ⟨mecha⟩ absolute mechanical value of as much of the sun's rays as fall on a unit of surface in a unit of time may be determined? The rate at which water in a non conducting vessel covered with a thin black lid would be heated above the temperature of the surrounding objects would do. Have any observations on these circumstances either in ordinary localities or on the tops of high hills been made? I was unfortunately not present at the meeting of the R. S. E. when your communication[1] was read, but the title of it suggested to me that you might be able to inform me on this point. I am thinking on communicating a paper[2] on the sources of the mechanical effect producible under the direction of man. I think that, with the exception of what might be got from tide mills, or the combustion of meteoric stones or other native metals, all is derived from the sun, and is merely a part of the mechanical value of the undulations which he has sent us from the epoch of the creation of plants.

Have you published your discoveries on light yet?[3] I am anxious to know more than from mere curiosity because I shall soon begin lecturing on the subject, and if your paper has been communicated to the R. S., or published in any way, I should like to refer to it. I heard from the printers that you were to be in Ireland about this time. Will you not give me a visit on your way back? There is an immense quantity of things I w^{d} like to talk about with you.

Your's always
WILLIAM THOMSON

1 Stokes (52), read to the Royal Society of Edinburgh on 5 January 1852.
2 Kelvin (58), read to the Royal Society of Edinburgh on 2 February 1852.
3 Stokes (42), read to the Royal Society of London on 27 May 1852.

90 KELVIN to STOKES, 15 January 1852
Stokes Collection, K54

2 College,
Glasgow
Jan 15, 1852

MY DEAR STOKES

Many thanks for your letter which I have just received. I wrote to you on Monday addressing to Cambridge, in the hope that my letter might be forwarded to your address in Ireland w^{h} I did not know. Now that I know it I write a line to ask if you will not take Glasgow on your way to Cambridge & spend a few days with me.

There are a great many things I w^{d} like to talk to you about, although I am afraid I cannot promise to shew you anything in the way of apparatus, worth coming to see.

I hope I shall have an opportunity in May of seeing your experiments.[1]

Your's very truly
WILLIAM THOMSON

1 Apparently, the missing letter from Stokes to Kelvin reported that Stokes (42) would be read to the Royal Society in May.

91 KELVIN to STOKES, 16 February 1852
Stokes Collection, K55

2 College, Glasgow
Feb 16, 1852

MY DEAR STOKES

Do you think it possible that the velocity of sound is affected by some solid elasticity of air existing during the rapid vibrations of sound, but not existing at all, or not appreciably, in any *statistical* circumstances of air? I have (in common with Joule) a very strong suspicion that 1.41 is a good deal greater than the true ratio of the specific heats. If Mayer's[1] hypothesis that the heat given out by the compression of air at constant temperature is = the thermal equivalent of the work spent, be as approximately true as it seems ⟨likely⟩ to be according to experiments by Joule, & certain theoretical considerations, for ordinary atmospheric temperatures, I think k cannot be greater than 1.38. Joule has been trying the sp[ecific] h[eat] of air under constant pressure & finds it to be .292 (in his first series of experiments. I have not heard whether he has confirmed this nor what confidence he puts in it.) This, with Mayer's hypothesis, & with $J = 1390$ (the mech^{l} equiv^{t} of the thermal unit centigrade) leads to $k = 1.31$. De la Roche & Berard's ⟨sp⟩ value[2] (about .265 I think) for the sp. h. under cons^{t} press. leads in the same way to $k = 1.352$. The value .2531 for the sp. h leads to $k = 1.375$. I do not think the sp. h. can be small enough to give $k = 1.41$. The question is between a great deviation from Mayer's hypothesis (1130 for the work required to develope [*sic*] a unit of heat by the comp^{n} of air, if Joule's sp. h. of air be right, instead of the mechanical equivalent 1390) and 'solid elasticity' in air as far as regards the propagation of sound.

Your's very truly
W. THOMSON

1 Julius Robert Mayer (1814–78) was a German physician who is best known for his claims in the 1840s, initially based on physiological observations, relating to the doctrine of the conservation of energy. In the 1860s, Tyndall championed Mayer in a priority dispute involving Joule, Kelvin, and Tait. The controversy drew several articles from both sides; see, for example, Kelvin and Tait (1); Tyndall, 'Remarks on Professor Tait's last letter to Sir David Brewster', *Phil. Mag.*, **26** (1863), 65–7 (which includes a letter from Mayer to Tyndall); and Tait, 'On the conservation of energy', *Phil. Mag.*, **26** (1863), 144–5 (in which Tait said that, unlike Kelvin, he had not yet 'declined to take part in personal controversy with [Tyndall]').

2 See J. E. Bérard and F. Delaroche, 'Mémoire sur la détermination de la chaleur spécifique des différens gaz', *Annales de Chimie*, **85** (1813), 72–110 and 113–82. Jacques Étienne Bérard (1789–1869) studied in Paris before returning to his birthplace, Montpellier, in 1813. He took his M.D. there in 1817 and became professor of chemistry. François Delaroche (1743–1812) took his M.D. at Edinburgh and was a medical practitioner in Paris. Bérard and Delaroche's joint paper – their only one listed in the *Royal Society Catalogue of Scientific Papers* – won a prize offered by the First Class of the Institute in 1811.

92 KELVIN to STOKES, 20 February 1852

Stokes Collection, K56

2 College, Glasgow
Feb. 20, 1852

MY DEAR STOKES

I troubled you some time ago regarding the possibility of 'solid elasticity' affecting the vibrations of air in sound & increasing the velocity of its propagation. I believe there will be no necessity of making the hypothesis that anything of the kind exists, when the true value of either sp. h. of air is known. There is a general relation of this kind between N & kN the two sp. h^s

$$kN - N\langle = EH \text{ h}^t \text{ of hom}^s\rangle$$

$$= \frac{E^2 \times \text{h}^t \text{ of hom[ogeneou]s atm. at } 0^\circ}{\mu(1 + Et)}$$

$$= \frac{E^2 \times 26215}{\mu(1 + Et)}$$

Now according to the gaseous laws for the $dens^y$ of sat[urate]d ⟨steam⟩ vapour at $diff^t$ $temp^{res}$, & $\frac{1}{1693.5}$ for its value at 100° Cent, we have

$$\frac{\mu(1 + Et)}{E} = \begin{matrix} 1357 \text{ when } t = 0 \\ 1369 \text{ when } t = 10 \end{matrix}$$

⟨&⟩ or if Mayer's hypoth.[1] were true, we should have, for all $temp^{res}$,

$$\frac{\mu(1 + Et)}{E} = \quad 1390.$$

Hence, if the value of kN were what De la Roche & Berard[2] make it (about .267) we should have $k =$ about 1.37. Or if, as Joule thought he had ma⟨k⟩de out experimentally ⟨ab⟩ a week ago, the value of kN were .29, we sh^d have $k = 1.30$ or thereabouts. But he has been repeating his experiments & he finds ⟨$k =$⟩ $kN = .234$ or a little more.

Now if $k = 1.41$ the preceding equations give

$$\left.\begin{matrix} kN = .2431 \text{ for } t = 0 \\ kN = .2410 \text{ [for] } t = 10 \end{matrix}\right\} \quad \begin{matrix} \text{acc}^g \text{ gaseous laws,} \\ \& = \text{for steam.} \end{matrix}$$

or $kN = .2374$, acc^g Mayer's hypoth. I believe therefore that the quantity k of w^h the value derived from the theory of sound & obs^{ns} on its $veloc^y$, &c, is 1.41, is exactly the ratio of the one spec. heat to the other.

Your's very truly
WILLIAM THOMSON

1 See the previous letter, note 1.
2 See the previous letter, note 2.

93 KELVIN to STOKES, 8 March 1852
Stokes Collection, K57

2 College, Glasgow
March 8, 1852

MY DEAR STOKES

Can you tell me whether Darker is living or dead or what has become of him? I have been writing letter after letter to him, without receiving any answer, and I have been most excessively put about by not receiving a quantity of apparatus which he promised to let me have, I can't tell you how long ago. I am now in very great straits for want of it, & I have written again & again asking him to let me have an immediate answer and to send me without any delay such of the apparatus as he has ready. I shall really be much obliged if you will tell me what you know of him now, & if you will, if you see him or write to him about your own business, ask him to communicate with me. I cannot tell you how much inconvenience I am suffering just now nor how impatient I am under Mr Darker's treatment which is such as I have never experienced from any instrument maker, or any body else, before. I am very sorry to trouble you by writing about it, but I do so as a forlorn hope of getting Mr D. stirred up.

I am just about to begin optics, and I am quite at sea about experiments as yet.

I sh^d have thanked you before for your letter in answer to my difficulties regarding the theory of sound. You should have received one from me about the same time that I received yours in which I told you of Joule's experiments by which it appears that probably the value of *k* derived from sound is after all the true ratio of the sp. heats. Do you not think that Challis's discovery[1] is really 'an aerial bore?' It is rather presumptuous in me to hazard such a conjecture, as I have not read his paper on this subject, but I am strongly impressed with the idea of *bore* connected with various discoveries of Challis's both in hydrodynamics & optics. Do you see that Potter's aerial bore has been published in French in the Annales de Chimie,[2] and answered by Bravais?[3] Is it not impossible that disturbances depending on any of the terms omitted in the ordinary theory can be propagated, even in the noise of a cannon, to distances from the locality of the disturbance which are great compared with its dimensions?

Your's very truly
WILLIAM THOMSON

1 Challis had published several papers on sound during the previous few years, the most recent being 'On the theory of the velocity of sound', *Phil. Mag.*, 1 (1851), 405–8.
2 R. Potter, 'The solution of the problem of sound, founded on the atomic constitution of fluids', *Annales de Chimie et de Physique*, **33** (1851), 327–32, and *Phil. Mag.*, 1 (1851), 101–4. Richard Potter (1799–1886) was sixth wrangler in 1838 and was professor of natural philosophy and astronomy at University College, London, from 1841 to 1865, except for one year.
3 A. Bravais, 'Note sur la vitesse du son', *Annales de Chimie et de Physique*, **34** (1852), 82–9. Auguste Bravais (1811–63) was professor of physics at the École Polytechnique from 1845 to 1857.

94 KELVIN to STOKES, 8 June 1852
Stokes Collection, K58

26 Lincoln's Inn Fields[1]
London
June 8, 1852

MY DEAR STOKES

I have taken lodgings at 32 Duke Street, St James', for myself, including a sitting room and bed room, which I get for 25/ a week. I saw a very nice bedroom in a house on the other side of the way, which could be had for 21/ a week and I have no doubt but that it could be had for 3/ or from that to 4/ a night. If you send me a line I shall take either that or some other hard by, for what ever time you want it.

Will you not come up in time for the R.S. meeting on Thursday? Make a point of ⟨coming for⟩ being here on Saturday so as to be able to go to Lord Rosse's[2] soiree.

If you post a letter for me tomorrow evening address it to 26 Lincoln's Inn Fields. If later, to the Oxford & Cambridge Club, Pall Mall.

Your's always
WILLIAM THOMSON

I hope you will be able to make some meaning out of the above, which is written in the dark, with Cayley & Sylvester talking about invariants all the time.

1 According to the *London Post Office Directory* for 1852, this was the address of the mathematician James Joseph Sylvester (1814–97), who was second wrangler in 1837 but, because he was Jewish, did not receive a degree until 1872. He held professorships at University College, London, and the University of Virginia, before entering the Inner Temple in 1846 and being called to the bar in 1850. Between 1855 and his death, he held a series of professorships at Woolwich, Johns Hopkins, and Oxford.
2 The Third Earl of Rosse (1800–67), a noted astronomer, was president of the Royal Society from 1848 to 1854.

95 KELVIN to STOKES, 29 June 1852

Stokes Collection, K59

32 Duke Street, St James',
June 29, 1852

MY DEAR STOKES,

When are you to be here on your way to Kew? Call here if you can. If you do not find me in, you will find a small prism (of quar[t]z I suppose) Darker sent here for you, and a pair of gloves which I believe are your's. Spottiswoode[1] would like you to dine with him on Thursday at 12 James Street, Buckingham Gate, at 7 p.m. if you are in town at that time. Cayley, Sylvester, and I are to be there. He asked me to communicate the invitation, as I told him I was going to write to you to ask you about your movements.

Your's very truly
WILLIAM THOMSON

1 The mathematician William Spottiswoode (1825–83) took his B.A. at Oxford in 1846 and was president of the Royal Society from 1878 to 1883, during the latter part of Stokes's term as secretary. There are some seventy letters from Spottiswoode to Stokes in the Stokes Collection, dating mostly from their time as fellow officers of the Royal Society.

96 KELVIN to STOKES, 31 July 1852

Stokes Collection, K60

2 College, Glasgow
July 31, 1852

MY DEAR STOKES

Sometime, probably early, in September, I am going to be married, to a Miss Crum.[1] I cannot describe her exactly to you, but I am sure that is unnecessary to ensure your good wishes at present, and when you come down to see us in Scotland, I am sure you will be glad to make her acquaintance.

This may alter my plans for the meeting in Belfast.[2] ⟨but⟩ I shall write to Col. Sabine as soon as I know for certain what I shall do. In the meantime I would not like you to mention to any one (except Col. Sabine if you chance to see him, or to be writing to him) the possibility of my not being present. ⟨as I⟩ Are you living at Kew now? I hope you have been keeping well since I saw you. Have you begun the pendulum experiments? I address to Cambridge as the safest.

Your's always sincerely
WILLIAM THOMSON

1 Margaret Crum (c. 1830–70) was Kelvin's cousin from Thornliebank, near Glasgow. Her father, Walter Crum (1796–1867), first cousin to Kelvin's father, was a fellow of the Royal

Society. She was in weak health throughout the marriage. (see, for example, letters 98 and 157.)

2 The British Association for the Advancement of Science met in Belfast in September 1852, and Kelvin attended as president of section A. Sabine was president of the Association for 1852.

97 Kelvin to Stokes, 21 December 1852

Stokes Collection, K61

2 College, Glasgow
Dec 21, 1852

My dear Stokes,

I wish to get some very light hollow metal balls, as truly spherical as possible, of dimensions of from 2 in to 6 in. diameter for electrical experiments, and as I believe you have had some things of the kind made for your pendulum experiments, you will perhaps be able to tell me of some maker by whom I could get what I want well done. I should be very glad also if you could give me some idea of the expense.

I am getting an apparatus constructed for the purpose of measuring electrical potentials in absolute units part of which will consist of equal and equally electrified metal balls repelling one another. I cannot quite determine what sizes will be best, and I may perhaps have to try various sizes; but if very accurately finished balls are much more expensive than ones made without great care as regards sphericity, I sh[d] probably first try various sizes of common balls and when I know what size w[d] give the most convenient forces, I might get accurate ones made of the right size.

Have you been making any more pendulum or other experiments lately? Your investigations in light have, as Fischer tells me, ⟨dr⟩ [?] excited much interest in Germany. But Sir David [Brewster] is yet incredible as to the 'change of refrangibility'.[1]

I have not had time to do almost anything since I saw you in Belfast. I hope however soon to know something definite from experiment regarding the thermal effects of electric currents in unequally heated conductors.

Where are you spending your vacation? I shall have to go for two or three days next week to Manchester to help in the air experiments Joule & I have undertaken. The rest of the holidays, which only last 10 days, I look forward to spending delightfully at Largs (sea side).

Your's very sincerely
William Thomson

1 See Stokes (42).

98 KELVIN to STOKES, 20 February 1854

Stokes Collection, K62[1]

Partially printed in Stokes's *MPP*, IV, 367.

2 College, Glasgow
Feb 20, 1854

MY DEAR STOKES

It is a long long time since I have either seen you or heard from you, and I want you to write to me about yourself & what you have been doing since ever so long. Have you made any more revolutions in science? or done any of the exp^{l} research on the friction of air? I saw a notice of your lecture at the R.I.[2] Tell me any new discoveries you have made &c. However I do not mean to impose upon you by demanding all this, but if there is anything short & good you can tell me I shall be glad to hear it.

I want to ask you about artif^{l} lights & the solar dark lines. (1) Is there any other substance than soda that is related to D? Are bright lines corresponding to it to be seen where soda is not present? Have any terrestrial relations to any other of the solar dark lines been discovered (or to the dark lines of any of the stellar spectra)? Are all artificial li⟨nes⟩ghts subject to dark[3] lines? I should be greatly obliged by your telling me in a word or two what is known on these questions, which I suppose you will easily do.

Is Mr Darker in existence? I bought & paid for a hydroelectric machine,[4] which I saw in its packing box last June year but it has never come. I am going to put it into a solicitor's hands to recover my property, unless I see my way of getting it otherwise, as I have sent Mr Darker letters without getting answers till I am tired.

I had a delightful tour through Malta Sicily Italy &c in summer with my wife and only arrived about the time of the meeting of the British Association to w^{h} I therefore felt scarcely up to going to.[5] I ⟨sh⟩ intend however to to [*sic*] go to the next, at Liverpool. Shall you be there? Is there a possibility of your coming this way at Easter going to or coming from Ireland? If you do visit Ireland at that time you *must* give us a few days here. I hope by that time my wife who is now recovering from a severe illness which has kept her in Edinburgh since Christmas will be well & strong. We intend to pass the summer in Arran w^{h} you would find not much out of your way to L'pool. I have been as busy as other occupations have permitted at the air exp^{ts} in Manchester & thermo-electric exp^{ts} here, since our return from the Continent.

Yours always truly
WILLIAM THOMSON

1 K63 in the Stokes Collection is a copy of K62 in neither Kelvin's nor Stokes's hand. It apparently was a copy which Larmor had made after Stokes's death and which he then sent to

Kelvin for his comments. (See Larmor's note in Stokes's *MPP*, IV, 367.) In any case, notes 3 and 4 below were added to K63 by Kelvin.

2 Stokes (49), delivered at the Royal Institution on 18 January 1853.

3 Kelvin has added a footnote to K63 reading, '"dark" was a "lapsus pennae" for "bright". See letter T. to S. March 2 1854'. The letter is letter 100.

4 Kelvin has added a footnote to K63 reading, 'I saw Darker some time later in Paradise Row Lambeth where he lived, and he told [me] that it, (or one like it, I forget which) had burst and killed a workman'.

5 In 1853 the British Association met in Hull in September.

99 STOKES to KELVIN, 24 February 1854

Kelvin Collection, S366

Partially printed in Stokes's *MPP*, IV, 368–9.

Pembroke Coll. Cambridge
Feb 24th 1854

MY DEAR THOMSON,

It is certainly a long time since we have had any communication with each other, hardly since your marriage. I hope to see you at Liverpool at the meeting of the British Assocn, but the Easter vacation I shall probably spend at Cambridge, as my lectures come on immediately afterwards.

Now for your questions. I am not aware that there is any pure substance known to produce the bright line D except soda. See end *. It would be extremely difficult to prove, except in the case of gases or substances volatile at a not very high temperature, that the bright line D, if observed in a flame, was not due to soda, such an infinitesimal quantity of soda would be competent to produce it. It is very common in ordinary artificial flames (such as a candle &c) but I think in such cases it may be attributed with probability to soda. In a spirit lamp I feel satisfied it is derived from the wick, for I find that alcohol burnt in a clean saucer does not give it, except perhaps a flicker now and then. Miller[1] told me (and I have verified the observation) that it is not found in an oil lamp, and I find that when the wick of a candle is cut short, so as to be surrounded by gas, and not to project into the luminous envelope where the combustion goes on, D disappears.

Sir D. Brewster states (Brit. Ass. Rep. 1842 p 15 of the 2nd part)[2] that the flame of deflagrating nitre contains bright lines corresponding to the dark lines A and B of Fraunhofer, and implies that other of the bright lines of this flame correspond to dark[3] lines of the solar spectrum. I saw somewhere a statement, I think by Sir D. B., that the flame, I think of burning potassium, certainly some flame in which potash was concerned, gave 7 bright lines corresponding to the dark lines forming Fraunhofer's group *a*. These are the only cases I know of in which identification has been[4] established, but the subject has hardly[5] yet been attacked. A vast deal of measurement has yet to be gone through. I think it likely that very interesting results will come out.

You will find in Moigno's Répertoire d'optique moderne (part III p. 1237)[6] much information on the subject of your questions.

You ask 'Are all artificial lights subject to dark lines?'[7] No, it is quite the exception. When there are lines of any kind it is usually *bright* lines. The flame of nitrate of strontia (i.e. a flame coloured red by nitrate of strontia) shows some[8] dark lines in the red, but then these same dark lines are found in the spectrum of common light transmitted across the flame, so that they appear to be due, not to the non-production of light of definite or almost definite refrangibility, but to its absorption by a certain gas or gases produced by combustion.

Darker had a sad accident about a hydroelectric machine, whether yours or not I am not sure. He was trying it, I think with a pressure less than that to w[h] it had previously been exposed, when it burst. Darker himself was knocked on the head, but not much hurt, but one of his best workmen was struck, and though apparently not severely hurt, he died of it the next day.[9]

Last summer I spent on the continent like yourself. Adams, Fischer, Wilson,[10] and I formed the party. We had a pleasant tour of it in Switzerland. Alas for the pendulums! I hope to swing them in the summer. As to other matters, I have had a good deal of correspondence with M[r] Schunck[11] on the subject of madder. I have convinced him of the existence of purpurine, which he did not before believe in. It has highly distinctive optical characters. I have thought of an explanation of certain very curious appearances seen with common light about the optic axes of biaxal crystals which are at the same time evidently doubly absorbing (axinite, epidote, iolite.) I wanted to compare theory and observation, but I have hitherto been stopped by the difficulty of procuring good crystals. Herschel (Light)[12] speaks of them as as rare as the most precious gems. However, M. de Senarmont[13] has lately found that the property of double absorption may sometimes be conferred on crystals by means of foreign colouring matters. The case he mentioned as most striking was that of nitrate of strontia crystallized out of a strong infusion of logwood turned violet by a little ammonia. He expressly mentioned the optic axes phenomena as shown by these crystals.

*Moigno mentions (p. 1244)[14] that Foucault in experiments with the electric pile, as well as Miller in observations on the flame of a spirit lamp to which various salts had been added, was struck with the constant occurrence of the bright line D.

Yours very truly
G. G. STOKES[15]

1 William Hallowes Miller (1801–80) was fifth wrangler in 1826 and professor of mineralogy at Cambridge from 1832 until his death.

2 David Brewster, 'On luminous lines in certain flames corresponding to the defective lines in the sun's light', *Brit. Assoc. Rep.* (1842), part 2, p. 15.
3 Larmor's printed version of the letter inserts *the* before *dark*.
4 Larmor's version inserts *yet* before *been*.
5 Larmor's version reads *barely* instead of *hardly*.
6 L'Abbé Moigno, *Répertoire d'Optique Moderne ou Analyse Complète des Travaux Modernes Relatifs aux Phénomènes de la Lumière*, 4 vols. (Paris, 1847–50).
7 See the previous letter, note 3.
8 Larmor's version reads *such* instead of *some*.
9 See the previous letter, note 4.
10 William Parkinson Wilson (1826–74) was senior wrangler in 1847, professor of mathematics at Queen's College, Belfast, from 1849 to 1854, and professor of mathematics at Melbourne from 1854 to 1874.
11 Henry Edward Schunck (1820–1903) studied chemistry at Berlin and Giessen before entering business in Manchester and becoming independently wealthy as textile merchant and calico-printer.
12 John Frederick William Herschel, 'Light', in Edward Smedley, Hugh James Rose, and Henry John Rose (eds.), *Encyclopaedia Metropolitana*, 20 vols. (London, 1845), IV, 341–586. Son of the astronomer William Herschel (1738–1822), John Herschel (1792–1871), as mathematician, physicist, philosopher of science, and especially as astronomer, was the one man who most clearly represented science for early Victorians. (See Walter F. Cannon, 'John Herschel and the idea of science', *Journal of the History of Ideas*, **22** (1961), 215–39.) An extensive correspondence between Herschel and Stokes is preserved in the Herschel papers at the Royal Society and in the Stokes Collection.
13 See H. de Sénarmont, 'Expériences sur la production artificielle du polychroisme dans les substances cristallisées', *Comptes Rendus*, **38** (23 January 1854), 101–5.
14 See note 6.
15 Kelvin has written the following two notes on opposite sides of the envelope. Kelvin's hand in the undated note suggests it was written at about the same time as the dated one.

Very valuable K[elvin] Dec 4, 03

Stokes Feb 24, 1854 Bright D not in lamp flame when wick short. Sir D. Brewster dark & bright lines. Moigno Cauchy. Other most interesting matter. Darker's accident. &c &c

100 KELVIN to STOKES, 2 March 1854

Stokes Collection, K64[1]

Partially printed in Stokes's *MPP*, IV, 369–70.

2 College, Glasgow
March 2, 1854

MY DEAR STOKES

Many thanks for your most satisfactory answers to all my questions. It was by a 'lapsus pennae' that I wrote *dark* lines instead of bright lines, when I asked if all artif[l] sources are subject to them.

I think it is really a splendid field of investigation, that of the relations betw. the bright lines of artif[l] light & the dark lines of the solar spec[m]. Don't you think any one who takes it up might find a substance for almost each one of the principal dark lines, by examining the effects of all salts on the flame of burning alcohol, or on other artif^{l} lights? I think it will lead us to a qualitative analysis of the sun's atmosphere. Do you think the supposed tendency to vibrate D of other salts besides common salt mentioned by Moigno[2] sufficiently established to want no confirmation. I am much disposed to doubt it, from what you have told me. I have tried a spirit lamp behind a slit through which sun light is coming & the effect is most decided. If I remember right the magnifying power I used was sufficient to divide D; but certainly whatever it was the light of the sp. lamp corresponded as exactly as could be observed with the dark of the solar D. It was curious to observe, the dark line not sensibly illuminated by the full light of the sp lamp coming through it, (the brightness on each side was so great) but a line of light above & below the solar spectrum appeared as an exact continuation of the dark line D. Will you not take up the whole subject of spectra, of solar & artif^{l} lights, since you have already done so much on it. I am quite impatient to get another undoubted substance besides vapour of soda in the sun's atmosphere. What you tell me looks very like as if there is potash too. I think copper & E[3] would be hopeful. The galvanic arc shows rather broad bands of green for copper than fine lines. Salts of copper sh^{d} be tried on flame. Do try iron too. There must be a great deal of that about the sun, seeing we have so many iron meteors falling in, & there must be immensely more such falling in to the sun. I find the heat of combustion of a mass of iron w^{d} be only about $\frac{1}{34000}$ of the heat derived from potential energy of gravitation, in approaching the sun. Yet it w^{d} take 2000 pounds of meteors per sq. foot of the sun, falling annually to account for his heat by gravitation alone.

I find that the emission of heat from a square foot of the sun is equivalent in energy to about 7 or 8000 horse power[4] – four times the heat that such an engine as that of the Duke of Wellington (w^{h} very likely actually works at double its nominal horsepower) could make by hammering alone. If the Duke of W. converts ⟨about⟩ $\frac{1}{16}$ of the heat of its furnace into energy in the engine (& I should think it does not do more) then the whole heat of its furnace w^{d} be about four times the heat from a sq. foot of the sun. So four sq. feet of the sun emit as much heat as the burning coals in the largest marine furnace – not altogether inconceivable, & quite in accordance with what we know of the excessive intensity of the sun's heat.

The mechanical value of the sun's heat per sq. foot at the earth's distance is about 84 foot lbs per second. If you calculate from this according to any tenable supposition as to the velocities of the vibrating particles (e.g. circularly polarized light with $\frac{1}{100}$ of the veloc. of prop^{n}, as the velocity of the particles), you will be astounded with the greatness of the density of the luminiferous medium required to produce the mech^{l} eff^{ts}. If for instance there were only 1 ft lb per sec^{d} produced per square foot, & if the mean vel. of the particles be $\frac{1}{100}$ of the veloc. of light w^{h} is about 1 000 000 000 ft per sec^{d}, the mass of ⟨a⟩ vib^{g} particles in a cub. f ⟨f⟩^{t} would have to be $\frac{32.2}{10^{23}}$ lbs. Now the density of the air in interpl. space, *if the temperature were uniform from the surf. of the earth upwards*, would be only some $\frac{1}{10^{230}}$ of the dens at the surface of the earth. What is the lum^{s} medium then? ⟨I suppose⟩ There must be matter in interplanetary space perhaps 10^{200} times as dense as the air w^{d} be on that hypoth.

I am much taken up with thermo elec^{c} currents in crystals now. I think I can make up a solid of

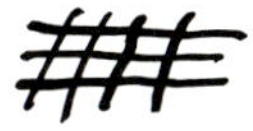

oblique gratings of iron & brass wires in w^{h} the thermo. el. axes will not be rectangular. If this be true a hollow cylinder cut out of it in a dir^{n} as shewn & heated inside while the outside is kept cool would have currents excited in

concentric circles. Although such an action is improbable & perhaps impossible, as you remark, as regards the conduction of heat, I believe it is possible in the therm. el. currents. Still, real crystals w^{d} probably have their axes rectangular.

Yours very truly
W. THOMSON

1 K65 in the Stokes Collection is a copy of K64 in neither Stokes's nor Kelvin's hand.

2 See the previous letter, note 6.

3 Larmor's printed version reads *&c* instead of *& E*. Fraunhofer's dark line E in the solar spectrum is in the green part of the spectrum, and it seems clear that Kelvin was hoping for a correspondence between copper and E similar to the correspondence he had just mentioned between sodium and D.

4 See Kelvin (73).

101 Stokes to Kelvin, 7 March 1854

Kelvin Collection, S367

Partially printed in Stokes's *MPP*, iv, 370–1.

Pembroke Coll. Cambridge[1]
March 7th 1854

My dear Thomson,

There are one or two points which occur to me with reference to your last letter which I wish to mention.

Miller told me of an experiment of his which he performed many years ago for testing the coincidence of the dark double line D of the solar spectrum with the bright line in the light of a spirit lamp. The sun's light was introduced by a slit, and refracted by 3 good prisms, and then viewed through a telescope with a pretty high mag. power. The two lines forming the line D were 'like that' (as he said holding two of his fingers about 3 inches apart) and he counted *six fixed lines between them*. The whole apparatus was left untouched till dusk, and then a spirit lamp was placed behind the slit. This gave two bright lines coinciding, as near as measurement could give, with the two dark lines D.

Miller seemed disposed to regard this as an accidental coincidence. It seemed to me that a plausible physical reason might be assigned for it by supposing that a certain vibration capable of existing among the ultimate molecules of certain ponderable bodies, and having a certain periodic time belonging to it, might either be excited when the body was in a state of combustion, and thereby give rise to a bright line, or be excited by luminous vibrations of the same period, and thereby give rise to a dark line by absorption.

But we must not go on too fast. This explanation I have not seen, so far as I remember, in any book, nor do I know a single experiment to justify it. I am not aware that any absorption-bands seen in the spectrum of light transmitted across any vapour that has been examined have been identified with D.

I gave you Moigno's statement[2] about Foucault's expert but I confess I am sceptical. I want more explicit proof of the absence of soda in some shape.

If I remember right chloride of copper on pack thread, put into the flame of a spirit lamp, gives two broadish green luminous bands (besides two more refrangible) which are each resolved, even by the naked eye with a highly dispersive prism of 60°, into a system of lines. Metallic copper in the voltaic arc is I believe somewhat different.

Your results respecting the density of the ether have been extremely interesting to me. I had indeed before contemplated the possibility of establishing a connexion between the density of the ether and the amplitude of the vibrations by means of the mechanical equivalent of heat, but I had no numerical data, respecting the effect of radiant heat, to go upon. As to the density $\left(\frac{32.2}{10^{23}} \text{ lbs per cubic foot}\right)$ which you deduce on a particular hypothesis respecting the vel^y, I confess I do not think there is anything unreasonable in it. I am altogether sceptical about the existence of air in the planetary spaces, but if it do exist I have no confidence in the truth of Boyle's law when pushed to such limits.

But what principally struck me was the magnitude of the excursions. Take your supposition of a vel^y of 10^9 feet per 1^s and suppose the circularly pol^d light to make 500 billion vibrations (5×10^{14}) in 1^s. If r be the radius of the circle described

$$2\pi r \times 5 \times 10^{14} = 10^9 \text{ feet}$$
$$= 12 \times 10^9 \text{ inches}$$

Putting 6 for 2π we have

$$r = \frac{2 \cdot 10^9}{5 \cdot 10^{14}} = \frac{4}{10^6} = \frac{1}{250\,000}$$

or say $\frac{1}{6}$th of ⟨a⟩ the length of a wave of light. This may be much too great, but it cannot be very enormously diminished without assuming an unreasonably large density. Now two views have hitherto been taken respecting the amount of the displacement of the ether relatively to the particles of ponderable matter, some supposing that it must be treated in the investigation as very small, others as very large. I think these considerations show that the latter view can alone be adopted.

Yours very truly
G. G. STOKES

1 Kelvin has written the following two notes on the envelope. They appear to date from the same time – 1903 – as those on letter 99. (See letter 99, note 15.)

Stokes 1854 March 7 Detailed description of Miller's exp^t ('many years ago') on solar dark D & spirit-lam[p] bright D

Density of ether Excursions of vib^g ? small or large relatively to size of atoms? suggests *large* as perhaps probable

2 See letter 99, note 6.

102 STOKES to KELVIN, 8 March 1854
Kelvin Collection, S368
Partially printed in Stokes's *MPP*, IV, 372.

Senate-House Cambridge[1]
March 8th 1854

MY DEAR THOMSON,

There was one thing more which I remembered just after I closed my letter last night, but as it was past 12 I did not think it worth while to open it. Being now engaged at a Bell Scholarship Examn I have plenty of leisure for writing.[2]

Ought not the $\frac{32.2}{2Vv^2}$ in your question to be $\frac{32.2}{Vv^2}$? The vol. of ether which is employed in heating a unit of surface in a unit of time is V; its mass $V\varrho$ ϱ being the density; its vis viva $V\varrho v^2$. The work necessary to produce this motion, and therefore the work it is capable of giving out, is $\frac{1}{2}V\varrho v^2$, and as much more work is given out in consequence of the distortion of the medium, giving altogether ϱVv^2.

Miller is now in the Senate-house examining for the Natural Sciences Tripos.[3] I spoke to him about the fixed line D. I find he has not published the observation. I find I did him wrong in supposing that he regarded the ⟨theory⟩ coincidence as accidental; he supposes that there is some cause for it.

Yours very truly
G. G. STOKES

1 Kelvin has written the following two notes on the envelope. The first is in pencil and probably dates from 1854. The second is in ink and probably dates from 1903. (See letter 98, note 1, and letter 99, note 15.)

Miller & Sodium Light Stokes Mar 9, 1854

Corrects error in previous letter saying Miller thought the coincidence accidental.

2 Bell scholarships were Cambridge University scholarships intended primarily for children of clergymen of the Church of England.

3 The natural sciences tripos was first held in 1851. (See letter 26, note 5.)

103 KELVIN to STOKES, 9 March 1854
Stokes Collection, K66[1]
Partially printed in Stokes's *MPP*, IV, 371–2.

2 College, Glasgow
March 9, 1854

MY DEAR STOKES[2]

It was Miller's experiment (wh you told me about a long time ago) which first convinced me there must be a physical connection between agency going on in and near the sun, and in the flame of a sp. lamp with salt on it. I never doubted after I learned Miller's experiment that there *must* be such a connection, nor can I conceive of any ⟨w⟩one knowing Miller's experiment, and doubting. There is I suppose something in Miller's mind inconceivable to me. You told me too your mechanical explanation, which struck me at the time, & for years has been taking a deeper & deeper hold [on] me ⟨m⟩. If it could only be made out that the bright line D never occurs without soda, I sh^{d} consider it as perfectly certain that there is soda or sodium in some state in or about the sun. If bright lines in any other flames can be traced as perfectly as Miller did in his case, to agreement with dark lines in the solar spectrum, the connection w^{d} be equally certain, to my mind. I quite expect a qualitative analysis of the sun's atmosphere by experiments like Miller's on other flames. Could you make anything decided[3] of your mechanical theory? ⟨in the way of⟩ Can you investigate mechanically the undulatory theory of radiant heat. E.G. A hot black ball, in the centre of a hollow black sphere, each of given temperature. What is the wave length, or lengths, of the undulations. The wave lengths as experiment shows are less the higher the temperature of the hot body. It is a splendid subject for mathematical investigation.

I chose $\frac{1}{100}$ of the vel. of propagation for the vel. of the particles, so that the excursions might be somewhat small fractions of the wavelengths. I almost think that, ⟨instead of⟩ besides being (as I never doubted they are) enormous[4] compared with the intervals betw. the atoms of matter, they must be sometimes much greater[5] than the wave length. What do you think of the radiation from the sun? If I remember right 4 000 000 or so f^{t} lbs per sec^{d} from each sq. foot: I know it is 7900 horse power from ea. sq. foot; as much as 30 lbs of coal burning per minute w^{d} produce. This is founded on Pouillet's[6] estimate .06th unit cent. per sec^{d} per sq. foot at the earth, & Joules m. eq. 1390, w^{h} gives 83 f^{t} lbs per sq. foot at the earth. This may be two or three times too much or too little, but still, how enormous must the truth be, either enormously dense ether or very large excursions of the vibrating particles. How can you think the air stops? Boyle's law need not hold of course. I have never seen or heard of any valid reason for supposing the air to stop.

Yours always truly
W. THOMSON

P.S. You did not tell me what you think of the gravitation theory of solar heat? I have been trying to make out what share of meteors the earth w^{d} take, if the sun gets enough to produce his heat, & I think it possibly reconcileable [*sic*] with what we have of falling stars &c.

1 K67 in the Stokes Collection is a copy of K66 in neither Stokes's nor Kelvin's hands. Notes 2, 4, and 5 below were added to K67 by Kelvin.

2 Kelvin has added a footnote to K67 reading, 'Yes K[elvin] Nov 30 1903'.
3 Kelvin has written above the line here, 'mathematical investigation'.
4 Kelvin has added a footnote to K67 reading, 'footnote of Dec 1, 1903. I now believe they are enormously *small* compared with the centres of neighbouring atoms in solids, liquids, and gases. In 1854 I believed sizes of atoms, and the smallest distances between centres of atoms in solids and liquids, to be vastly smaller than we now know them to be. K. Dec 1, 1903'.
5 Kelvin has added a footnote to K67 reading, 'I cannot think how I *ever* imagined the excursions, in the most intense light or radiant heat, to be greater than *very* small in proportion to the wave length. K Dec 1, 1903'.
6 See C. S. M. Pouillet, 'Memoir on the solar heat, on the radiating and absorbing powers of the atmospheric air, and on the temperature of space', Taylor's *Scientific Memoirs*, **4** (1846), 44–90. Claude Servais Mathias Pouillet (1790–1868) was assistant professor of physics at the Faculty of Sciences in Paris from 1826 to 1838 and professor of physics from 1838 to 1852.

104 STOKES to KELVIN, 28 March 1854

Kelvin Collection, S369

Partially printed in Stokes's *MPP*, IV, 372–3.

Pembroke Coll. Cambridge[1]
March 28th 1854

MY DEAR THOMSON,

Since I wrote last I have been somewhat of an invalid. Between the sailing of the Baltic fleet one day, Westminster Abbey the next, and one or two other things of a similar nature, I caught cold which produced a severe inflammation in my left eye. I was confined to my room for a week, and spent some days in darkness, not however making experiments on invisible light. My eye is now well, so that I can read and write as usual.

I certainly think that Foucault's &c. observations as to the general occurrence of the fixed line D require confirmation.[2] I am disposed to suspect contamination with sodium, or some compound of sodium. At the same time, although the compounds of sodium do produce this bright line in flames, I do not think that we can necessarily infer the presence of sodium or any compound of sodium from the appearance of this bright line. I will explain my reasoning presently.

As to the gravitation theory of the heat of the sun, I do not know of any objection to it. I never could bring myself to believe in the luminous atmosphere that Herschel talks about.[3] I have been in the habit of regarding the sun as an enormous body in a state of intense heat, emitting continually a portion of its original heat; as in fact 'growing dim with age', but at a rate not to be measured by the lives of us mortals. According to the gravitation theory the heat might be kept up, but there would be a progressive change of another kind, namely an increase of the sun's mass. This would of course produce a secular augmentation of the Earth's and retardation of the Moon's mean motion. I have not put the thing in numbers, but it would be requisite to attend to this point, to see if the augmentation of mass be sufficiently small.

You ask me how *can* I believe in the finiteness of the atmosphere. I answer, How *can* you believe in the community of atmosphere of the bodies of the solar system? In the first place, what is gained? Nothing that I know of, except the removal of the trifling difficulty of having to contemplate the free surface of an elastic fluid. I call this a trifling difficulty, for such I regard it. Of course if Boyle's law were true however small the density, and the temperature did not decrease *indefinitely* in ascending, there could be no such thing. But what do we know of the relations of pressure and density in such extreme limits of tenuity? Why absolutely nothing. If with Joule we suppose the elasticity of air to be simply an effect of motion, we get a very considerate velocity, about that acquired by a heavy body falling in vacuo from a height of $7\frac{1}{2}$ miles.

Now consider some of the difficulties of the supposition. Suppose the atmosphere at rest, and of a uniform temperature. Then p will be some function of ϱ (not necessarily $k\varrho$ if Boyle's law do not hold in case of extreme tenuity) and we shall have

$$\int \frac{dp}{\varrho} = \frac{m}{r} + \frac{m'}{r'} + \ldots$$

m, m' ... being the masses of the Earth, Sun, Jupiter &c and r, r' ... the distances of their centres.

For a point near the Earth we need only attend to the potentials of the Earth and sun, and we have

$$\int \frac{dp}{\varrho} = \frac{E}{r} + \frac{S}{r'}$$

Now $S = 356000^4\, E$ about, $r' = 95\ 000\ 000$ miles about and at the surface of the Earth

$$r = 4000 \text{ miles about}$$

$$\therefore \quad \frac{\text{pot}^{\text{l}} \text{ of sun}}{\text{d[itt]o earth}} = \frac{356354 \times 4000}{95\ 000\ 000}$$

$$= 15 \text{ about}$$

Hence the variation of the sun's potential due to the eccentricity of the Earth's orbit would be about $1\frac{1}{2}$ times the whole pot^{l} of the Earth at the surface of the Earth. Consequently, if we suppose the earth to start in its usual state from perihelion, long before it arrives at aphelion the barometric pressure would be as much reduced as it would have been by ascending hundreds of miles from the surface of the Earth. For whatever be the functional connexion of p with ϱ, the variations of $\int \frac{dp}{\varrho}$, and therefore those of p, will be the same whether a given variation of

$$\frac{m}{r} + \frac{m'}{r'} + \ldots$$

be produced by altering its first or its second term.

Again, Venus, not to mention Mercury, must be circulating in an atmosphere of tremendously great density; and therefore offering an enormous resistance.

In the face of such perfectly extravagant results, will any reasonable supposition respecting the temperature or motion of this supposed common atmosphere set matters to rights?

As to the augmentation of the refrangibility of the emitted rays when the temperature of a body is more and more raised, I would refer you to some exper^{ts} of Draper's, in case you should not have seen them. They were published a few years ago in the Phil. Mag.[5]

As to the cause of this result, I can state nothing positive, as you may[6] conceive. It falls in very well with certain conjectures which I have made in my paper On the Change of Refrangibility of light (art 230 &c)[7] but these are only conjectures.

I was greatly struck by[8] the enormous length of the spectrum which I obtained with my quartz train with the powerful battery belonging to the Royal Institution. It far surpassed in length the solar spectrum, I mean even taking in that highly refrangible invisible[9] part which a quartz train is required to show. In the case of metallic points this spectrum consisted of isolated bright lines. I cannot help thinking that decompositions of a very high order may be going on in such an arc (the voltaic arc I mean) and that a careful examination of these lines may lead to remarkable inferences respecting the bodies at present regarded as elementary. There is nothing extravagant in this supposition: few chemists I imagine believe that the so-called elements are all really such.

Now it is quite conceivable that chemically pure metals should agree with compounds of sodium in giving the bright line D. If this were made out, I should say that perhaps these metals were compounds of sodium, but more probably they and sodium were compounds of some substance yet more elementary.

Yours very truly
G. G. Stokes

P.S. I forget whether I told you that when I saw Darker 2 or 3 weeks ago your hydroelectric machine was very nearly ready.

1 Kelvin has written the following on the envelope, probably in 1903 (see letter 98, note 1, and letter 99, note 15): 'Stokes Mar 28, 1854. Grav[l] Theory of Sun heat. Sodium & D dark or light. ?D without Sodium. Finiteneness [*sic*] of Earth's atmosph &c'.

2 See letter 99, note 6.

3 This apparently referred to William Herschel's idea (accepted also by his son) of the sun consisting of a hidden, earth-like core surrounded by a luminous shell. (See, for example, A. M. Clerke, *A Popular History of Astronomy during the Nineteenth Century* (London, 1908), 54–5, and J. F. W. Herschel, *Outlines of Astronomy*, 4th edn (London, 1851), section 389.)

4 Stokes has written above the line, '356354 (Pratt)'. See J. H. Pratt, *The Mathematical Principles of Mechanical Philosophy, and Their Application to the Theory of Universal Gravitation* (Cambridge, 1836), 362.

5 J. W. Draper, 'On the production of light by heat', *Phil. Mag.*, 30 (1847), 345–60.

6 Larmor's printed version reads *can* instead of *may*.

7 Stokes (42).

8 Larmor's version reads *with* instead of *by*.

9 Larmor's version omits *invisible*.

105 KELVIN to STOKES, 21 March and 20 April 1854

Stokes Collection, K68

2 College, Glasgow
March 21, 1854

MY DEAR STOKES

I thought I was right in the matter of the mechanical energy of light, even after I received your letter, not thinking much about it but always intending to write & defend myself, till yesterday, when on a little consideration, ⟨when⟩ I saw that you were right & I was wrong.

I have been thinking some more on my theory of thermo-elec^{y} in crystals and I find very simple illustrative structures of a variety of kinds. I have just ordered three ⟨equal similar⟩ [?] bars each composed of thin pieces of copper & iron soldered together constituting equally thick alternate layers of the two metals. One of the bars, consisting of ⟨transverse pieces⟩ layers perpendicular to its length, will when put lengthwise in a circuit, have a thermo-el. character betw. copper & iron but much nearer ⟨copper⟩ iron, the ⟨better⟩ worse conductor of heat. Another, consisting of longitudinal layers, will have a character much nearer that of ⟨iron⟩ copper the ⟨worse⟩ better conductor of electricity. A third, consisting of oblique layers, will have a character as regards the effect of diff^{ce} of temp^{res} betw. its ends, intermediate betw that of the other two. If heated on ⟨the⟩ one side & cooled on the opposite, a current will tend from end to end, in w^{h} respect it possesses a new property. The electromotive force in each case due to a diff^{ce} of temp^{s} from end to end is independent of the length. Th⟨at⟩e long^{l} electromotive force in the oblique bar due to difference of temp^{res} at two sides will be directly proportional to its length & independent of its breadth for given diff^ce^ of temp^res^. The bars have come, and show the effects perfectly.

[One illegible word deleted] ⟨exp^ts^⟩ [?] [one illegible word deleted] ⟨succeeded perfectly in their⟩ [first part of word illegible] ⟨ations, which⟩ [two or three illegible words deleted]

April 20, 1854

Dear Stokes

Just a month ago I began to write to you, one day in bed, and was prevented from going on by a variety of hindrances. You see how far I got. I intended to go on to say that I could prove, both for thermo-electricity, and for the conduction of heat, the possibility of making a structure ⟨having three⟩ not symmetrical with reference to three axes ⟨not⟩ at right angles to one another. A homogeneous solid arranged in ⟨infinitely thin⟩ layers connected by oblique pillars would, if ⟨*AB*⟩ kept with a surface (*AB*) parallel to the layers, at one temperature, and another parallel surface at another, ⟨occasion⟩ experience no ⟨transverse⟩ conduction of heat except in a direction perpend^r^

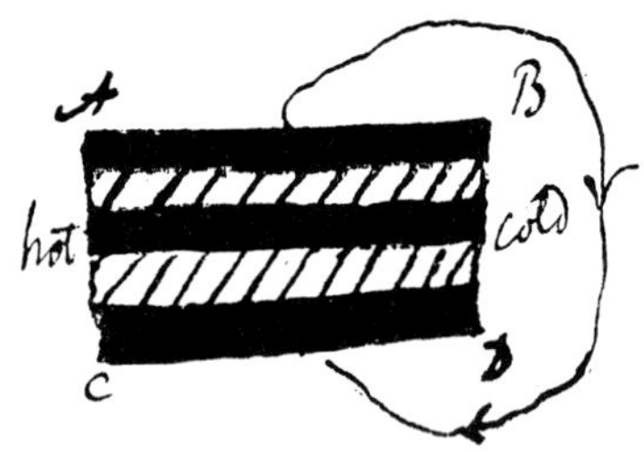

to the isoth^l^ surfaces. If on the other hand, the isothermal surfaces are perpendicular to these, there would be some conduction ⟨along⟩ perpendicular to them, or ⟨if they⟩ [?] to make the⟨m⟩ ⟨be no⟩ conduction be solely along them, the isothermal surfaces would have to be not exactly perpendicular to them, but inclined between the perp^r^ to them and the directions of the pillars. A crystal with a structure comparable to this, would have the wonderful conductive properties you show to be ⟨w^h^ equ⟩ [?] possessed by a body for w^h^ the 9 const^s^ are not reducible to 6.

If the pillars be of a different metal from the layers, there will be a corresponding obliquity in the thermo electric properties. If *AB* is connected with *CD* by means of a homogeneous equally heated conductor, ⟨no⟩ current would be produced and sustained by keeping *AC* and *BD* at two different temperatures: but if *AC* & *BC* be connected by a homog^s^ equally heated cond^r^

no curr^t would be occasioned by keeping *AB* and *CD* at two diff^t temperatures. I have got a piece of wire cloth woven of brass and iron wire, in the two directions; & by pulling it oblique

I shall have a representative of thermo electric crystalline obliquity. If a solid possessing this property be cut into a cylindrical shape, and the axis hollowed & heated, while the outer surface is kept at a uniform temperature, circular currents will be produced, & it will for the time act like a bar magnet.

I was in bed the whole week, in which I commenced this, and before the end of it I received your letter telling me you too had been unwell. I should have written at the time to condole with you, if I had known I was to be kept so long from writing to you on other matters. I was not very well for some time after, but was able to return to my work, and I am nearly quite re-established now. I hope you have long been quite well again.

I was very much struck with what you said about the occupation of the interplanetary spaces, and I wish very much to write to you about it, but I must stop now as I have scarcely time to get ⟨all re⟩ away to catch an omnibus for the country. I still believe in the continuity of atmosphere through space, and I have no doubt but that the difficulty you show regarding the earth's atmosphere when at diff^t distances from the sun will be explained by taking into acc^t the centrifugal force due to the revolution round the sun of a portion of it carried round with the earth, or else, by considering, w^h is very probable, ⟨by considering⟩ the whole interplanetary atmosphere to be revolving round the sun. I am much disposed to go back to the Vortices, differing only from DesCartes in being dragged round by the planets instead of drag[g]ing them round. I think I can prove that the sun's light is due to parts of the zodiacal light (which is merely a whirling cloud of stones, ⟨)⟩ acc^g [to] Herschel) falling in.[1] I communicated a paper on the subj^t to the ⟨R.S.E. on Monday. If the sun be a burning mass, he would burn out in 4 or 5⟩ R. S. E. on Monday.[2] If the sun were a burning mass, he would burn out in 4000 years at the present rate; (& 4000 years ago must have been of double diameter;) unless supplied with fuel from without. He is therefore not a ball of gun cotton burning off round the out side but must have supplies from without, of ⟨matter (if)⟩ fuel, if he be really a burning mass. But he can't have supplies of any kind of matter from without, without getting at least a thousand times as much heat from them by gravitation as they could possibly give by burning either intrinsically or with any bodies they might meet. The sun would become dark in a few minutes I suppose if he were merely a primitive heated mass of any kind of matter resembling in physical properties any thing we know.

A ⟨bright⟩ star suddenly appearing is merely a dark body perhaps as large as the sun moving into a cloud of dust in space. In a few seconds its surface might become incandescent by the impacts.

Yours very truly
WILLIAM THOMSON
In great haste

1 See J. F. W. Herschel, *Outlines of Astronomy*, 4th edn (London, 1851), section 897. In his (73), Kelvin cited the 'last edition' of 'Herschel's *Astronomy*'. This would have been the fourth edition of Herschel's *Outlines*, the first edition of which in 1849 was an expansion of his *Treatise on Astronomy* (London, 1833). The fifth edition appeared in 1858.

2 Kelvin (73), which was read to the Royal Society of Edinburgh on 17 April 1854.

106 STOKES to KELVIN, 24 April 1854
Kelvin Collection, S370

Pembroke College Cambridge[1]
April 24th 1854

MY DEAR THOMSON,

I am sorry to find that you have been ailing like myself, or rather unlike myself, for I had only an inflamed eye which did not affect my general health, and which was well again in a week.

I am still not satisfied about the rotation of heat.[2] I will endeavour to explain myself.

In the first place however I would observe that the propagation of heat, or more correctly the 'successive distribution' of heat, in a crystal depends only on *six* arbitrary constants; and with respect to this successive distribution there always exist three rectangular planes of symmetry, whose intersections constitute a system of what may be called 'thermic axes' (Camb. & D. M. J. N° 27 p. 220).[3] I am inclined to think that the most scientific way of treating the subject would be to banish the notion of a flux, and consider only the successive distribution of heat in a solid.

If however we please to introduce the notion of a flux, we get 9 constants instead of 6, the 9 entering into the eqn of successive distribution of heat in such a manner as to reduce themselves to 6. Now the 3 extra constants correspond to an idea necessarily connected with a flux, but not necessarily connected with the varying distribution of heat, namely, that of the *motion* of heat. (C. M. J. XXVII, 222)[4]

Now it seems to me that we can attach no idea to the motion of heat without entering into some speculation as to the nature of heat. I mean, that we cannot conceive as different from one another two modes of transfer of heat which would, both one and the other, bring the temperature of a solid from its state at the end of the time t to its state at the end of the time $t + dt$.

If then we make no supposition respecting the nature of heat, we are at liberty to introduce into our *definition* of the flux of heat that the mode of transfer is to be such as to make the constants D_1, E_1, F_1 (p. 220)[5] vanish.

If however we make some supposition respecting the nature of heat, it may be that our supposition necessarily involves, or at least leads to, the notion of a flux of heat. In such a case we are not at liberty to make the constants D_1, E_1, F_1, vanish. Rather, we may make these constants vanish if we please, but if we do we must distinguish the resulting analytical expressions

$$-A\frac{du}{dx}, \quad -B\frac{du}{dy}, \quad -C\frac{du}{dz},$$

from fluxes of heat.

If we suppose heat to be material, and conceive each ponderable element to be continually emitting in all directions a quantity of heat proportional to some function of its temperature, the heat so emitted being in turn absorbed by the elements among which it passes, the idea of a flux does not instantly present itself, but we are naturally lead [*sic*] to it by considering the difference between the quantities of heat which in a given short time pass, in opposite directions, across a given small plane. According to the theory of molecular radiation, as commonly understood, the constants D_1, E_1, F_1 vanish.

If we suppose heat to be material, but conceive no motion of heat to take place except what arises from differences of temperature, and instead of assuming the mutual intrapenetration of particles of heat suppose that the whole heat moves continuously in the manner of an elastic fluid (I speak only of the motion geometrically possible) we are lead instantly & unavoidably to the notion of a flux.

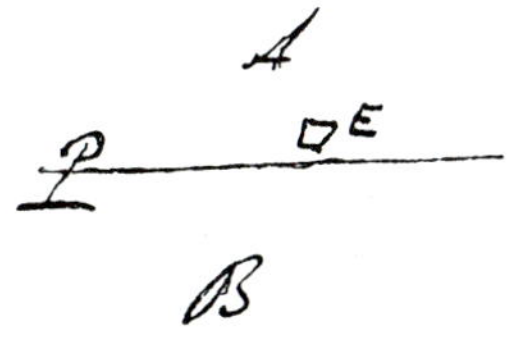

If we adopt the dynamical theory the notion of a flux does not immediately present itself. Yet we may give a precise definition to 'flux', at least if we introduce some simple and natural suppositions respecting the constitution of matter. Suppose a body, whether solid ⟨of⟩ [?] or fluid, to be divided by an infinitesimal mathematical plane P. If the body be at rest (not counting the disturbances w[h] constitute heat) let the plane P be fixed, otherwise let it move with the part of the ⟨s⟩ [?] body in which it is situated. Let dt be an increment of the time, which is regarded as infinitesimal, but nevertheless very great compared with the period of the molecular disturbances. ⟨Le⟩ During the time dt the adjacent portions of A and B may cross the plane P a little backwards & forwards, but this is no matter. Let E be an element of A situated at a distance from P less than the distance (δ) at which the molecular forces become insensible. During the time dt, we may conceive the rest of A as doing work on ⟨A⟩ E, and E as doing work on B. The difference of these will correspond to the change of vis viva (or it may be also compression or distortion) of E, and will introduce only insensible terms, if the distance δ be insensible. Let $dt \times dW$ be then the work done by A on E or by E on B, and $\bar{W} = dt \int\int\int dW$, the integral extending through a right cylinder having P (of which I will suppose the area to be equal to dS) for base, and having a height on the side A at least equal to δ. Then the quantity of heat w[h] is the equivalent of the quantity of work

$$\frac{\bar{W}}{dt\, dS}$$

may be defined as the flux of heat across the plane P.

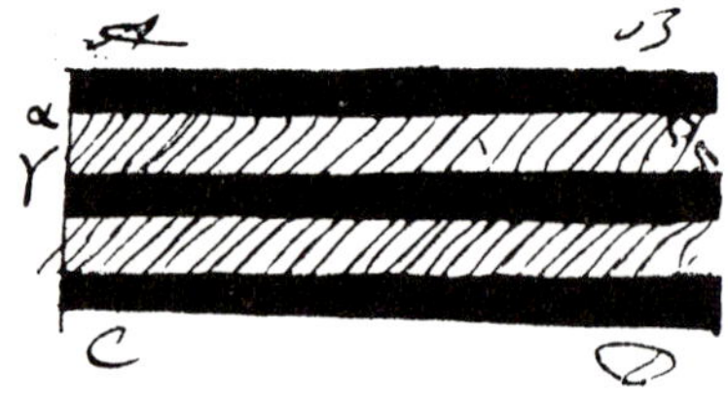

But whatever notion we adopt of a flux, I cannot see on what ground you say that in your solid if the surfaces AB and CD were kept up to different constant temperatures there would be no conduction of heat except in a direction $\perp$ to these surfaces. Suppose for simplicity the pillars to be infinitely narrow & close compared with the distance from AB to $\alpha\beta$ and from $\alpha\beta$ to $\gamma\delta$, and suppose that there is no lateral conduction from pillar to pillar. According to any notion I can form of flux, the flux would take place $\perp^{ly}$ to AB in the solid parts, and elsewhere along the pillars. Therefore if a be the breadth of one of the solid parts such as $AB\alpha\beta$, b the brea[d]th of one of the open spaces such as $\alpha\beta\gamma\delta$, i the inclination of the pillars to the normal, the direction of the flux would be inclined to the normal by the angle.

$$\tan^{-1}\left(\frac{b \tan i}{a+b}\right)$$

I think you mistook what I said about the sun. I stated, as that which had been my notion, not that the sun was in a state of combustion, but that it retained original heat, heat perhaps of condensation. Even if the numerical results of this hypothesis, made on the supposition of the sun's being solid, were not tenable, that objection would not be fatal. For we may well suppose the sun to be in a state of fusion, in which case if the surface cooled a little, and so became denser, the outer portions would descend, and be renewed from beneath.

I do not want to argue in favour of this view; I only say that I do not see it to be absurd on the face of it.

Yours very truly
G. G. STOKES[6]

1 Kelvin has written the following on the envelope, probably in 1903 (see letter 98, note 1, and letter 99, note 15): '⟨Stokes⟩ Rotational (? Magnetic) quality in conduction of heat. Stokes Ap 25, 1854'.
2 Kelvin has added (probably in 1903) a footnote here reading, 'See Stokes C & D Math Journal 1851, (or his MPP Vol III). See also T's MPP Vol I pp 280, 281, 286, 288'. The references are to Stokes (44) and Kelvin (62).
3 Stokes (44).
4 *Ibid.*
5 *Ibid.*
6 Kelvin drew the following diagrams below Stokes's signature in connection with the next letter.

107 KELVIN to STOKES, 26 April 1854

Stokes Collection, K69

2 College, Glasgow
April 26, 1854

MY DEAR STOKES

I have just received your letter, of Monday and I write to apologize for the haste of my last. I did not think of you either holding or at all inclining to either of the theories of solar heat that I objected to. I perfectly remembered your mentioning the notion that the sun is an enormous body in a state of intense heat emitting a portion of its original heat and 'growing dim with age,' but when I wrote last I was only impressed with what you had said about the gravitation theory which from not being unfavorable I had got into the habit of considering as favorable during the month after I got your former letter. It was during this time, principally during the last week of it when I was writing a paper w^h I had first intended merely to show some numerical relations as to heat of combustion, gravitation &c, that my own conviction became so strong as to the truth of the gravitation theory. I quite forgot (what I now see on reading your letter again) that you ⟨mentioned⟩ mentioned the other as one you had been in the habit of holding, and I did not consider myself as arguing with you at all when I put down that & the combustion theory with so little reason shown. I had always inclined to the primitive heat theory till rather more than two years ago, when I became convinced that neither as solid nor fluid the sun (of his own finite dimensions) could have given out heat as he has done for 6000 years with no source but primitive heat to draw upon. The quantity emitted is, according to Pouillet's data,[1] about

$.06 \times \frac{95\,000\,000^2}{441\,000} = 4639$ thermal units cent. per sec^d from each square foot – equivalent to 3 869 000 ft lbs. of energy per sec^d or about 7000 horse power.

The number of cubic feet in a pyramid to the centre of the sun [f]rom a sq. foot of his surface being

$$\frac{1}{3} \times 441\,000 \times 5280$$

the abstraction of heat from the sun per sec^d per cub. foot of his whole mass, would be about

$$\frac{1}{3} \times 528\,000$$

or per pound (if 60 lbs be taken as the w^t of a cub. f^t of the Sun, w^h is not far from the truth)

$$\frac{1}{10\,560\,000}$$

or per pound, per annum

$$\frac{31\,400\,000}{10\,560\,000} = \text{about } 3.$$

That is, the loss of heat per annum, per pound of the Sun's mass, is about 3 times the qu^{y} requ^{d} to raise the temp^{re} of a pound of water 1° Cent. We know no substance of greater thermal capac. than water, all metals, earths &c are of far less. Suppose the sun's mass to have the same th capac. as water w^{t} for w^{t}. Then the mean reduction of temperature of his whole mass per annum w^{d} be 3° Cent, or 18 000° in the 6000 years we know he has been giving out heat at something like the present rate. I think this makes it certain (or as nearly certain as any thing in w^{h} we consider possi⟨bi⟩l⟨ity⟩[2] properties of matter out of our reach) that the primitive heat theory cannot be true. I think I can show with nearly equal certainty that the combustion theory is insuff^{t} without a supply of fuel from ⟨with⟩ external space. Fuel coming from external space cannot come without giving from gravit^{n} more than 1000 times the heat of combustion, and therefore I conclude the grav^{n} theory. In my theory I anticipate the objections of physical astronomy as to the Sun's mass, ⟨in my paper⟩.[3] The matter that is falling in has probably for ages been inside Mercury's orbit & is now seen as the zodiacal light. ⟨In⟩ By its falling in, the force of grav^{n} on the planets towards the Sun's centre is not sensibly ⟨increased⟩ altered. A mass of meteors equal to the earth's mass would ⟨give a⟩ fall in in 50 years or so according to my form of the gravitation theory. I have not yet[4] considered what amount of effect such an add^{n} to the Sun's mass if coming from extra planetary space would ⟨pre⟩ produce on the planetary motions but I shall try directly. Mercury falling in w^{d} give 10 years heat. If he falls in as a whole, ⟨w^{h}⟩ how will the luminiferous ether stand it? Of course the whole surface of the earth would be instantly scorched. Mercury will probably fall in before the earth has varied very much from its present orbit; but the solidity of his mass would not so far as I can see be a guarantee that he w^{d} fall in as a whole. It is just possible that he might burst or evaporate into zodiacal light before getting to the region of intense resistance (ignition) and not cause any increase in the sun's heat, but I think it more probable that he would make[5] a great increase. I believe that if the solar system is ⟨perp⟩ permitted to go on long enough fulfilling the laws of matter which we know, the surface of the earth will in a few seconds or minutes be scorched & melted by Mercury falling in to the Sun. Till very lately I ⟨was in doubt⟩ saw no decided mechanical reason for anticipating that the Earth would first cease to be habitable for man by the Sun becoming cold or by itself becoming scorched by getting too near the Sun.

Taking 389551 *E* acc^{g} [to] Herschel[6] (very diff^{t} from Pratt's[7] by the way) or

400000 E as the Sun's mass we find $\frac{1}{400000}$ of his own, as the increase in 100 years, on the extra planetary meteoric theory, w^h^ requires only half the matter that mine does, or $\frac{1}{\langle 4 \rangle 8000}$ in 5000 years. That would give an accel^n^ of angular motion (since $\omega^2 r$ varies as the mass & $\omega^{1/2} r$ is cons^t^)

$$\frac{2}{3} \times \frac{1}{\langle 4 \rangle 8000} \quad \text{or} \quad \frac{1}{1 \langle 4 \rangle [?] 2000}$$

in 5000 years. The effect of this would be $\frac{5}{\langle 12 \rangle 24}$ of a year in that period, w^h^ I suppose is quite inconsistent with facts. If so, the extra planetary meteoric theory must be false, & we may conclude that the sun's heat is generated ⟨w⟩by the falling in of meteors which for ages have been far inside the Earth's orbit.

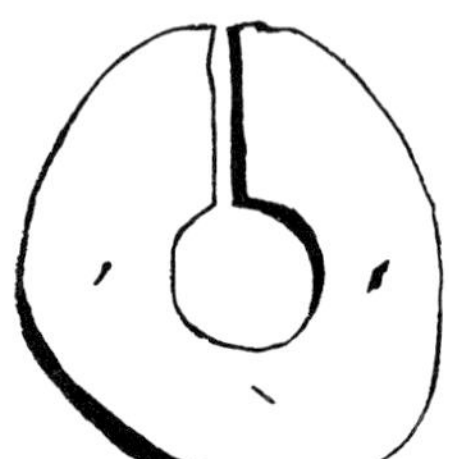

As to the rotation of heat, the supposed obliquity which I mentioned will certainly not in general affect the succession of distributions. But it would ⟨in⟩ [?] do so in some cases. Thus ⟨If⟩ if a cylindrical piece be bored out along its axis of figure, & slit as shown, ⟨the⟩ with a non conductor inserted into the slit, the ⟨ap⟩ symmetrical application & withdrawal of heat ⟨from⟩ by the two cylindrical surfaces will cause a difference of temperatures to subsist on

the two sides of the non conductor in the slit. One way of making a structure having that property is to take a number of parallel slips or bars cut obliquely from a real crystal, & lay them parallel to one another, & connect them into a⟨n⟩ rectangular net work by another set of bars similarly obtained, turning them so that the principal crystalline axes of the two sets are parallel respectively to the lines AA' &c indicated in the diagram. When I said that if the surfaces AB CD are kept at diff^t^ cons^t^ temperatures there would be no

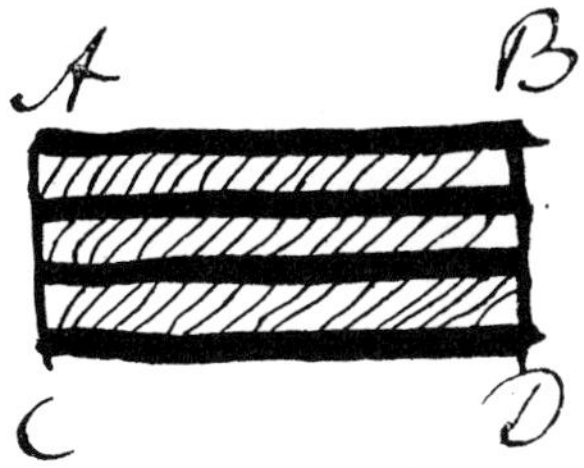

conduction of heat except perp^y to the isoth^l surfaces, I meant that to ⟨keep⟩ make the isoth^l surfaces parallel to *AB* & *CD* throughout the body, no heat would have to be given or taken ⟨from⟩ by the ends or sides *AC*, *BD*. This I think you will see is true, provided the dimensions *AB*, *CD* &c be infinite compared to the thicknesses & distances of the successive layers. But on the contrary if *AC*, *BD* be kept at diff^t temperatures, either the isoth^l surfaces between them will be oblique & curved, or heat must be given to *CD* and taken away from *AB*.

Whether or not this conductive obliquity exists in any homogeneous crystalline substance, the mere possibility of it for ⟨a⟩ an artificial structure makes it not to be excluded from the possibilities of crystalline or quasi-crystalline natural bodies, and ⟨I think⟩ the effects of it, although generally *latent*, yet may (if I am right in the illustration above) give rise to a remarkable phenomenon. Here is a simple enough structure possessing it, and at the same time having its principal conductivities equal ⟨pa⟩ [?] (i.e. its conductivities parallel to your princ. thermic axes equal) (the thin lines forming the sides of the square sh^d not have been drawn). The way of making it up from oblique crystalline bars would also give principal conductivities equal if the bars of the two sets be cut from similar positions in the same substance. It is in fact equivalent to the elementary structure here shown.

The black squares & thick bars all one substance

I was much concerned with what you told me about Darker's misfortune in the matter of my hydro electric machine. I have not written to him since you told me of the accident and of course I shall not urge him to do anything more about it. The last I had heard from him was last May, when he spoke of some delay having been occasioned, but said that he would send it almost immediately. I am sorry now that I had not time to go to see him as I wished to do on my way to⟨o⟩ or from the continent as I should have heard of what had happened.

When we meet or have time to write, I would like a great deal more talk with you on inter planetary & extraplanetary atmosphere; but in the meantime must pull up. I am glad you got over the illness so soon and completely as what you told about your eye made me afraid you might have had more trouble.

Yours very truly
W. THOMSON

1 See letter 103, note 6.
2 Kelvin obviously intended to change *possibility* to *possible*.
3 Kelvin (73).
4 Kelvin has written above the line '*See below'.
5 Kelvin has written the following between the lines and deleted it: '⟨(I feel it presumptuous to say he will, as⟩'.
6 See J. F. W. Herschel, *Outlines of Astronomy*, 4th edn (London, 1851), section 360, where the number given is actually 354936.
7 See letter 104, note 4.

108 KELVIN to STOKES, 3 May 1854

Stokes Collection, K70

Thornliebank, Glasgow,
May 3, 1854

MY DEAR STOKES

I have been thinking I made a mistake in estimating the acceleration the earth's motion w[d] experience from an augmentation of the central mass. The true relations, instead of the ones I gave are

$$\omega^2 a \propto \frac{m}{a^2}$$

$$\omega a^2 \text{ constant}$$

(ω being the mean angular vel. & a the mean dist[ce]). These give

$$\omega^2 a^3 \propto m$$

$$\omega^2 a^4 \text{ const}$$

$$\therefore \quad a \propto \frac{1}{m}$$

$$\omega \propto m^2 \text{ (instead of } m^{2/3} \text{ as I said before)}$$

This will make the true effect be three times what I said, if I made no other mistake. The result is I think quite conclusive against the extra planetary meteoric theory of Solar heat, and convinces me of the other form of the meteoric theory. The more I think of it the more probable it appears. I am only at a loss to account for the dark spots. Perhaps they are hurricanes in the sun's atmosphere keeping the stones off a locality by centrifugal force.

I am quite anxious to hear your opinion, and I shall be very glad if you will write & tell me especially if you see any objections.

Yours always truly
WILLIAM THOMSON

109 STOKES to KELVIN, 8 May 1854
Kelvin Collection, S371

Pembroke Coll. Cambridge[1]
May 8th 1854

MY DEAR THOMSON,

I delayed answering your last letter till I should have time to go more fully into it, which my lectures just commencing left me hardly leisure enough to do at the time. Particularly, I wanted to consider further what you said about heat. But as you seem to wish to have my further opinion on the Sun question I may as well say at once that I do not know any solid objection to your theory. I think the extra-planetary meteoric theory, from the numerical results which you have given, cannot be maintained. The theory seems to account more simply than any other I know for the fluctuations of temperature of the Earth in past ages which geologists consider we have evidence of. Geologists never stick at a trifle in the way of upheavals of continents &c., and indeed we have evidence that great change must have taken place in this respect during the Earth's history. But some changes which they assume seem almost too great to be produced by mere upheaval & sinking, and steady altering the state of the Earth as to configuration of land and water, currents &c. One is rather startled at first to give the sun such an appetite as to devour 10 Mercurys in a century. Still, we do not know that he is more abstemious. As to the spots, I have long thought they might be due to matter falling on the sun converted into vapour. It had not occurred to me that falling *stones* were to be the source of the sun's heat. But even still may not the spots be due to masses of more easily vaporizable matter falling in (water mercury &c?). Might not the volatilization of such masses be sufficient to use up, mainly by converting into work done in lifting the Sun's atmosphere, not only the heat due to their own impact, but also a portion of heat derived from the Sun in the neighbourhood where they fell? Though *comparatively* dark (which a Drummond's[2] lime ball looks against the Sun) they may notwithstanding be at a good high temperature.

We still do not agree about heat. I will take your pattern. Let the perimeters

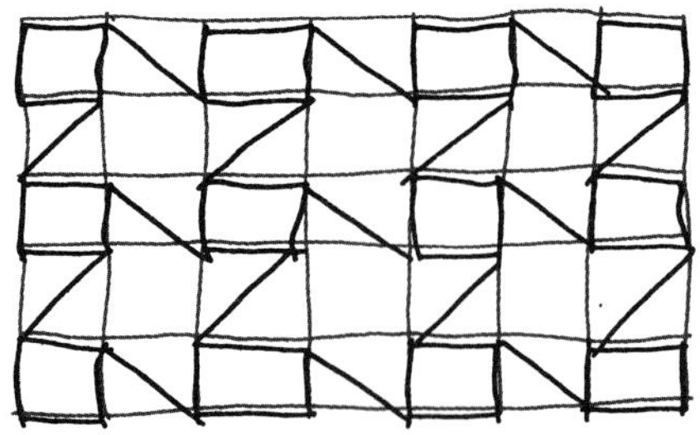

of the squares & the slant lines filled in with ink be infinitely narrow conducting rods, the rest being non-conducting.[3] Let heat be applied so that the isothermal surfaces are horizontal. Let t, u, v, w be the temp^{res} at the corners A, B, C, D of any square in one hor^{l} rank. The flux along ⟨BA⟩ AB

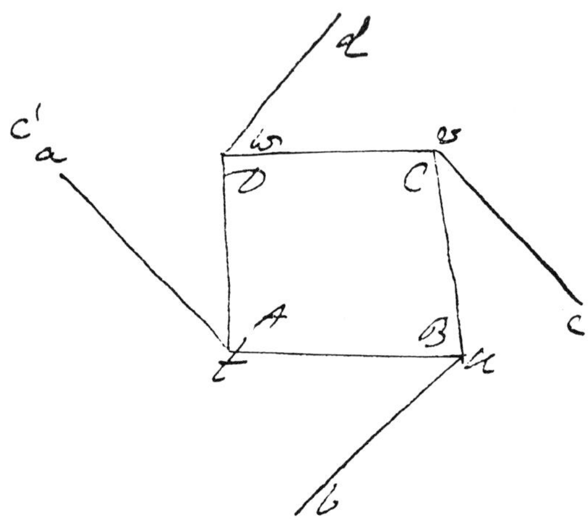

will be proportional to $t-u$, and may be represented by $t-u$. The flux along Aa will be represented by $\dfrac{t-v}{\sqrt{2}}$ or say $n(t-v)$, which will take in the case in which the slant bars are of different conducting power from the others. We have

$$\begin{array}{lll} t-u+t-w+n(t-v) & =0 & 1 \\ u-v+u-t+n(u-w_{-1}) & =0 & 2 \\ v-w+v-u+n(v-t) & =0 & 3 \\ w-v+w-t+n(w-u_{+1}) & =0 & 4 \end{array}$$

the -1, $+1$ referring to the ranks one below, above. Adding & subtracting (1) and (3) we find

$$t+v=u+w$$
$$t=v$$
$$\therefore \quad t-u=v-u=w-v=w-t \text{ \&c.}$$

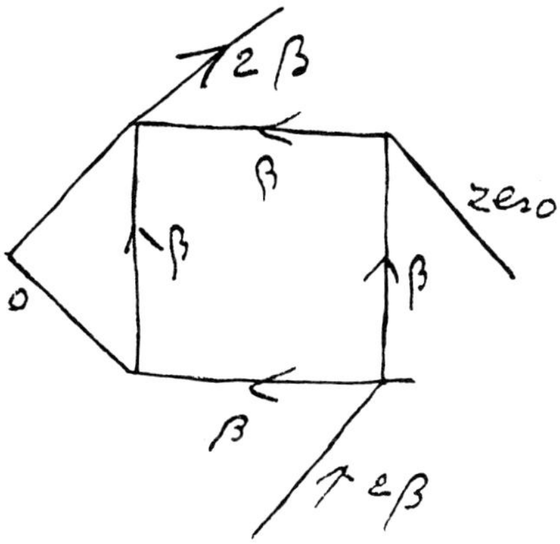

To save the post I will merely put down the result as to the flux which indeed is evident.[4] Hence the heat does not travel sideways, but (the squares being regarded as inf^y small) straight ahead, whence it follows (since the thermic ellipse (two dimensions) in this case is a circle,) that there is no rotation of heat, since if there were there could not fail to be an obliquity in the propagation. It is ⟨if⟩ evident if only one set of lines be laid down, & the isothermal surfaces be as before, that whether we have squares or circles, full or empty, the temp^res at the points such as *aa′* must be equal, and therefore = the temp^re at *a″* and therefore we may lay down the additional lines such as *a′ a″* without altering the flux, and therefore the perpendicularity of the flux to the isothermal surfaces.[5]

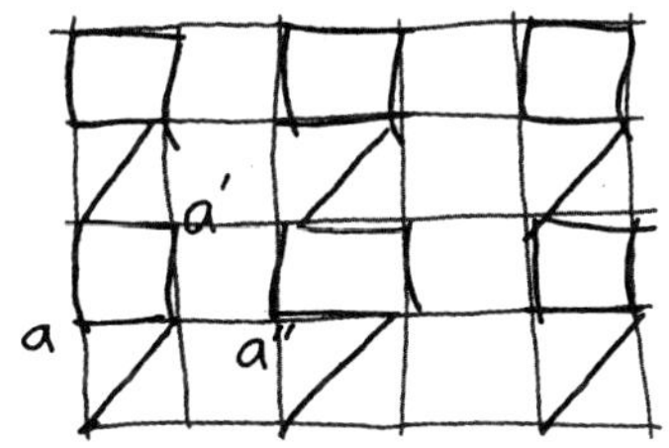

A sort of instinct seems to tell me that no structure which would not give a rotation could be made to do so by the mere change of right angles into obtuse or acute.

As you do not say whether your absence from Glasgow be more than temporary I think it best to direct to the College.

Yours very truly
G. G. Stokes

1 Kelvin wrote the following on the envelope, probably in 1903 (see letter 98, note 1, and letter 99, note 15): 'Stokes May 8, 1854. Grav^l Theory of sun heat. Rotational conduction of heat condemned'.

2 Thomas Drummond (1797–1840) developed his new source of light to ease work on the

ordnance survey, with which he was involved. His discovery attracted much attention and was later tried in lighthouses. See Drummond's 'Description of an apparatus for producing intense light visible at great distances', *Edinburgh Journal of Science*, 5 (1826), 319–23.

3 Stokes drew the figure by first making a grid of squares in pencil. Then in ink he darkened some of the squares and added the slanted lines.

4 In the figure, Stokes has tried to delete the slanted line which goes from near O to the upper left-hand corner of the square.

5 Stokes has drawn the final figure in the same manner as the first.

110 STOKES to KELVIN, 23 May 1854

Kelvin Collection, S372

Pembroke Coll. Cambridge[1]
May 23[d] 1854

MY DEAR THOMSON,

In the 20th livraison of the first vol. of the Abbé Moigno's Cosmos is an account of a memoir which Regnault has just communicated to the Academy on the sp. heat of air (and other gases) at constant pressure and constant volume.[2] From his exper[ts] he concludes that the ratio of the sp. heats, instead of being equal to 1.41, as results from the theory of sound, or ⟨1.7⟩ 1.37, as was deduced from experiments which have been made on direct compression, is sensibly equal to unity, from whence (according to him) Laplace's explanation of the vel[y] of sound can no longer be maintained. A closed vessel, capacity 1 litre, filled with air at a pressure of 10 atmospheres, was enclosed in a vessel, cap[y] 10 litres, exhausted of air and closed. The whole was immersed in a water bath. On establishing a communication there was no development of ⟨heat⟩ cold. Now this is just as it should be, and does not in the least invalidate Laplace's theory. Regnault found results inexplicable on the supposition of the substantiality of heat in the course of his experiments, but perfectly in accordance with the supposition that heat is vis viva; but the latter view is no doubt new to him, and he does not apparently perceive the full bearing of the dynamical theory of heat. I think it would be well, after you have seen the account of Regnault's experiments, that you should send a short paper to the Phil. Mag. or Annales de Chimie to point out that Regnault's results are not in the least inconsistent with Laplace's theory of sound.[3]

I hope to swing my pendulums this summer. I think it possible that something interesting may turn up from the value of the index of friction, which is as it were the door through which vis viva passes from the mechanical state (observable motion) to the molecular state (heat).

Yours very truly
G. G. STOKES

1 Kelvin sent Stokes's letter to Joule, writing to him on the envelope of Stokes's letter:

Please return the enclosed when you have read it! You have answered Regnault long ago. Will you not write a note to that effect to the ⟨Ann. de C⟩ Academy. ⟨to be⟩ If you do it would be immediately published in the Co[m]ptes Rendus.

On 29 May 1854, Joule returned Stokes's letter to Kelvin, writing:

Such a conclusion as he [i.e., Regnault] arrived at about the ratio of sp heats is virtually an assertion that 10 = 10 + 4 and ⟨with⟩ any one who believes in it must be ignorant of what yourself Rankine & I have written & of the experiments we have made. I think it is not worth reply until we see it elsewhere than in 'Cosmos.' [Kelvin Collection, J182.]

2 The report on Regnault was part of a report on the meeting of the Académie des Sciences on 15 May 1854 in *Cosmos*, **4** (19 May 1854), 597–8. Henri Victor Regnault (1810–78), after studying at the École de Polytechnique and the École des Mines, became professor of chemistry at the École de Polytechnique in 1840 and professor of physics at the Collège de France in 1841. He was, above all, a talented experimenter, especially regarding the thermal and physical properties of gases. During his 1845 visit to Paris, Kelvin attended lectures by Regnault and worked in his laboratory at the Collège de France. Also, Regnault wrote a testimonial supporting Kelvin for the Glasgow professorship in 1846. (Kelvin Collection, Tm22.)

3 So far as I have found, Kelvin did not write such a paper.

111 KELVIN to STOKES, 31 May 1854
Stokes Collection, K71

Invercloy, Brodick
by Ardrossan
⟨Ju⟩ May 31, 1854

MY DEAR STOKES

I have delayed for a long time writing to thank you for your letter of the 8th, because I was very much occupied getting some notes &c written for a paper[1] on sun-light, and wished to have some time to consider the heat question farther. You have certainly saved me from committing myself to a mistake, or rather from publishing & recanting a mistake, in my supposed rotatory structures. I am afraid I must give up all my illustrations, but I have not made out yet to my satisfaction that no such structure is possible. I am almost inclined to think not from the failure of a variety of cases, but I would like to have it proved one way or other. The impossibility of making up from non crystalline elements a structure having the rotatory property would not (?) absolutely prove the impossibility of a crystal having it. Any crystal which has ⟨an ay⟩ [?] an axis with polarity (a crystal of Iceland spar with a magnetized steel needles [*sic*] stuck in it parallel to its axis); ⟨the⟩ a section of which ⟨parallel to⟩ perpendicular to the axis would have a difference as to its sides, of which a geometrical type is a plane spiral; would have the kind of property required for the rotatory conductivity. I am not aware whether any such properties have been discovered in crystals. I still think I am right in the matter of thermo-electricity but I have been so completely occupied with

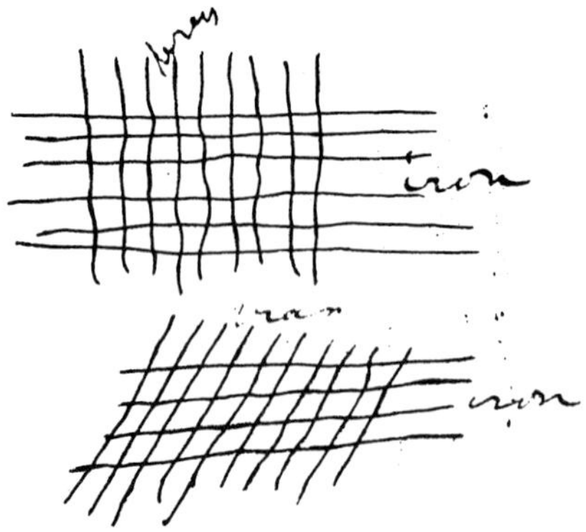

other subjects since I received your letter that I have not been able to make myself quite confident that there may not be a flaw. You will however I think see that a rotatory property may actually be induced by a change of right angles into acute, if you consider a wire cloth, with different metals for warp & woof. When pulled so that the squares become oblique. Thus if you look from this side of the paper, ⟨at the⟩ from a⟨n obtuse angle⟩ point of intersection along the shorter diagonal of a square, you have always brass on your right and iron on your left. Look at the same structure in the same way, from the other side, & you have always iron on your right & brass on your left. This shows that such an oblique structure is a type at least possessing the rotatory characteristic. As to the electricity, if you consider a stream passing from *AB* to *CD*, without getting out at all at the sides *AC* & *BD*, you will see that it will pass from brass to iron ⟨to brass most towards⟩ at positions nearer *AC* and from iron to brass ⟨to iron⟩, at positions nearer *BD*, than the straight lines of average conduction. Hence if *AC* be kept hot & *BD* cold, a current would pass

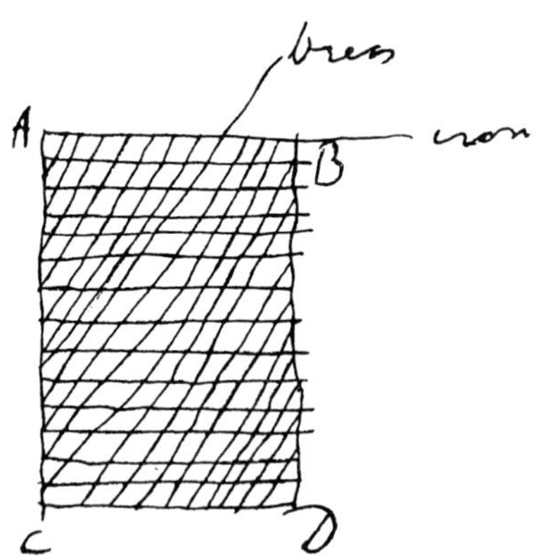

from *AB* through the structure to *CD* provided there is an independent uniform conductor connecting *AB* & *CD* externally. On the other hand a current from *BD* to *AC* would ⟨keep⟩ pass entirely by the iron, & ⟨I think⟩ therefore no such current could possibly be produced [b]y a difference of temperatures between *AB* & *CD*, when *AC*, *BD* are connected by an independent conductor. The *two* agencies of electric conduction, & thermo-electric force are essentially involved here. I am very anxious to make out whether or not with one single agency such as electric or thermal conductivity, there can be in any structure or in any crystalline substance, such an obliquity.

Just before I received your letter I had ⟨had⟩ added to my paper on the Sun, that probably *all* the meteors evaporate long before falling in; that the heat and light are produced in a stratum of intense frictional action between a great vortex of meteoric vapour (revolving at the rate of a satellite very nearly) & the portion of atmosphere & vapours held back by the sun's surface; and that the dark spots are great hurricanes checking the influx of vapours ⟨to⟩ temporarily over great tracts of the Sun's surface.[2] I think both the luminous atmosphere, & the cloudy stratum below assumed by Herschel[3] (showers of condensed vapours) are made very probable by the meteoric theory. The intense frictional action will be checked through a larger space than the showers below, & hence the 'penumbra' of the spots. The dark centres will be, just as Herschel said, the dark body of the Sun, seen where the clouds have cleared away (fallen, according to this theory, instead of blown upwards as conjectured by Herschel). It always appeared to me however that there was a mistake in the way the dark surface of the Sun is spoken of in giving Herschel's explanations. Under the luminous atmosphere the Sun's surface must be sensibly at the same temperature as the luminous atmosphere itself; but a few seconds or minutes after the luminousness ceases in any locality of the atmosphere, the matter in & below it will become comparatively black (although perhaps, as you say, still a bright red heat) in consequence of the loss of heat by radiation.

I am much obliged to you too for your last letter. Regnault gave a short account[4] last April year of his researches on the specific heats of gases which showed that he did not quite take up the bearings of the dynamical theory. The special case of expansion which you mention as experimented on by him, was experimented on by Joule long ago and the same result (as Regnault has found) is published with complete explanation, in the Philosophical Magazine as early as 1845.[5] Regnault knew of this paper long ago and has I rather think referred to it. It is very unfortunate that he does not understand English at all. The experiments which Joule & I have been making and are continuing are for the purpose of more rigorously testing that result. We find that there is in reality a slight cooling, although much too slight to be discovered by such an experiment. I think we shall show the bearings of it on the case Regnault has tried, as soon as we get the details of his experiment. The whole matter of the two specific heats, & the bearing on ⟨th⟩ Laplace's explanation of the velocity of sound have been very fully given in various published papers. It is quite certain that 1.41 is the ratio in the case of air, not 1.37. Joule has remarked that the writer in the Enc. Brit.[6] who made a more perfect experiment of Gay Lussac's kind, got (1.333) more nearly unity (Gay Lussac's number is 1.354) because more mechanical effect was dissipated in sound. In fact in a perfect experiment of this kind, a vacuous space would be suddenly opened wide and then closed again (for instance the end suddenly taken from a cylinder from which air had been exhausted, & then quickly replaced) and there would be no heating or cooling effect left after the noise of the explosion ceases.

I have been thinking a good deal on the fall of Mercury, and I believe that if the Sun is still warm as at present when he approaches, he will evaporate before he falls in.

Yours very truly
WILLIAM THOMSON

1 Kelvin (72).
2 The additions to Kelvin (73) are dated 9 May 1854. (See Kelvin's *MPP*, II, 16–22.)
3 J. F. W. Herschel, *Outlines of Astronomy*, 4th edn (London, 1851), section 389, where he gives this as William Herschel's theory. In the fifth edition of 1858, Herschel added section 389a on new research on sun spots.
4 H. V. Regnault, 'Recherches sur les chaleurs spécifiques des fluides élastiques', *Comptes Rendus*, **36** (1853), 676–87.
5 J. P. Joule, 'On the changes of temperature produced by the rarefaction and condensation of air', *Phil. Mag.*, **26** (1845), 369–83.
6 See the articles 'Hygrometry' and 'Sound' in *Encyclopaedia Brittanica*, 7th edn, 21 vols. (Edinburgh, 1830–42).

112 KELVIN to STOKES, 7 June 1854
Stokes Collection, K72

Invercloy, Brodick
by Ardrossan
June 7, 1854

MY DEAR STOKES

I find now that it is impossible to compose a structure possessing the obliquity as regards thermal conductivity, (or rotatory conductivity) out of materials which do not possess it, and I can demonstrate the impossibility thus.

If a ball or cylinder composed of elements each either crystalline or non crystalline but not possessing rotatory conductivity be placed filling a hollow space in an infinite homogeneous solid, ⟨and exposed to⟩ containing any sources of heat, the distribution of ⟨temperature⟩ flux of heat within it will follow the same law as the distribution of magnetization in a ⟨certain⟩ sphere or cylinder composed in a certain way of possible crystalline or non-crystalline elements and subjected to the influence of magnetic poles (North or South) in situations corresponding to the sources of heat (or cold) in the other case. But if the sphere or cylinder be made to revolve slowly round a fixed axis, the work done on it by the magnets acting on the induced magnetization must ⟨b⟩amount to nothing, & therefore the ⟨constants⟩ coefficients of inductive magnetization must be reducible to 6 in the regular way, & there can be no obliquity or rotatory property in the resultant magnetization. Hence there can be no such property as regards thermal conductivity in the corresponding case for heat; & therefore no such property can be obtained by putting together elements which do not possess it.

Yours very truly
WILLIAM THOMSON

P.S. The comparison of the magnetic & thermal circumstances is made on the understanding that the communication between the elements of the structure is purely conductive. I suppose your demonstration will ⟨make it[1] include⟩ show that the same conclusion is applicable even for structures of which the elements radiate to one another.

1 Kelvin has written the following above the line and deleted it: '⟨the demon⟩'.

113 STOKES to KELVIN, 16 June 1854
Kelvin Collection, S373

Pembroke Coll. Cambridge[1]
June 16th 1854

MY DEAR THOMSON,

I confess I have not been convinced of the possibility of producing rotation of heat from materials which have not the property even when thermo-electricity is called in, but this is a subject of which I know so little that I would not be positive. Of course there is no difficulty in conceiving the existence of that particular kind of want of symmetry without which rotation would be impossible. A pencil of parallel lines of magnetic force for instance (though it is not a structure) has the required want of symmetry, as is shown by the action of currents on magnets, though not by the action of magnets on each other.

I feel satisfied that the oblique wire cloth will not give rotation. This may be shown thus, where it will be sufficient to consider two dimensions. First let the wires be placed rectangularly: then there can be no rotation on account of symmetry. If then heat be applied at one spot of the gauze (regarded as extending indefinitely) the lines of motion of heat, or rather what those lines become in the limit when the number of wires in a given space is increased without limit, are curves starting from the source and passing off to infinity, which are symmetrically situated with respect to two rectangular axes passing through the source. Now let the wires be pulled diagonally so as to become oblique. No change will be made in the lines of motion of heat except that they will be pulled out in the same way as the wires, remaining as they do fixed relatively to the wires.

⟨B⟩ [?] But this pulling out cannot make a line of motion take one, much less an infinite number of turns round the source as you go along the line from the source to an infinite distance. Yet this is necessary in the case of rotation as I have shown near the end of my paper in the C. M. J.[2]

Yours very truly
G. G. Stokes

1 Kelvin has written the following on the envelope:

June 16, 1854
rotatory thermal conduction questions
adversely

The date and the following four words are written in pencil and probably date from 1854. The word *adversely* is in ink and probably dates from 1903 (see letter 98, note 1, and letter 99, note 15).

2 Stokes (44).

114 Stokes to Kelvin, 16 October 1854

Kelvin Collection, S374

Pemb. Coll.
Cambridge
Oct 16th 1854

My dear Thomson,

Am I right in attributing the finiteness, and even (for such experiments) considerable magnitude, of the time concerned in the phenomena described by Faraday Phil. Mag. March 1854 p. 197[1] relative to the charging &c. of a long (100 miles) telegraph wire coated with gutta percha and immersed in water to the following two causes?

Cause 1

The water wire (supposing for simplicity the further end to be unconnected) acts as an enormous Leyden Jar, and therefore drinks in an immense quantity of electricity before it attains the full charge corresponding to the state of tension of the battery. To supply so large a quantity, the chemical action of the battery must work for an appreciable time. The battery being connected with the earth at one end, say the negative, and insulated at the other, remains inert; but as soon as a connexion is formed between the $+^{ve}$ end of the battery and the nearer end of the water wire chemical action is resumed. The circuit is in a certain sense closed: $+^{ve}$ electricity flows into the wire, and as the water becomes $-^{ve}$ at the inner surface (that in contact with the gutta percha) by induction, $+^{ve}$ elec^{y} flows into the water and earth regarded as the common reservoir which supplies the $+^{ve}$ elec^{y} which has to flow in at the $-^{ve}$ end of the battery.

Cause 2

Suppose the $+^{ve}$ end of the battery to be connected with the nearer end of the water wire, and then suddenly disconnected before the effect has time to travel near to the end. For the instant the state of the wire may be represented approximately by a nearly constant tension over the first say 10 miles, a tension falling gradually off for the next 20, and insensible for the last 70. Then the electricity is very nearly in statical equilibrium. Any elementary slice of the wire has its free elec^{y} ($+^{ve}$) nearly all on the surface, and the corresponding slice of the water has a very nearly equal quantity of $-^{ve}$ elec^{y} on its inner surface and very little free elec^{y} elesewhere [*sic*]. Hence the electromotive forces are very small; but the quantity of electricity to be differently distributed is by no means small, and in the change of distribution resistance has to be overcome not only in the wire but in the water, because the induced elec^{y} as well as the elec^{y} in the wire has to be differently distributed.

In the actual experiments causes 1 and 2 would both come into operation.

I should be obliged to you to write me a line to say whether you think this the true view. I have not mentioned a possible expenditure of time in producing possibly a change in the molecular state of the dielectric gutta percha, because such a cause would be altogether hypothetical.

Believe me
Yours very truly
G. G. STOKES[2]

1 M. Faraday, 'On electric induction – associated cases of current and static effects', *Phil. Mag.*, 7 (1854), 197–208.

2 Kelvin has written the following on the reverse of the letter. See the next letter.

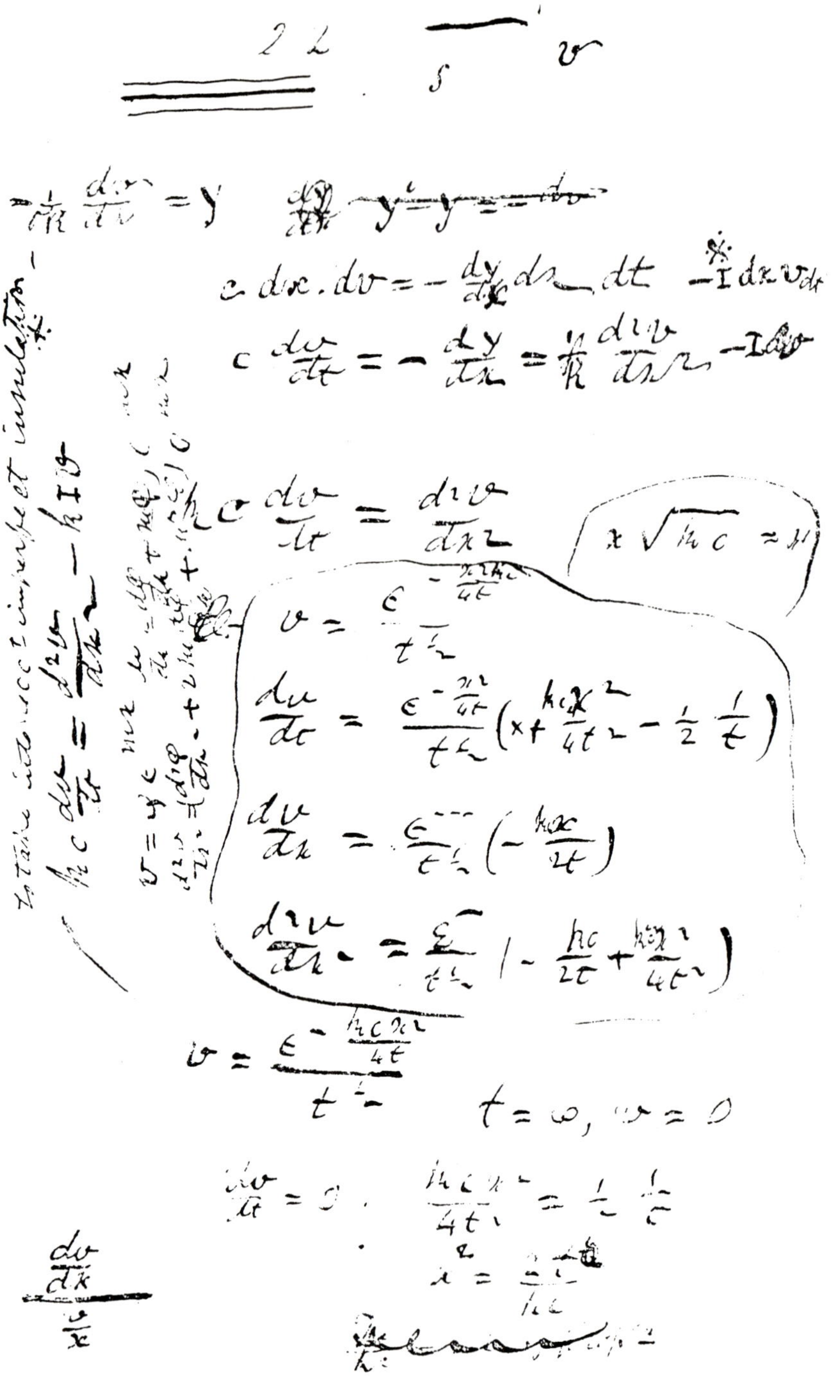

22

$\overline{s}\ v$

$-\frac{1}{k}\frac{dv}{dx} = y$

$c\,dx.dv = -\frac{dy}{dx}dx\,dt$ $- I\,dx\,v\,dt$

$c\frac{dv}{dt} = -\frac{dy}{dx} = \frac{1}{k}\frac{d^2v}{dx^2} - Iv$

To take into acct imperfect insulation

$kc\frac{dv}{dt} = \frac{d^2v}{dx^2} - kIv$

$kc\frac{dv}{dt} = \frac{d^2v}{dx^2}$

$x\sqrt{kc} = u$

$v = \frac{\epsilon^{-\frac{x^2kc}{4t}}}{t^{\frac{1}{2}}}$

$\frac{dv}{dt} = \frac{\epsilon^{-\frac{x^2}{4t}}}{t^{\frac{1}{2}}}\left(x + \frac{kcx^2}{4t^2} - \frac{1}{2}\frac{1}{t}\right)$

$\frac{dv}{dx} = \frac{\epsilon^{---}}{t^{\frac{1}{2}}}\left(-\frac{kcx}{2t}\right)$

$\frac{d^2v}{dx^2} = \frac{\epsilon^{-}}{t^{\frac{1}{2}}}\left(-\frac{kc}{2t} + \frac{k^2c^2x^2}{4t^2}\right)$

$v = \frac{\epsilon^{-\frac{kcx^2}{4t}}}{t^{\frac{1}{2}}}$ $\quad t = \infty,\ v = 0$

$\frac{dv}{dt} = 0 \ . \quad \frac{kcx^2}{4t^2} = \frac{1}{2}\frac{1}{t}$

$x^2 = \frac{2t}{kc}$

$\frac{\frac{dv}{dx}}{\frac{v}{x}}$

115 KELVIN to STOKES, 28 October 1854

Stokes Collection, K73

Partially printed in Kelvin's *MPP*, II, 61–6.[1]

Largs, Oct 28, 1854

MY DEAR STOKES

I should have written to you sooner in answer to your letter of the 16th, had I not been engaged all my working time in getting over some arrears of correspondence, but on the whole devoting myself as much as possible to open air & the sea, in preparation for the 'session' about to commence.

The two reasons you assign for the finite intervals occupied by the ⟨phenomena relative to⟩ agencies concerned in the charging &c of the telegraph wire under water are inseparable & I believe contain the ⟨whole⟩ explanation of ⟨t⟩ [?] all the perceptible features of the phenomena as observed by Faraday. Your 'Cause 1' appears to have been fully appreciated by Faraday himself, but your 'Cause 2' I think I have not seen explicitly stated except in your letter & must be taken into account to get a complete investigation of the action. In taking up your letter this morning to answer it, I find that the whole may be worked out definitely as follows.

Let c be the capacity (as a Leyden phial) per unit of length, of the wire: that is, let c be such that clv is the quantity of elecy required to charge a length l of the wire up to potential v. In a Note[2] added (as 'Note V') to ⟨a⟩ an old paper of mine recently republished in the Philosophical magazine (last June I think) but accidentally omitted in the printing (to be published yet I believe) I proved that the value of c is $\dfrac{I}{2\log\dfrac{R'}{R}}$, if I denote the specific inductive capacity of the gutta percha, & R, R' the radii of ⟨the⟩ its inner & outer cylindrical surfaces.

Let k denote the 'galvanic resistance' of the wire in absolute electrodynamic (or electromagnetic) measure (See a paper 'On the Application of the Principle of Mechanical Effect to the Measurement of Electro motive Forces, and Galvanic Resistances'. Phil. Mag. Dec. 1851)[3]

Let γ denote the strength, at time t, of the current, in absolute electromagnetic measure, at a point P of the wire at a distance x from one end which may be called O.

Let v denote the potential in electrostatic measure, at the same point P, at time t.

The potential at the outside of the gutta perch[a] ⟨will⟩ may be taken as at each instant be [*sic*] rigorously zero. (the resistance of the water, certainly if the wire be extended as in a submarine telegraph, being incapable of ⟨preventing⟩ ⟨causing⟩ ⟨any appreciable accumulation⟩ preventing the inductive action from ⟨takin⟩ being completed instantaneously round each part of the wire. If the wire be closely coiled the resistance of the water may produce sensible ef⟨f⟩fects).

Hence the quantity of electricity on a length dx of the wire at P will be $cdx \cdot v$.

The quantity (in electromagnetic units) that leaves it in time dt will be

$$dt\,\frac{d\gamma}{dx}\cdot dx,$$

or in electrostatical units,

$$\frac{1}{\sigma}\,dt\,\frac{d\gamma}{dx}\,dx,$$

if σ be the number of electromagnetic units of electrical quantity in the electrostatic unit. Hence we must have

$$-cdx\,\frac{dv}{dt}\,dt = \frac{1}{\sigma}\,dt\,\frac{d\gamma}{dx}\,dx \quad \dots\dots\dots\dots\dots\dots \quad (1).$$

But the electromotive force, in electrostatic units, at the point P is $-\frac{dv}{dx}$ or in electromagnetic units, $-\sigma\frac{dv}{dx}$ (See Phil. Mag. June 1853 'On Transient Electric Currents' equation (21))[4] and therefore at each instant

$$k\gamma = -\sigma\frac{dv}{dx} \quad \dots\dots\dots\dots\dots\dots\dots\dots \quad (2).$$

Eliminating γ from (1) by means of this we have

$$ck\sigma^2\,\frac{dv}{dt} = \frac{d^2v}{dx^2} \quad \dots\dots\dots\dots\dots\dots\dots\dots \quad (3),$$

which is the equation of electrical excitation in a submarine telegraph wire perfectly insulated by its gutta percha covering.

This equation agrees with the well known equation of the linear motion of heat in a solid conductor, and various forms of solution which Fourier has given are perfectly adapted for answering practical questions regarding the use of the telegraph wire. Thus first, suppose the wire infinitely long & communicating with the earth at its infinitely distant end. Let the end O be suddenly raised to the potential V, (by being put in communication with ⟨one⟩ the positive pole of a galvanic battery of which the negative pole is in communication with the ground; the resistance of the battery being small, say not more than that of a few yards of the wire) kept at that potential for a time T, and then put in communication with the ground (i.e. suddenly reduced to, and ever afterwards kept at, potential zero). The solution of the equation ⟨on⟩ for this case is

$$v=\frac{V}{2\pi}\int_{-\infty}^{\infty} dn\epsilon^{-zn^{1/2}}\frac{\sin[2nt-zn^{1/2}]-\sin[(t-T)2n-zn^{1/2}]}{n}\ \ldots\ldots\ (4),$$ [5]

where, for brevity,

$$z=x\cdot\sigma\sqrt{kc}\ \ldots\ldots\ (5).$$

This solution may be put under the following form

$$v=\frac{V}{\pi}\int_{t-T}^{t} d\theta\int_{-\infty}^{\infty} dn\epsilon^{-zn^{1/2}}\cos(2n\theta-zn^{1/2})\ \ldots\ldots\ (6),$$

which, if T be infinitely small, becomes

$$v=\frac{V}{\pi}\cdot T\cdot\int_{-\infty}^{\infty} dn\epsilon^{-zn^{1/2}}\cos(2n\langle\theta\rangle t-zn^{1/2})\ \ldots\ldots\ (7)$$

This last expresses the effect of ⟨an infinitely short period⟩ putting ⟨the pos. pole of the battery in communication with⟩ the end O of the wire for an infinitely short time ⟨, and then suddenly⟩ in communication with the battery and ⟨then suddenly uninsulating it⟩ immediately after, with the ground. It may be reduced at once to finite terms by the evaluation of the integral, which I believe you will find to be as follows

$$\int_{-\infty}^{\infty} dn\epsilon^{-zn^{1/2}}\cos(2nt-zn^{1/2})=\frac{\pi^{1/2}z}{2t^{3/2}}\ \epsilon^{-\frac{z^2}{4t}}\ ;$$

so that we have

$$\left\langle v=\frac{VT}{2\pi^{1/2}}\right\rangle$$

$$v=\langle TV\rangle T\frac{Vz}{2\pi^{1/2}t^{3/2}}\ \epsilon^{-\frac{z^2}{4t}}\ \ldots\ldots\ (8);$$

or, by (6), when T is not infinitely small

$$v=\frac{Vz}{2\pi^{1/2}}\int_{t-T}^{t}\frac{d\theta}{\theta^{3/2}}\ \epsilon^{-\frac{z^2}{4\theta}}\ \ldots\ldots\ (9),$$

or, which is the same,

$$v=\frac{Vz}{2\pi^{1/2}}\int_{0}^{T}\frac{d\theta}{(t-\theta)^{3/2}}\ \epsilon^{-\frac{z^2}{4(t-\theta)}}\ \ldots\ldots\ (10).$$

It ought to be remarked that in (6), ⟨and⟩ (9) and (10) the limits of the integral must be taken 0 to t, (instead of $t-T$ to t, or 0 to T) if it be desired to express the potential at any time t between 0 & T.

These last forms may be obtained synthetically from the following solution

$$v = \frac{\epsilon^{\frac{-z^2}{4t}}}{t^{1/2}} \cdot \frac{Q}{\langle 2 \rangle \pi^{1/2}} \quad \ldots\ldots\ldots\ldots\ldots\ldots \quad (11)$$

which expresses the potential in the wire consequent upon instantaneously communicating a quantity Q of electricity to it at O and leaving this end insulated. For if we suppose the wire to be continued to an infinite distance on each side of O and its infinitely distant ends to be in communication with the earth, the same equation will express the consequence of instantly communicating a quantity 2Q at O. Now suppose at the same instant a quantity $-2Q$ to be communicated at a point O′ at ⟨an infinitely small⟩ distance ⟨α⟩ $\frac{\alpha}{\sigma\sqrt{kc}}$ on the negative side of O. The consequent potential at any ⟨p⟩time t, ⟨in any part⟩ at a distance ⟨z⟩ $\left\langle \frac{z}{\sigma\sqrt{kc}} \right\rangle$ $\frac{z}{\sigma\sqrt{kc}}$ along the wire from O will be

$$v = \frac{Q}{\pi^{1/2}} \left\{ \frac{\epsilon^{-\frac{z^2}{4t}}}{t^{1/2}} - \frac{\epsilon^{-\frac{(z+\alpha)^2}{4t}}}{t^{1/2}} \right\} \quad \ldots\ldots\ldots\ldots\ldots\ldots \quad (12)$$

and if α be infinitely small this becomes

$$v = \frac{Q\alpha \cdot \langle z \rangle [?]}{2\pi^{1/2}} \; \frac{z\epsilon^{-\frac{z^2}{4t}}}{t^{3/2}} \quad \ldots\ldots\ldots\ldots\ldots\ldots \quad (13)$$

which, with positive values of z, expresses obviously the effect of ⟨touching⟩ communicating the point O ⟨or⟩ [?] with the positive pole for an infinitely short time, and then instantly with the ground.

The strength of the current at any point of the wire, being ⟨, as given by equation⟩ equal to $-\frac{\sigma}{k}\frac{dv}{dx}$, as shown above, in equation (2), will vary proportionally to $\frac{dv}{dx}$, or to $\frac{dv}{dz}$. The time at which the maximum electrodynamic effect of impulses such as those expressed by (11) or (13) will be found by determining t so that $\frac{dv}{d\langle x \rangle z}$ is a maximum in each case. Thus from (11) we find

$$t = \frac{z^2}{6} = \frac{\sigma^2 kcx^2}{6},$$

⟨and from (13) $t =$ ⟩

as the time at which the maximum electrodynamic effect of connecting the battery for an instant at O and then leaving this point insulated, is experienced at a distance x.

In these cases there is no regular 'velocity of transmission'; but on the other hand, if the potential at O be made to vary regularly according to the simple harmonic law (sin $2nt$), the phases are propagated regularly at the rate $2\sqrt{\frac{n}{\sigma^2 kc}}$, as is shown by the well known solution

$$v = \epsilon^{-zn^{1/2}} \sin(2nt - zn^{1/2}) \quad \ldots\ldots\ldots\ldots\ldots\ldots \quad (14)$$

The effects of pulses at one[6] end, when the other is in connection with the ground, and the length finite, will be most conveniently investigated by considering a wire of double length, with equal positive and negative agencies applied at its two extremities. The synthetical method, founded on the use of the solution (11) appears perfectly adapted for answering all the practical questions that can be proposed.

To take into account the effects of imperfect insulation (which appear to have been most sensible in Faraday's experiments) we may assume the gutta percha to be uniform & the ⟨loss⟩ flow of electricity across it to be proportional to the difference of potentials at its outer & inner surfaces. The equation of electrical excitation will then become

$$\sigma^2 kc\frac{dv}{dt} = \frac{d^2v}{dx^2} - hv \quad \ldots\ldots\ldots\ldots\ldots\ldots\ldots \quad (15)$$

and, if we assume

$$v = \epsilon^{-\frac{h}{\sigma^2 kc}t}\phi \quad \ldots\ldots\ldots\ldots\ldots\ldots\ldots\ldots \quad (16)$$

we have

⟨α⟩ [?]

$$\sigma^2 kc\frac{d\phi}{dt} = \frac{d^2\phi}{dx^2} \quad \ldots\ldots\ldots\ldots\ldots\ldots\ldots\ldots \quad (17)$$

an equation to the ⟨solution⟩ treatment of which the preceding investigations are all applicable.

It would be most interesting to consider numerical estimates in connection with Faraday's results, but, as I have not his paper[7] here, I cannot do so, and indeed I shall have to leave the subject altogether just now and for some time. Just one thing more I wish to say. It is that the quantity σ, ⟨m⟩ by which electrostatical & electrodynamic units are compared, may be determined by finding the velocity of propagation of a regular periodical effect (sin $2nt$); and I believe it may be actually estimated (roughly) from what Faraday has already done.

Yours, (in great haste)
WILLIAM THOMSON

1 Kelvin (78), which is a greatly revised version of the manuscript letter, the most substantial modification being an extra paragraph inserted after equation (5).
2 Kelvin (88) was published in the *Philosophical Magazine* in 1855 as an additional note to Kelvin (2) and (19), originally published in 1843 and 1846, respectively, and republished in the *Philosophical Magazine* in June and July, respectively, of 1854.
3 Kelvin (52).
4 Kelvin (67).
5 Kelvin's brackets.
6 Kelvin has written above the line and deleted, '⟨ applie ⟩',
7 See the previous letter, note 1.

116 KELVIN to STOKES, 30 October 1854

Stokes Collection, K74

Partially printed in Stokes's *MPP*, II, 66–9.[1]

Largs Oct 30, 1854

MY DEAR STOKES

An application of the theory of the transmission of electricity along a submarine telegraph wire, which I omitted to mention in the haste of finishing my letter on Saturday, ⟨is to⟩ shows how the question raised by Faraday as to the practicability of sending distinct signals along such a length as the 2000 or 3000 miles of wire that would be required for America, may be answered. The general investigation will show exactly how much the sharpness of the signals will be worn down, and will show ⟨how long the⟩ what maximum strength of current through the apparatus in America, ⟨may⟩ would be produced by a specified battery action on the end in England, with wire of given dimensions &c

The following form of solution of the general equation

$$\sigma^2 kc \frac{dv}{dt} = \frac{d^2v}{dx^2} - h \left\langle \frac{d}{dx} \right\rangle v$$

which is the first given by Fourier, enables us to compare the times until a given strength of current shall be obtained, with different dimensions &c of wire

$$v = \epsilon^{\frac{-ht^2}{\sigma^2 kc}} \left\langle \frac{ht}{\text{[illegible deletion]}} \right\rangle \sum A_i \sin\left(\pi \frac{ix}{l}\right) \cdot \epsilon^{-\frac{i^2\pi^2 t}{\sigma^2 kc \cdot l^2}}$$

If l denote the length of the wire, and V the potential at the end communicating with the battery, the ⟨permanent⟩ final distribution of potential in the wire will be expressed by the equation

$$v = V \frac{\epsilon^{(l-x)\sqrt{h}} - \epsilon^{-(l-x)\sqrt{h}}}{\epsilon^{l\sqrt{h}} - \epsilon^{-l\sqrt{h}}}$$

$$v = \left\langle V - V\frac{x}{l}\right\rangle V\left(1-\frac{x}{l}\right)$$

and the final (maximum) strength of current at the remote end,

$$\gamma = \frac{\sigma V}{kl}\frac{2l\sqrt{h}}{\epsilon^{l\sqrt{h}} - \epsilon^{-l\sqrt{h}}}$$

Hence if we assign A_i so that ⟨when $t=0$⟩

$$\sum A_i \sin\left(\pi\frac{ix}{l}\right) = -\langle(\rangle\, V\left(1-\frac{x}{l}\right),$$

when $x>0$ & $<l$, the equation

$$v = V\left(1-\frac{x}{l}\right) + \sum A_i \sin\left(\pi\frac{ix}{l}\right)\epsilon^{-\frac{i^2\pi^2 t}{\sigma^2 kcl^2}}$$

will express the actual condition of the wire at any time t after the one end is put in connection with the battery, the other being always kept in connection with the ground.

We may infer that the time required to reach a stated fraction of the maximum strength of current at the remote end will be proportional to kcl^2. We may be *sure* before hand that the ⟨a⟩ American telegraph will succeed, with a battery sufficient to give a sensible current at the remote end when kept long enough in action; but the time required for each deflection will be 16 times as long as would be with a wire a quarter the length, such for instance as in the French submarine telegraph to Sardinia & Africa.

One very important result is that by increasing the diameter of the wire & of the gutta percha covering, in proportion to the whole length, the distinctness of utterance will be kept constant. For k varies inversely as the square of the diameter; and c (the electrostatical capacity of the unit of length) is unchanged when the diameter of the wire & the covering are altered in the same proportion.

Hence when the French submarine telegraph is fairly tested, we may make sure of the same degree of success in an American telegraph by increasing all the dimensions of the wire in the ratio of the greater distance to which it is to extend, to that for which the French one has been tried.

The insulation will probably be better with the thicker covering, (& therefore the coeff[t] h diminished) so that the advantage, under these conditions, will probably be rather in favour of the greater.

The solution derived from the type $\epsilon^{-z^2/4t}/t^{1/2}$, regarding which I wrote to you on Saturday, may be applied to give the condition of the wire when one end, E is kept connected with the ground, and the other, O, is operated on so

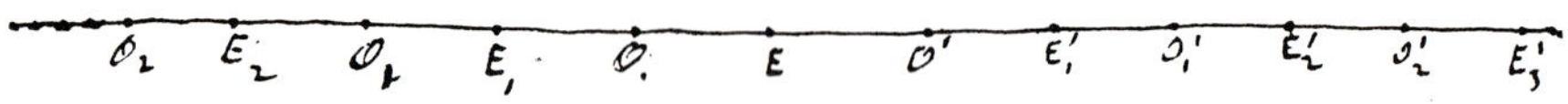

that its potential may be kept varying according to a given arbitrary function of the time: only this which I omitted to mention must be attended to. Instead of merely considering sources (so to speak) at O and O' (the latter in an imaginary continuation of the wire) we must suppose sources at O, O_1, O_2, &c & O'_1, O'_2, &c, arranged according to the general ⟨pont⟩ [?] principle of successive images, so that the potential at ⟨O⟩ E, ⟨E_1, E_2 &c E'_1, E'_2, &c⟩ may be zero, and that at O may be uninfluenced by all the other sources except the ⟨single⟩ source at O itself. ⟨If⟩ Taking O_1, O_2, . . . O'_1, O'_2, . . . equidistant we have only to suppose ⟨a⟩ equal sources, each represented by the type

$$\frac{z\epsilon^{-\frac{z^2}{4t}}}{t^{3/2}}$$

to be placed at each of the points O, O_1, O_2, . . . O'_1, O'_2, &c. For the effects of O_1 & O' will balance [*sic*] ⟨at O.⟩ on the potential at O will balance one another.

So ⟨So⟩ [?] will those of O_2 & O'_1
O_3 & O'_2
&c

And again O & O' ⟨will keep⟩ would alone keep the potential at E, zero

So would O_1 & O'_1
O_2 & O'_2
&c

Hence if we denote $2l\sigma^2 kc$ by a, for brevity, the general solution is

⟨$v = A\{$⟩

$$v = \langle A\rangle \int_0^t \frac{d\theta F\theta}{(t-\theta)^{3/2}} \{\ldots (z+2a)\epsilon^{\frac{-(z+2a)^2}{4(t-\theta)}} + (z+a)\epsilon^{\frac{-(z+a)^2}{4(t-\theta)}}$$

$$+ z\epsilon^{\frac{-z^2}{4(t-\theta)}} + (z-a)\epsilon^{\frac{-(z-a)^2}{4(t-\theta)}} + (z-2a)\epsilon^{\frac{-(z-2a)^2}{4(t-\theta)}} + \ldots\}$$

where $F(\theta)$ is an arbitrary function such that $F(t)$ varies as the potential sustained at O by the battery. (One of the equations in my letter of Saturday will show what the numerical coeff^t is by w^h $F(t)$ must be multiplied ⟨⟩ to give the⟩ It is in fact the limiting value of

$$\int_{0}^{t} \frac{d\theta \cdot F\theta}{(t-\theta)^{3/2}} z\epsilon^{-\frac{z^2}{4(t-\theta)}}$$

when z = 0.)

The corresponding solution of the equation

$$\sigma^2 kc \frac{dv}{dt} = \frac{d^2 v}{dx^2} - hv$$

is

$$\langle v = \epsilon^{-\frac{ht}{\quad}} \rangle$$

$$\langle v = \epsilon^{\frac{-ht}{\sigma^2 k[?]c[?]}} \int_{0}^{t} d\theta F\theta \epsilon^{\frac{-[?]h\theta}{\sigma^2 [?] kc}} \rangle$$

$$v = \epsilon^{\frac{-ht}{\sigma^2 kc}} \int_{0}^{t} \frac{d\theta \cdot \epsilon^{\frac{-h\theta}{\sigma^2 kc}} \cdot F\theta}{t-\theta} \sum_{-\infty}^{\infty} \{(z+ia)\epsilon^{\frac{-(z+ia)^2}{4(t-\theta)}}\},$$

by which the effects of imperfect insulation may be taken into account.

I remain here, except for two days, till Monday next, when I go to Glasgow to commence my class. So address Largs, Ayrshire, if you write in the meantime.

Yours always truly
WILLIAM THOMSON

1 Kelvin (78), which is a greatly revised version of the manuscript letter.

117 STOKES to KELVIN, 4 November 1854
Kelvin Collection, S375

49 New Bond S[t], London
Nov 4th 1854

MY DEAR THOMSON,

Here I am and expect to remain till the end of March, except going up to Cambridge for the Smith's Prize. I was too much engaged in preparing for my lectures, especially my first lecture,[1] to go into your letters when I received them. Even now I have not gone properly into them, so that at present I write mainly to thank you for them.

Faraday no doubt was perfectly familiar with the electrical appetite of the water wire as compared with the ⟨earth⟩ air wire, and in all probability regarded the finite rate of supply as at least one cause of the finiteness of the times observed. I did not refer to what he or any one else had done, for I wrote simply for my own information, taking for granted that the subject was all ABC to you. I merely endeavoured to put the question clearly as it occurred to my own mind. It seemed to me that both the causes I mentioned were true ones, but I did not know but that the effect of one might be numerically insignificant compared with that of the other. I was led to think of the subject from answering one of Sir William [Rowan] Hamilton's son's questions.[2] Hamilton's question, obvious though the answer in many respects seemed, led me to think of the subject, and thereby see some things in a clearer light, and my questions seem to have led you, at any rate to ⟨have⟩ [?] work out analytical results bearing on an important practical question, possibly even to see the principles in a clearer light than before. The mental translation of propagation of electricity into propagation of heat which your equations show to be permissible puts the mode of transfer in a clear light.

Yours very truly
G. G. STOKES

1 Stokes apparently meant his lectures at the Government School of Mines in London, not his lectures as Lucasian professor.
2 Kelvin explained how Stokes happened to be answering Hamilton's son's questions in his (307), in *MPP*, III, 486. Hamilton had two sons, but the one here is probably William Edwin (b. 1834), who received honours in science at Trinity College, Dublin, and became a civil engineer.

118 STOKES to KELVIN, November 1854

Printed in Stokes's *MPP*, IV, 61–2, and in Kelvin's *MPP*, II, 69–71.[1]

In working out for myself various forms of the solution of the equation $\frac{dv}{dt}=\frac{d^2v}{dx^2}$ under the conditions $v=0$ when $t=0$ from $x=0$ to $x=\infty$; $v=f(t)$, when $x=0$ from $t=0$ to $t=\infty$, I found that the solution with a single integral only (and there must necessarily be this one) was got out most easily thus:–

Let v be expanded in a definite integral of the form

$$v=\int_0^\infty \varpi(t,\alpha)\sin\alpha x\,dx,$$

which we know is possible.

Since v does not vanish when $x=0$, $\frac{d^2v}{dx^2}$ is not obtained by differentiating under the integral sign, but the term $\frac{2}{\pi}\alpha v_{x=0}$ must be supplied,[2] so that (observing that $v_{x=0}=f(t)$ by one of the equations of condition) we have

$$\frac{d^2v}{dx^2}=\int_0^\infty\left\{\frac{2}{\pi}\alpha f(t)-\alpha^2\varpi\right\}\sin\alpha x\,dx.$$

Hence

$$\frac{dv}{dt}-\frac{d^2v}{dx^2}=\int_0^\infty\left\{\frac{d\varpi}{dt}+\alpha^2\varpi-\frac{2}{\pi}\,\alpha f(t)\right\}\sin\alpha x\,dx,$$

and the second member of the equation being the direct development of the first, which is equal to zero, we must have

$$\frac{d\varpi}{dt}+\alpha^2\varpi-\frac{2}{\pi}\,\alpha f(t)=0,$$

whence

$$\varpi=\epsilon^{-\alpha^2t}\int_0^t\frac{2}{\pi}\,\alpha f(t)\,\epsilon^{\alpha^2t}\,dt,$$

the inferior limit being an arbitrary function of α. But the other equation of condition gives

$$\varpi=\epsilon^{-\alpha^2t}\int_0^t\frac{2}{\pi}\,\alpha f(t)\,\epsilon^{\alpha^2t}\,dt=\left(\frac{\pi}{2}\right)^{-1}\alpha\int_0^t\epsilon^{-\alpha^2\overline{t-t'}}f(t')\,dt',$$

therefore

$$v=\left(\frac{\pi}{2}\right)^{-1}\int_0^\infty\int_0^t f(t')\,\alpha\epsilon^{-\alpha^2\overline{t-t'}}\,sin\ \alpha x\,d\alpha\,dt'.$$

But

$$\int_0^\infty\epsilon^{-a\alpha^2}\cos b\alpha d\alpha=\tfrac{1}{2}\left(\frac{\pi}{a}\right)^{1/2}\epsilon^{-\frac{b^2}{4a}},$$

therefore

$$\int_0^\infty\epsilon^{-a\alpha^2}\sin b\alpha\cdot\alpha d\alpha=-\frac{d}{db}\left\{\tfrac{1}{2}\left(\frac{\pi}{a}\right)^{1/2}\epsilon^{-\frac{b^2}{4a}}\right\}$$

$$=\frac{\pi^{1/2}b}{4a^{3/2}}\,\epsilon^{-\frac{b^2}{4a}},$$

whence writing $t-t'$, x, for a, b, and substituting, we have

$$v=\frac{x}{2\pi^{1/2}}\int_0^t(t-t')^{-3/2}\,\epsilon^{\frac{x^2}{4(t-t')}}f(t')\,dt'.$$

Your conclusion as to the American wire follows from the differential equation itself which you have obtained. For the equation $kc\frac{dv}{dt}=\frac{d^2v}{dx^2}$ shows that two submarine wires will be similar, provided the squares of the lengths x, measured to similarly situated points, and therefore of course those of the whole lengths l, vary as the times divided by ck; or the time of any electrical operation is proportional to kcl^2.

The equation $kc\frac{dv}{dt}=\frac{d^2v}{dx^2}-hv$ gives $h\propto l^{-2}$ for the additional condition of similarity of leakage.

1 Kelvin (78). The original letter does not survive. This extract from Stokes's letter was initially published by Kelvin along with his two previous letters to Stokes (letters 115 and 116).
2 The footnote in the printed version reads, 'according to the method explained in' Stokes (21).

119 KELVIN to STOKES, 1 December 1854

Stokes Collection, K75

2 College, Glasgow
Dec 1, 1854

MY DEAR STOKES

I should be much obliged if you would not mention to any one what I wrote to you regarding the remedy for the anticipated difficulty in telegraphic communication to America, at present, as Rankine has suggested that I should join with him in applying for a ⟨way⟩ patent[1] for a way of putting it in practice, & improving conductors in ordinary circumstances, to which I have agreed. In a very short time I believe there will be no necessity of keeping any part of our plan secret. In the mean time there can be no harm in publishing the theory, & if I can get time to write it out, I shall try to do so soon. As I did not preserve any memoranda of what I wrote to you, I would be greatly obliged by your letting me have my letters until I get a copy of them made, after w^h I shall return them if you care.[2]

Thanks for your two letters on the subject, of w^h the last was particularly interesting.

I hope you find your lecturing & class satisfactory.

Yours always truly
WILLIAM THOMSON

1 This was the first of Kelvin's many patents. (See Thompson, II, 1275.)
2 Kelvin (78) included revised extracts from letters 115 and 116 from Kelvin to Stokes and an extract from letter 118 from Stokes to Kelvin.

120 Stokes to Kelvin, 2 December 1854

Kelvin Collection, S376

49 New Bond S[t]
London
Dec 2nd 1854

My dear Thomson,

I send you your letters as you wish. There is one point which I do not feel altogether certain about, that the resistance of the water may be neglected, although I believe that it may. Still I am not satisfied altogether. I have begun to make some calculations about it with a view to satisfy my doubts, but as I have several things now to attend to I am not sure whether I can well go on with it. That being the case I may as well state to you my doubts; perhaps rather probably you can see through them at once.

First suppose electricity to be diffused outwards from a metallic sphere whose radius $=a$ into an infinite mass of water, the supply being constant, and the tension uniform throughout the sphere. The resistance in crossing the shell whose radii are r and $r+dr=\frac{a^2}{r^2}\,dr\times$ the resistance in passing across an equal thickness dr close to the sphere, so that if dx be the ⟨height⟩ inc[remen]t of the height of a cylinder having a base equal to the surface of the sphere and giving the same resistance as the shell whose radii are a and r

$$dx=\frac{a^2}{r^2}\,dr$$

$$\therefore \qquad x=\int_a^r \frac{a^2}{r^2}\,dr=a^2\left(\frac{1}{a}-\frac{1}{r}\right)$$

{I am supposing the diffusion to be what in hydrodynamics would be called 'steady'}

As r increases x soon approaches a its limit when $r=\infty$.

If however we perform the same calculation for a cylinder we find

$$dx=\frac{a}{r}\,dr, \quad x=a\log\frac{r}{a},$$

which increases indefinitely with r. If the motion, though not steady, be nearly so, the resistance is liable to become very great if the superior limit r be ⟨ver⟩ infinite. If now the tension when $r=a$ be a variable function of the time, the flux for $r=a$ is capable, like any arbitrary function, of expression in the form

$$\int_0^\infty \phi(\alpha, r)\cos \alpha t\,d\alpha+\int_0^\infty \varpi(\alpha, \langle t\rangle\, r)\sin \alpha t\,d\alpha$$

and the part of the resistance which corresponds to small values of α is liable to be large.

I have merely expressed hastily and vaguely a source of doubt which probably would clear itself up on finishing the calculation. This I have not time to do before despatching your letter. When the motion is steady in the wire the motion in the water is of course not steady but null, so that probably the resistance would turn out not to be very great.

Yours very truly
G G STOKES

121 KELVIN to STOKES, 25 December 1854
Stokes Collection, K76

Largs, Dec 25, 1854

MY DEAR STOKES

I thought the best way of answering your last letter was to give every moment of my time that I could to correcting the proofs[1] I had in my hands. I had been doing this from Monday in consequence of a letter I received on that day from Dr Sharpey,[2] and I despatched the last instalment on Thursday.

I regret very much that the delay I made has been so inconvenient. Still, I do not wish you to think that I have been wilfully negligent in the matter, and I think you will see that there could not have been a *much* earlier return of the whole proof corrected, when I tell you that it was only on the Monday preceding that one that I received the principal part of the proof which consisted of 27 quarto pages. I saw the printer's date Dec 4 upon them, but they had been sent to Joule & kept by him for 3 days, and then posted for me. I only received them on Monday the 11th on coming up from this place where I had been since the Saturday. Then I had a particularly busy week with other matters w^h I had to work off before undertaking so heavy a piece of work as the proofs, and on ⟨the⟩ coming up from Largs on the next Monday (the 18th) determined to work at the proofs without letting anything else interfere, I received Dr Sharpey's letter telling me that it had been expected to get the part out on the preceding Thursday (just 3 days after I had received the 27 pages of proofs) and that my corrections were anxiously waited for. This of course only confirmed my resolution, & made me determine not to lose a post in returning them when corrected. I wrote to Dr Sharpey, but did not mention particulars as to dates. Perhaps if you see him you would tell him something [o]f what I have written to you. By the bye I sh^d add that the first 16 pages were longer in my hands by about a week (or perhaps more) bearing the printer's date Nov 25) but I could scarcely return them without having ⟨s⟩ first seen the sequel, as some of the passages were accidentally transposed, owing to the manuscript having been supplied partly from me & partly from Joule (although the order was clearly described on the MSS.) I asked the printers for a revise, without w^{h} our paper could scarcely appear safely. I w^{d} rather even have it kept back for the next part than let it appear without us seeing a revise. However as I expect to have to make very few if any corrections, I think the revise will occasion but little delay.

You will see why I have been so slow in writing in reply to your letters on the elec. telegraph matter. I shall write as soon as I have had time to consider fully what you wrote to me. In the mean time, I am still confident that the inductive electrification of the outside of the insulating tube must proceed without sensible loss of time, at least in all cases when the wire is ⟨laid⟩ extended out under water. For consider a mile of the wire, twenty miles from the battery. The inside of the insulating tube gets its elec^{y} by conduction along twenty miles of the wire. ⟨The outside gets it, over e⟩ Each inch of the outside gets it from the earth and water all round causing, (if the inside charge be positive) a flow of electricity outwards from the outside of the gutta percha tube in all directions, to the surface of the sea, or round through the earth & water to the negative pole of the battery, or of any other battery that chances to be working or any source of negative elec^{y} anywhere in the earth.

Excuse this hasty letter & believe me

Yours always truly
W. THOMSON

P.S. I was very glad to see your name as Sec. R. S., and I am glad now to find you entering office.

I don't know whether you have yet received my paper on Th. elec. currents[3] on w^{h} I enclose two corrections, but if not, you shall have it as soon as I can get a copy. I have got a wave machine I w^{d} like to describe to you, if you w^{d} think on getting one made. It shows (with actual motion) 1st a plane polar^{d} wave wave [*sic*] in hor^{l} plane

2nd d[itt]o in vertical plane

3^{d} The two superimposed, each in motion and one advanced any fraction of a phase on the other. It is the same as Wheatstone's except that it shows the ⟨waf⟩ waves in motion not me⟨e⟩rely an instantaneous photograph of their forms. It is to Wheatstone's like a running automaton to a statue of a figure running. I had my model (a very rough one) in Watkins & Hill's[4] hands for a long time to let them make something after it, but I believe they never made anything of it, and probably did not understand it. It w^{d} be very easy in London to get a good model made on the same principle, & I think would be most convenient for lecturing purposes. WT

1 Kelvin and Joule (4).
2 William Sharpey (1802–80) was professor of anatomy and physiology at University College, London, from 1836 to 1874 and secretary of the Royal Society from 1853 to 1872.
3 Apparently, Kelvin (101).
4 Watkins and Hill were opticians and philosophical instrument makers in London. They are not listed in the *Post Office Directory* for either Glasgow or Edinburgh.

122 Kelvin to Stokes, 12 February 1855
Stokes Collection, K78

2 College, Glasgow
Feb 12, 1855

My dear Stokes

I duly received Tyndall's M.S.S.[1] from the R. S. and I shall be glad to read it carefully and write my opinion as requested. I w^{d} like to know how long I may keep it: that is *when* it would be more convenient for you to have it returned than a little later. I should arrange my plans accordingly so as ⟨if possible⟩ not to cause inconvenience by keeping it longer than the shortest time in which I could manage to get through it.

I have been reading again and carefully considering your letters on the resistance to the electric flow required for the inductive charge outside the gutta percha tube of a telegraph wire, and I am still convinced that it must be wholly inappreciable. In your letter of the 16th Dec. you say that looking at it in a common sense way it seems to you as if the resistance would be comparable with, but a good deal smaller than that w^{h} would oppose the return current if the ends of the wire were merely dipped some way into the water & not connected with metallic plates. I believe it w^{d} be certainly considerably smaller, & perhaps comparable, as you say, if for instance a hundred yards, or a quarter or a half mile of the end⟨s were⟩ was dipped in the water. The resistance in this case w^{d} be certainly insensible, since it is found in practice, that if one pole of the battery & the remote end of the telegraph wire, be connected with a plate sunk in earth or water the whole resistance is sensibly the same as if the remote end were brought direct to the uninsulated pole: and a hundred yards of the copper wire would conduct into water as well as a copper plate of equal superficial area.

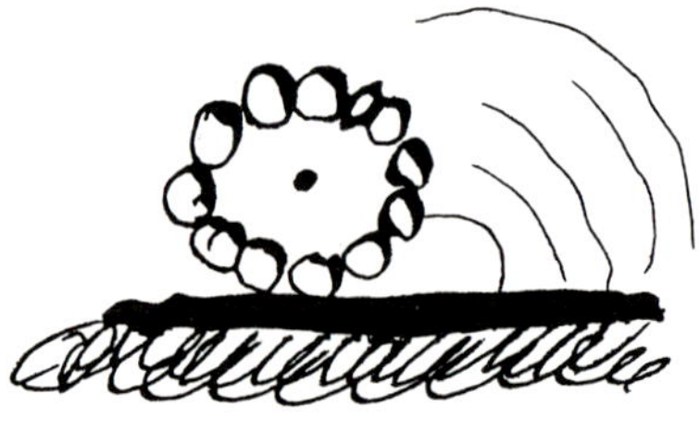

However when you consider the iron sheathing w[h] surrounds the gutta percha, resting on the bottom and taking its charge partly from contact with the ground, & partly from the ground through the water in curved lines of motion & partly perhaps downwards & inwards from all sides, from the mass of water itself, ⟨&⟩ I think you cannot but say that any foot of the outside of the insulating covering gets its charge with infinitely less resistance than a foot of the copper wire inside it, w[h] can only get its charge by conduction through miles of the copper wire itself.

The philosophy of electrical conduction through water is very obscure: so much so that it would be impossible [to] say exactly what the condition of the water round a gutta percha tube ⟨solid⟩ [?] (with no iron or bottom or surface near it) must be when the surface of the gutta percha commences to receive an inductive charge which an internal electrified body forces it to do. If it were a conducting solid, e.g. metal which surrounded the gutta percha instead of water, there would certainly be, to supply pos. elec. to the gutta percha, first a partial vacuum of pos. elec. in the solid immediately round it, then in those parts a little farther out, & so on: i.e. there would be a regular propagation of negative electricity outwards, according to a law no doubt expressed by the equ[n]

$$\frac{d^2 V}{dr^2} + \frac{1}{r}\frac{dV}{dr} = q\frac{dV}{dt}$$

In the case of water an electro polarization, & perhaps decomposition, of the elements w[d] accompany the process, in a very mysterious way. The resultant effect as regards conduction would very possibly be the same as if a solid of constant conducting power were substituted for the liquid: but (I believe) certainly a foot of the outside of the gutta percha covering would receive its charge more easily if there were nothing but water round it, than it would if coated with tinfoil & put in communication with the uninsulated pole of the battery by a thin copper wire (⟨of such ext⟩ [?] several miles long. I sh[d] be very glad to hear from you if you consider this & what I said before on the subject satisfactory.

I am anxious to get something brought out upon it: either a set of ⟨abstracts⟩ extracts from the letters that have passed between us upon it (commencing with your's of the 16th Oct) or an abstract of their substance, published without delay.[2] I have no hope of being able to write a satisfactory & complete paper upon it for at least a year. I think I shall have to keep myself quite clear of all writing next summer, as I felt a good deal knocked up at the end of last, and am only now (a most unusual result of winter ⟨direction⟩ of variation)[3] beginning to feel about as well as usual again. If you think anything such as I propose could appear in the Proceedings, I sh[d] therefore be very glad to have it so, & with little delay. One form which perhaps w[d] answer w[d] be one or more letters to you of future dates, in w[h], so far as necessary I could refer to the letters that passed between us formerly.

I have got a good many curves described, showing the potential & current at different parts of a tel. wire for successive instants of time: also for showing the same at the receiving end continuously as the time advances. These I shall send you, with short descriptions: also an estimate, founded on the retardation $\frac{1}{10}$ of a sec[d] observed between Greenwich & Brussels & some data Airy has supplied me with regarding the wires, by which I find that there cannot be more than 419 000 000 nor less than 104 000 000 electrostatic units in the electrodynamic, & consequently that the ⟨force⟩ attraction between the separated electricities flowing from the decomposition of a grain of water ⟨cannot⟩ if concentrated in points 1 foot apart cannot be less than 10 tons nor more than 42: and that the ⟨force⟩ attraction between two parallel discs each a squ. foot area, & $\frac{1}{100}$ foot apart, when connected with the two poles of a Daniell's battery[4] of 100 cells (which I am now going to try & measure directly) cannot be less than 4 grains nor more than 71. There is not the slightest occasion now for any secrecy as to any of the letters I have been writing to you on this subject.

If you are busy with your lectures do not let me trouble you at all just now with this matter. A few lines about Tyndall's paper & any general idea on publication you may have, I w[d] be glad to have at your leisure.

Yours always truly
WILLIAM THOMSON

1 J. Tyndall, 'On the nature of the force by which bodies are repelled from the poles of a magnet; to which is prefixed an account of some experiments on molecular influences', *Phil. Trans.* (1855), 1–52. Letter 124 is Kelvin's report on the paper.
2 Kelvin (78).
3 Kelvin apparently intended also to delete *of variation*. He has interlined *result of*.
4 John Frederic Daniell (1790–1845) was an experimental natural philosopher with a special interest in instruments. He had already developed a hygrometer and pyrometer when he presented the idea for his improved battery to the Royal Society in 1836 in his paper, 'On voltaic combinations', *Phil. Trans.* (1836), 107–24, 125–30. The new battery won him the Royal Society's Copley medal in 1837.

123 KELVIN to STOKES, 14 March 1855
Stokes Collection, K79

2 College
Glasgow, March 14
1855

MY DEAR STOKES

By this post I send Tyndall's paper with some leaves of memoranda in one packet, and a letter[1] addressed to you under a separate envelope, each addressed to you as Secretary the R.S. Somerset House, where I suppose you will get it all right I trust in time.

I am thinking on sending a short article[2] by the evening post on variation of force in the magnetic field, & on the action exp^{d} by small ferromag^{s} or diamagnetics, *a propos* of some of the matters ⟨under⟩ discussed in M^{r} T's paper & my memoranda.

Yours truly
W. THOMSON

P.S. Many thanks for your last letter on the telegraph. Can you tell me whether Faraday's optical effect of magnetism has been reduced to elastic forces. The following is an action of the same type (*dipolarity*). A cylinder of elastic solid is subjected to equal external and internal pressures so as to produce no tension perp^{r} to its radius, or is subjected to internal pressure and equal external normal tension, so as really to experience nothing internal but ⟨tangential⟩ tension perp^{r} to the radius. In either case the outside will turn round the axis on the inside, one way or the other, provided coeff^{ts} of elasticity

$$(y, \xi), \quad (z, \xi)$$

(OX being the axis of the cylinder x, y, z referring to strains parallel to the axes, σ, ξ, η, z to distortions such as slipping in plane (y, z) parallel to x, &c). If these coeff^{ts} be equal one to the other with neg. sign, & if all the other coeff^{ts} are those of an isotropic solid, the body will be symm^{l} about the axis OX but that axis will have the dipolar property referred to.

P.S. Some time when you see Ferrers, w^{h} I suppose as co editor [you] will do, will you oblige me by giving him the paper[3] on thermoelectricity w^{h} reaches you with this? Any time will do.

1 See the next letter.

2 Kelvin (86), dated 15 March 1855 and published in the April number of the *Philosophical Magazine*, was a simplified restatement of his (27) and (48). Although (86) did not mention Tyndall specifically, Kelvin had on 12 March 1855 written Tyndall a letter – Kelvin (87) – which, also, was published in the April number of the *Philosophical Magazine*.

3 Kelvin (102), dated 10 March 1855. On Stokes as 'co-editor', see letter 80, note 4.

124 Kelvin to Stokes, [14 March 1855][1]
Royal Society of London, RR.2.253

My dear Stokes

I return Mr Tyndall's paper 'on the Nature of the Force by which Bodies are repelled from the Pole of a Magnet'[2] – with some leaves of memoranda hastily put down in reading it. As regards the publication of it in the Transactions, I think there is so much of important and curious experimental investigation in it – for instance the directional and absolute effects on woods, ⟨and⟩ the experiments on compressed powder of bismuth described in the Appendix, the numerical results showing the relative forces of repulsion on the same mass, held in different positions, the investigation in p 20 by which Mr Tyndall first showed (what I think no other experimenter has ever since done) ⟨the⟩ a case of magne-crystallic action depending on three principal axes at right angles to one another with three difft inductive capacities; and so much of interesting illustrations ⟨of⟩, such as in the experiments on the actions experienced by and reactions produced by diamagnetic bars (pp 41–71) as to fully entitle to a place in the Transactions. Still I think that (especially with reference to the Title of the paper) Mr Tyndall is frequently contending against an imaginary adversary. In fact except Von Feilitsch[3] whose 'theory' is founded on a mistake most obvious from the beginning and in my opinion not worthy of more notice than very shortly to point out that mistake, all Mr Tyndall's experiments and views are in perfect accordance with those indicated by Faraday from the beginning, and advocated by myself as early as 1846 in the C & DMJ.[4] uniformly on many different occasions of which short Reports have been published in the British Assocn Volumes,[5] the Philosophical Magazine,[6] & the Comptes Rendus.[7] The real question is 'whether are the phenomena presented by diamagnetics to be explained by a contrary magnetization to that of soft iron, or by a less magnetization than that of the medium (air or luminiferous ether) surrounding them'. Which ever be the truth the resultant action is undoubtedly the same as that which would be experienced by a small magnet with its axis *reverse*, and therefore Mr Tyndall's experiments, which amply confirm the view forced on us by Faraday's original investigations that the resultant action is such as that, do not at all contribute to a foundation of any theory in which one or the other alternative is essentially involved.

Impressed with this belief I could wish much modification to be made in the controversial parts of the communication, but should Mr Tyndall be disposed to make no change, I should advise its publication as it stands.

I remain, My dear Stokes

Yours very truly
William Thomson[8]

1 Presumably, this is the letter which was enclosed in the previous one.
2 See letter 122, note 1.
3 Fabian Carl Ottokar von Feilitzsch (1817–85) was professor of physics at the University of Greifswald. See his 'On the physical distinction of magnetic and diamagnetic bodies', *Phil. Mag.*, 1 (1851), 46–51.
4 Kelvin (27), actually published in 1847. This article and those mentioned in notes 5 through 7 were all reprinted together in Kelvin's *PEM*, 499–531, immediately preceding Kelvin (86) (*PEM*, 531–9).
5 Kelvin (55) and (56).
6 Kelvin (48).
7 Kelvin (70).
8 Stokes has written on the reverse of the letter, 'Report on Dr *Tyndall's* paper by Prof *Thomson*'.

125 KELVIN to STOKES, 26 March 1855
Stokes Collection, K80

2 College, Glasgow
March 26, 1855

MY DEAR STOKES

I hope to be able to send you in a few days an article[1] of an incomplete kind, but such as I would like to have published without much more delay, on the Electric Telegraph. I should think it would occupy about 10 pages of the Proceedings. I have got a good many curves traced illustrative of the propagation of electric pulses & of the linear motion of Heat, which I should wish to communicate sometime later to the R.S. when I shall have been able to write a more comprehensive paper[2] on the subject. In the meantime, if the incomplete article which I propose to send immediately be published in the Proceedings, could a diagram of curves, such as I enclose (of which I should provide a much better copy for the wood cutter or engraver) appear along with it? I should be obliged by your writing me a line as soon as you conveniently can in answer to this so that I may know how to adapt what is to be written, and at the same time please return the copy of the curves, which I send you as the shortest way of ⟨explaining⟩ showing the kind of diagram wanted.. I should also probably want a woodcut such as this which I suppose could readily be admitted into the Proceedings.

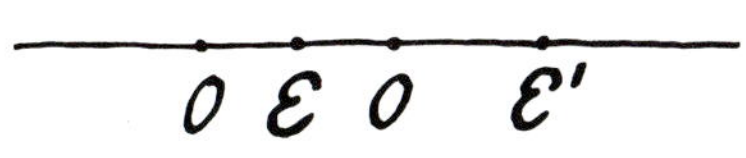

Could you also inform me regarding the Government Grant Committee, whether any conclusion was come to with reference to a Report, and an application for a farther allowance for experimenting, which I forwarded on the 31st of March last.

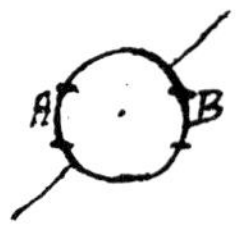

I have lately fallen on a new branch of thermodynamics which proves that two steel magnets ⟨become cold⟩ experience a cooling effect when drawn asunder against attr^{n}, and that a piece of soft iron experiences the same when drawn from a magnet, or a bar of soft iron when turned from being parallel to the dipping needle to a position at right angles to it. I am going to make a rotatory apparatus as an illustration in which a uniform ring of soft iron wire balanced horizontally ⟨and supported on a point by a pivot rigidly connected with⟩ and free to turn about its centre which is fixed, will turn (very slowly I suppose) in the direction of the hands of a watch when all of it, except two parts *A* and *B*, is kept hot by a lamp (the dotted line being the north & south magnetic line).

Yours very truly
WILLIAM THOMSON

1 Kelvin (78).
2 This intended paper was not sent to the Royal Society but was published in Kelvin (103) and as parts of Kelvin (243A). See Kelvin's introduction to the reprint of this material in his *MPP*, II, 41.

126 KELVIN to STOKES, 31 March 1855
Stokes Collection, K81

2 College, Glasgow
March 31, 1855

MY DEAR STOKES

Mr Joule and I had an intimation shortly after this time last year of the decision of the Grant Committee on our joint application to which you refer in your letter of Wednesday; but there was another application, made at the same time (the 31st of March 1854) by myself alone, accompanying a Report of my Electrical Experiments, to which I have not yet received any answer. My application was for an additional sum of £50 over and above the sum of £100 previously granted, to enable me to pursue certain researches described in the Report. I am very sorry to trouble you again, but I should be much obliged by your letting me know, or directing whoever a⟨sk⟩cts as Secretary to the Grant Committee, to let me know what conclusion was come to on that application. I am ready, on a few days notice, to send a farther Report, and an account of Expenditure ⟨(⟩ for all the amount already drawn (viz £100) of which £82 was accounted for this time last year.

I am much obliged for the information regarding the Proceedings, & I shall try to send the article[1] very soon.

Yours very truly
WILLIAM THOMSON

1 Presumably, Kelvin (78), read before the Royal Society on 24 May 1855.

127 KELVIN to STOKES, 12 May 1855
Stokes Collection, K82

Malvernbury House
Malvern, May 12,
1855

MY DEAR STOKES

I am much obliged to you for pointing out the fault in my formula. I am not sure whether or not it is a mistake, but it is certainly a useless mystification. It may be corrected simply by reducing the limits to 0 . . . ∞, and omitting *2* from the denominator – so that the expression stands thus[1]

$$v = \frac{V}{\pi}\int_0^\infty dn\, \epsilon^{-zn^{\frac{1}{2}}} \frac{\sin[2nt - zn^{\frac{1}{2}}] - \sin[2n(t-T) - zn^{\frac{1}{2}}]}{n}$$

I prefer this form to that in w[h] m^2 is substituted for n, because it is better adapted to show the relation with the primitive form

$$\epsilon^{-z\sqrt{\frac{\pi i}{T}}} \sin\left(2\pi i\frac{t}{T} - z\sqrt{\frac{\pi i}{T}}\right)$$

from which it is derived. This form you will see fulfils the diff[l] equ[n] $\frac{dv}{dt} = \frac{d^2v}{dz^2}$;

and the superposition of such terms may be effected to give an arbitrary function of t, when $z = 0$. The actual formula is adapted to make $v = 0$, when $z = 0$, except for values of t between 0 and T, and is verified by observing that

$$\int_0^\infty \frac{dn \sin(2An)}{n} = \pm\frac{\pi}{2},$$

according as A is positive or negative.

I do not know of any alteration to be made in the paper before going to press except to omit the words 'if successfully laid', after the mention which is made of the telegraph cable from Varna to Balaclava.[2] I would be much obliged by your looking at the copies of the diagrams before sending them to the wood cutter, & seeing if the references to the curves in the latter are distinctly indicated. I would like too something in the way of dots thus to be added to the first diagram to show that it represents a line extending to infinity in each direction.

I expect to be here for five weeks certainly, and for that period any communication from the R.S., proofs, &c. had better be addressed as above. It is likely I may remain till about the end of June.

Are you going to be one of the India examiners?[3] Have you made any farther investigations in light? You should try to include military projectiles in your experiments on the resistance of air. Do you think the resistance to a cannon ball, or other shot varies rigorously as the square of the veloc^y, & have you any idea of the absolute values of the coeff^{ts} for diff^t cases?

I only ask these questions out of curiosity, & do not wish you to be at the trouble of writing expressly to answer any of them. Any time you are writing, I shall be glad to hear of what you are doing or proposing to do. I reckon on seeing you in Glasgow at the B. Assoc^n.

Yours always truly
WILLIAM THOMSON.

1 Kelvin's brackets. Kelvin (78).
2 *Ibid.*, in *MPP*, II, 75.
3 See letter 130.

128 STOKES to KELVIN, 28 May 1855
Kelvin Collection, S377

Pembroke College Cambridge
May 28/55

MY DEAR THOMSON,

Whether your integral be right or not as it stands is a question about which I did not trouble myself since as you say it is at least a useless mystification to integrate between imaginary limits. Your integral stands thus:–

$$\langle V \rangle \; v = \frac{V}{2\pi}\int_{-\infty}^{\infty} dn\, \epsilon^{-zn^{\frac{1}{2}}} \frac{\sin(2nt - zn^{\frac{1}{2}}) - \&c.}{n}$$

I have verified the correctness of

$$v = \frac{V}{\pi}\int_{-\infty}^{\infty} dm\, \epsilon^{-zm} \frac{\sin(2m^2 t - zm) - \&c.}{m}$$

which you get by putting m^2 for n in the former and supposing the limits to be real $-\infty$ and $+\infty$. Putting

$$\int_{-\infty}^{\infty} f(m)\,dm = \int_{0}^{\infty} f(m)\,dm + \int_{0}^{\infty} f(-m)\,dm$$

the latter integral becomes

$$\frac{V}{\pi}\int_{0}^{\infty} \epsilon^{-mz}\frac{\sin(2m^2t - mz) - \&c.}{m}\,dm$$

$$-\frac{V}{\pi}\int_{0}^{\infty} \epsilon^{mz}\frac{\sin(2m^2t + mz) - \&c.}{m}\,dm$$

and not

$$\frac{2V}{\pi}\int_{0}^{\infty} \epsilon^{-mz}\frac{\sin(2m^2t - mz) - \&c.}{m}\,dm$$

so that the change you propose won't quite do. I think it best to send you the pages as they are and let you modify them for yourself. The rest of the paper[1] I am sending to Dr Sharpey.

As to the wood-cut I presume you don't mean the *small* squares to be retained. It would much increase the expense [Here a portion of the letter is torn away, deleting probably six to twelve words.] and would not I think look well. I suppose you want the large squares for measurement by the eye; otherwise I think the curves would look better without them. If they be retained I think they had best be retained complete and not merely inside the curves.

Please to send your answer to Dr Sharpey, correcting the sheets and saying what you would like done with the woodcut [Here is the reverse of the tear mentioned above.]

Yours very truly
G. G. STOKES

1 Kelvin (78).

129 KELVIN to STOKES, 31 May 1855
Stokes Collection, K83

Malvernbury House
Malvern, May 31, 1855

MY DEAR STOKES,

I have again tried the verification of the formula[1]

$$v = \frac{V}{\pi}\int_{0}^{\infty} dn\,\epsilon^{-zn^{\frac{1}{2}}}\frac{\sin[2nt - zn^{\frac{1}{2}}]\ \&c.}{n}$$

and I think it is quite right. It is necessary to add to what I wrote to you last, that it is completely convergent, and vanishes when $z=\infty$, to make the justification complete. One term of the integral you give is divergent which I think will account for the apparent discrepance [*sic*]. I have verified too (not in a most satisfactory way) that the integral

$$\int_0^\infty dn\,\epsilon^{-zn^{\frac{1}{2}}}\cos(2nt - zn^{\frac{1}{2}})$$

is equal to

$$\frac{\pi^{\frac{1}{2}}z}{4t^{\frac{3}{2}}}\,\epsilon^{-\frac{z^2}{4t}}$$

for all positive values of t, and to zero for all neg[ative values of t]. This last is curiously demonstrated [on] physical considerations by the fact that the solution is expressed by the equation ref⟨f⟩erred to.

I have sent the leaves of my paper with the requisite modifications, to Dr Sharpey and along with them, instructions regarding the diagram. As I have not his address ⟨here⟩ to refer to here, I address simply, Sec. Royal Society, which I suppose will find him without any inconvenient delay.

Yours very truly
WILLIAM THOMSON

P.S. I think I omitted to say that the extract from your letter[2] forming part of my paper was given with your permission. I shall make an insertion to that effect on the proof, if you have no objection.

Please ask Mr Weld[3] when you are at the R.S. for copies of the 1st & 2nd joint papers[4] by Joule & myself, from the Transactions, which the authors will be much gratified by your accepting. I have asked Mr W. to keep copies for you and for Cayley, Sylvester & some others, before sending the remainder to myself.

1 Kelvin's brackets. Kelvin (78).
2 Letter 118, which became part of Kelvin (78).
3 Charles Richard Weld (1813–69) was assistant secretary of the Royal Society from 1843 to 1861.
4 Kelvin and Joule (3) and (4).

130 STOKES to KELVIN, 2 June 1855
Kelvin Collection, S378

Pembroke College Cambridge[1]
June 2nd 1855

MY DEAR THOMSON,

I worked out my integral during an examination and took for granted without examination that it would be convergent on account of the fluctuation of the sine factor, which, had I examined it, I should have seen was not the case. I have verified your integral in its present shape by means of the differential equation and the eq^{ns} of condition. In its original shape your integral would I believe have been divergent: I mean the shape $\int_{-\infty}^{\infty} f(u)\,du$ instead of $2\int_{0}^{\infty} f(u)\,du$.

There is one sentence or part of a sentence which may as well I think be struck out of my letter as being too much of a private nature namely 'I have not worked at these subjects lately and I have not Fourier by me' but if you have returned the proof it does not signify.[2]

I got a letter this morning announcing my appointment as one of the math^{l} examiners for the East India Civil Service. The salary is £100. The exam^{n} is to take place in July.

Yours very truly
G. G. STOKES

1 Kelvin has written on the envelope:
June 2, 1855 Stokes
formulas for signal through cables
appointment as examiner for East India Civil Service

2 The phrase was omitted from the printed version of letter 118 in Kelvin (78).

131 KELVIN to STOKES, 28 August 1855
Stokes Collection, K84

Broomfields, Largs,
Ayrshire, Aug 28, 1855

MY DEAR STOKES

I hope you are coming to the British Association, and I have been looking forward all summer to it as an opportunity of seeing more of you than I have been able to do for a long time. I hope too that you are not engaged to stay with anybody during your visit to Glasgow, as my father-in-law, Mr Crum has commissioned me to say that he would be glad if you would take up your quarters in his house. He expects Liebig, Faraday, and some others whom I think you would like to meet, and I am sure they would be glad to meet you. I shall be there, as in consequence of being away all summer, and from Mrs Thomson's not being well enough to be present. I shall not be able to have my own house open.

I should be glad to hear from you as soon as you can let me know if we may expect you. If you write before Saturday, address as above. If later, *care of Walter Crum Esq, 51 Cochrane Street, Glasgow.*

Believe me, My dear Stokes
Yours very truly
WILLIAM THOMSON

132 Stokes to Kelvin, 24 October 1855
Kelvin Collection, S379

69 Albert Street Regent's Park[1]
London Oct 24th 1855

My dear Thomson,

I have been looking up Pasteur's researches lately,[2] and as I perhaps set you wrong about one thing it is right I should set you right. I said that solutions of right-handed and left-handed tartaric acid on being mixed developed an amount of heat sensible to the hand; but I find it is that when sufficiently concentrated solutions of the two acids are mixed, crystallized racemic acid is immediately thrown down with the development of an amount of heat sensible to the hand. Thus the heat given out is probably only the latent heat of liquefaction (solution) and does not therefore indicate chemical combination.

I don't think Pasteur's researches prove that racemates necessarily exist as such in solution. I think it may be that the solutions are merely ⟨a⟩ mixtures in equal quantity of a right and left tartrate and that the so-called racemates in a solid state are merely double salts right and left tartrates.

Yours very truly
G. G. Stokes

1 Kelvin has written on the envelope:
Stokes Oct 25 1855
Pasteur heat generated in crystallization of the [?]

2 Stokes was probably doing so in connection with the Royal Society's awarding Pasteur its Rumford Medal in 1856.

133 Kelvin to Stokes, 6 November 1855
Stokes Collection, K85

2 College, Glasgow
Nov 6, 1855

My dear Stokes

I am much obliged to you for your letter regarding the Bakerian lecture[1] (also for its predecessor) and if appointed I shall be very glad to choose the subject which is considered more appropriate. As for the rotatory motion, I shall try if I can get a paper written on it in time to be communicated this session, but I fear from having 'too many irons in the fire' I may be unable.

I think I could be present on either the 24th of January or the 28th of Feb. (both Thursdays & therefore I presume days of meeting) but the only month on the third Thursday of wh I could be present is March. I would not like (as I feel just now) to throw it so late as that, although I may possibly find myself not ready sooner. I would try to have it ready for the day of January, if you are disposed to give me that provisionally in the mean time, and I would be glad if you would tell me how long notice I ought to give, if I find I have to beg for delay.

I have received Joule's paper and I shall return it, with a report, in a few days.[2]

Believe me

Yours very truly
WILLIAM THOMSON

1 Kelvin (92), delivered 28 February 1856.

2 J. P. Joule, 'Introductory research on the induction of magnetism by electrical currents', *Phil. Trans.* (1856), 287–96. Kelvin's report, dated 16 November 1855, is in the Royal Society, RR.3.156.

134 STOKES to KELVIN, 10 November 1855
Kelvin Collection, S380

Royal Society,[1]
Somerset House,
Nov 10th 1855

MY DEAR THOMSON,

The Bakerian lecturer is appointed by the *new* council, which will not be elected till the 30th, so nothing can be fixed before the next council meeting after that, which will probably be on the 20th of December.

It is of exceedingly little consequence whether the 3d or some other Thursday be chosen, so we should suit your convenience supposing you are appointed.

The council not only appoint the lecturer but the day on which the lecture is to be delivered. It would be desirable therefore if a little before the 20th of December you would let us know whether you would prefer the 24th of Jany or the 28th of Feby.

Foucault has lately brought forward an experiment which seems to have created quite a sensation in Paris.[2] I don't think it would have created any sensation here, though it shows the result certainly in a very striking manner. He finds his gyroscope is very rapidly stopped by a magnet. If now the motion be continued mechanically while the magnet is still in presence the disk of the gyroscope is heated I think some 15° or 20° Centigrade. A very pretty result certainly but I apprehend nothing new in principle, at least new in England. I presume any Joulite would have regarded it & did regard it as a matter of course that the work spent in moving the conductor in a manner contrary to that in wh the mutual force of the magnet and induced currents would tend to move it was represented by the heat generated by the circulation of those currents in the conductor. I should feel obliged to you if you would refer me to something in print, the earliest that you happen to know, in which this is explicitly stated.

Yours very truly
G. G. STOKES

1 The following has been written on the envelope, not by Kelvin but presumably by his secretary: 'Rec[d] Sund Nov 11. answ[d] Nov 12 1855'.

2 H. Foucault, 'De la chaleur produite par l'influence de l'aimant sur les corps en mouvement', *Comptes Rendus*, **41** (1855), 450–2.

135 KELVIN to STOKES, 12 November 1855
Stokes Collection, K86

2 College, Glasgow
Nov 12, 1855

MY DEAR STOKES

It is only from ignorance of what Joule taught twelve years ago, that Foucault's experiment on the heat generated in his revolving mass has excited anything like a 'sensation'. I saw the account of it in the Comptes Rendus,[1] and *admired* but did not wonder at it. I knew well from Faraday how instantaneously a mass of continuous copper comes to rest between the poles of a strong magnet, no matter how rapidly it may have been revolving, and had frequently shown the phenomenon in my lectures. I do not think I ever showed it without saying at the same time that if the copper were kept rotating uniformly, an equivalent to the work spent would be produced in heat generated in the copper. I often thought of making an experiment to show the elevation of temperature of the copper which I anticipated could easily be got very considerable, but I have not yet put it in practice. (By the bye, one thing I had determined on before ⟨I saw⟩ Foucault's exp[t] was descr[d] was to make a large top of copper for another purpose, and I intended to show the effect of an electromagnet brought near it in causing sudden death: this however was illustrative of Faraday, not Joule.)

I have this morning looked for a reference to give you & I find in Joule's paper 'on the Calorific Effects of Magneto electricity and on the Mechanical Value of Heat' (commun[d] to the British Assoc[n] Cork, Aug. 1843, & published Phil Mag . . .) the following passage in p. 2⟨6⟩2 of the article itself. 'After the preceding experiments there can be no doubt ⟨but⟩ that heat would be evolved by the rotation of non magnetic substances, in proportion to their conducting power, D[r] Faraday having proved the existence of currents in such circumstances and that their quantity is proportional *ceteris paribus* to the conducting power of the body in which they are excited. I have not made any experiments on this subject, but in the next part we shall have occasion to avail ourselves of the good conducting power of copper, in conjunction with the magnetic virtue of the bar of iron, in order to obtain a maximum result from the revolution of ⟨the⟩ a metallic bar.'[2]

In the exp[t] of Part II to w[h] he alludes, an iron bar is coated with copper to a thickness of ⅛ inch & made to rotate betw. the poles of a magnet, by a measured application of force, & the heat developed is measured. One of the best ⟨of⟩ determinations (743 ft lbs to the un[it] Fahr) of the mechanical equivalent contained in that paper is so obtained. You will see from that I think that Joule in 1843 was rather ahead of France in 1855. By the bye I sh[d] mention that he opens his paper with an experimental proof that induced currents generate heat throughout their circuit, in the moving as well as in the fixed parts. His conductor being copper, the thing proved by exp[t] actually includes the phenomenon observed by Foucault.

Would it be worth while to write to the French Academy giving these references, for the credit of England? I hope if the matter is talked of in London Joule will be properly referred to.

Another phenomenon which he brings forward in the same paper is the heat evolved by the rotation of a bar of soft iron betw. the poles of a magnet. If you want to show the same thing without motion, purely by magnetization & demagnetization or reversal, take an iron tube into which you can put a thermometer. Put this inside a double brass tube ⟨to give a current of water⟩ and the whole inside a magnetizing helix. The iron tube should be well wrapped in cotton wool to keep it from thermal communication as much as possible, with the brass, & the inside of the iron tube stuffed with cotton round the thermometer, which may show its scale & column of mercury externally. If you pass a current continuously through the helix, ⟨the⟩ and keep water flowing from the town pipes through the double brass tube, the thermometer will not rise at all (although if you stop the flow of water, it will soon rise greatly, because of the heating effect of the electric current). If now, by a self acting break, or otherwise you rapidly break & make, or make in contrary directions, alternately, the current in the helix, the thermometer will rapidly & steadily rise. I showed it as a class exp[t] rising 4 or 5°. I think in a very few minutes. I intend to heat a little steel wire, (tempered brittle at first) by this process[3] till I find it soft, when I take it out.

Yours very truly
WILLIAM THOMSON

How sh[d] I address when you are in London?

P.S. I think I had better say 28th Feb., to be sure of being ready with my paper.[4] If before December I find I can be sure of having it ready in January, I shall write & let you know.

1 See the previous letter, note 2.
2 J. P. Joule, 'On the calorific effects of magneto-electricity, and on the mechanical value of heat', *Phil. Mag.*, **23** (1843), 263–76, 347–55, 435–43, and *Brit. Assoc. Rep.* (1843), part 2, p. 33.
3 Kelvin has written above the line, '(taking away the iron tube of course)'.
4 Kelvin (92).

136 Stokes to Kelvin, 12 November 1855
Kelvin Collection, S381

London, Nov 12th 1855[1]

My dear Thomson,

I may spare you part of the trouble I have been giving you, namely that relating to the transformation of the work done in overcoming the mutual force of a magnet and induced current into heat, as while engaged in looking for something else I met with Joule's paper[2] in which he determines the mechanical equivalent of heat in that way. I fear that I have been giving you a deal of trouble of late though as your lectures have commenced you have not much time to yourself.

Yours very truly
G. G. Stokes

1 Kelvin has written on the envelope, 'Stokes Nov 12, 55. Finds in Joule his question answered re work done by moving a magnet in neighbourhood of a conductor'.
2 See the previous letter, note 2.

137 Stokes to Kelvin, 13 November 1855
Kelvin Collection, S382

69 Albert St Regent's Park[1]
London Nov 13/55

N.B. this is my usual address for the winter & early part of spring

My dear Thomson,

I am sorry I should have put you to so much trouble. I felt certain that Joule & you were thoroughly imbued with the notion, and that it would be found explicitly stated in some of your numerous writings, but I thought you could much more readily refer me to the places. When Wheatstone first told me of the experiment[2] I was like you were disposed to ⟨wonder⟩ admire than wonder. In fact I had long been accustomed to recollect the way in which an induced current travelled by considering that it must be such that work was spent not given out, and I took for granted that the heating of the conductor was the representative of the work.

The Copley medal has been awarded to Foucault for his various experimental researches interference, pendulum, gyroscope &c. and the recent thing was mentioned among others. I said that in *that* particular I considered Foucault's merits were as nothing in comparison with Joule's.

I shall consider the 28th of Feby as the day on w^{h} you would be ready to deliver the Bakerian lecture[3] unless I hear in the mean time to the contrary.

As to writing to the Academy I think perhaps we may as well wait to see whether Lord Wrottesley[4] will say anything about it in his speech at the anniversary on the 30th.

Yours very truly
G. G. STOKES

1 Kelvin wrote the following on the envelope, probably in 1903 (see letter 98, note 1, and letter 99, note 15):
Stokes Nov 13, 1855
ref^{ce} to Joule's
{Joule
{Foucault (medal RS
(Copley
2 See letter 134, note 2.
3 Kelvin (92).
4 The Second Baron Wrottesley (1798–1867) was president of the Royal Society from 1854 to 1858. His presidential address on 30 November 1855 is in the *Proc. Roy. Soc.*, 7 (1854–5), 560–77. The section of the address listing Foucault's accomplishments (pp. 571–4) does not mention his experiments on heat produced by the influence of magnetism on moving bodies.

138 KELVIN to STOKES, 20 November 1855
Stokes Collection, K87

2 College Glasgow, Nov 20, 1855

MY DEAR STOKES

I have kept the enclosed report for several days because I wished to make some memoranda from the papers (Joule's M. El.) w^{h} ought to accompany it, but I have been so constantly occupied otherwise that I have not yet been able to do so. Rather than wait longer I send the Report, & I shall send the paper after, I trust two days later at the most.[1] I hope too to be able to send almost immediately the acc^{t} you asked for of the exp^{ts} in which I have been concerned, conducted with assistance from the R.S. fund. I hope it will reach you in time for the purpose for w^{h} it is wanted.

Yours truly
WILLIAM THOMSON
In great haste

1 See letter 133, note 2.

139 KELVIN to STOKES, 24 November [1855][1]
Stokes Collection, K87A

Thornliebank
Saturday Nov 24

MY DEAR STOKES

I enclose, ⟨explanations⟩ according to your request, explanations of the experiments in which I have been engaged with assistance from the R.S. fund. They are more detailed I sh^{d} think than Lord Wrottesl[e]y wants, but I hope they will supply him with the information ⟨he wants⟩ required. I am sorry that I have not managed to get them sent sooner.

Yours very truly
WILLIAM THOMSON

1 The year is determined by the similarity of the letter's content to that of the previous one and by the fact that in 1855 November 24th was a Saturday.

140 STOKES to KELVIN, 26 November 1855
Kelvin Collection, S383
Printed in Stokes's *MPP*, IV, 374.

69 Albert St Regent's Park[1]
London Nov 26, 1855

MY DEAR THOMSON,

I write to refer you to a paper[2] of Foucault's in which he finds that the voltaic arc *produces* by absorption the fixed line D in lights[3] in w^{h} it did not before exist, and that the *bright* line D is of constant occurrence in the arc.

l'Institut N° 788 Jan. or Feb. 1849.

Yours very truly
G. G. STOKES

1 Kelvin has written the following on the envelope:

1855 Stokes Nov 26
Foucault dark & bright D
Feb 1849

2 The relevant portion of Foucault's paper is translated in Stokes (69).

3 Larmor's printed version reads *light* instead of *lights*.

141 Stokes to Kelvin, 6 December 1855

Kelvin Collection, S384

Partially printed in Stokes's *MPP*, IV, 374.

69 Albert Street Regent's Park[1]
London Dec 6th 1855

My dear Thomson,

At the meeting of the Committee of papers of the Royal Society it was decided that Joule's paper[2] should be printed, being first returned to him for revision, and that you be requested to communicate with him. So now you are free to act directly.

I am much obliged to you for your account of the experiments, and I am now sorry I should have given you the trouble. I posted it to Lord Wrottesley the same evening, but somehow or other it miscarried, for he had not received it when he left Wrottesley Hall, though he received, I think a day before he left for Town, a letter which I did not post till the following morning. As it turned out the miscarriage was of no consequence, for it was ultimately decided not to attempt any account of the appropriation of the grant.

I have it I think on my memory that you ultimately fixed on the 28th of February as the day on which you would be ready to deliver the Bakerian lecture,[3] but I cannot find your letter. I have got the letter in which you mention the days in Jany or Feby and give the title. Please to write me a line, which I can receive *before* the 20th, to say which day you would finally prefer. It is on the 20th that the Bakerian lecturer is to be appointed and the time of delivery fixed.

I was speaking to Foucault about the artificial dark line D.[4] It is easily produced by arranging so that the coke pole shall be seen *through* the arc, the pole itself giving an uninterrupted spectrum in which the *dark* line D is seen by absorption while in the less bright spectrum formed by the arc itself is seen the *bright* line D as a continuation of the other.

Foucault left me several copies of his magnet & disk experiment,[5] of which I send you one.

Yours very truly
G. G. Stokes

1 Kelvin has written on the envelope, 'Stokes Dec 6, 1855. Foucault, bright & dark D in electric arc'.
2 See letter 133, note 2.
3 Kelvin (92).
4 Presumably, Stokes talked with Foucault on the occasion of the Copley Medal for 1855 being presented to him. See letter 137.
5 See letter 134, note 2.

142 KELVIN to STOKES, 14 December 1855
Stokes Collection, K88

2 College, Glasgow
Dec 14, 1855

MY DEAR STOKES

The day which would be most convenient for me to give the Bakerian lecture[1] on is Thursday the 28th of February.

I have written to Joule and heard from him regarding his paper.[2] He is going to make some changes; and to include some additional experiments, but may be prevented from this by wishing to cause no inconvenient delay. If I do not hear from you to the contrary I shall assume, in any farther correspondence I may have with him on the subject, that it is desirable to have the paper ready to put into the printers' hands as soon as possible.

I was much obliged to you for the information you gave me regarding Foucault's observations on the line D, & immediately looked for & read the paper[3] in the Institut to wh you referred me. What you mention too in your last letter, is most interesting. It appears as if sodium was not essential, but it is still possible that there may have been some of it ⟨among⟩ in the charcoal. However this is a minor point compared to the great fact of the same medium ⟨causing the generation⟩ giving rise to the light D when generated in itself & absorbing the same kind of light from light passing through it.

Have you seen any striking effect in the heating a metal by rotation near a magnet? I am thinking on spinning a copper top on an agate point, & showing its sudden death when an electromagnet is made in its neighbourhood.

Yours very truly
WILLIAM THOMSON

1 Kelvin (92).
2 See letter 133, note 2.
3 See Stokes (69).

143 STOKES to KELVIN, 17 December 1855
Kelvin Collection, S385

69 Albert Street Regent's Park
London Dec 17, 1855

MY DEAR THOMSON,

There is no particular hurry about Joule's paper.[1] It is indeed ordered for printing, but it might if need be stand over till the next Part. If the proposed changes are a decided improvement I am sure the Society would prefer having the paper in its improved form to having it a little sooner, and Joule no doubt would rather it should come out in its improved shape. The Part will be a fat one even without it. It will not however be published for some time. If there should be *extensive* and *material* changes in the paper perhaps the best plan would be for Joule, when the paper is ready, to ask leave of the Council to substitute the fresh paper, of which the title could then be read for form's sake, in order that it might take date correctly.

Yours very truly
G. G. STOKES

1 See letter 133, note 2.

144 STOKES to KELVIN, 20 December 1855
Kelvin Collection, S386

The Royal Society,
Somerset House,
Dec 20 1855

MY DEAR THOMSON,

You have just been appointed to deliver the Bakerian Lecture – Researches on the Electrodynamic ⟨Equiva⟩ Qualities of Metals – on the 28th of February 1856.[1] A happy Xmas to you.

Yours very truly
G. G. STOKES

1 Kelvin (92).

145 KELVIN to STOKES, 12 January 185[6][1]
Royal Society of London, RR.3.299

2 College, Glasgow
Saturday Jan 12, 1855 [*sic*]

MY DEAR STOKES

I have not seen Joule's paper[2] since he made his alterations, but I have had a good deal of correspondence with him both before and after, and I think it probable that in its present shape it will be on the whole satisfactory. I expect to see him here in about a fortnight when I could discuss the subject farther with him if necessary, but if you are satisfied with the paper as you have it, I think it would not be necessary for me to see it again⟨, and it could⟩.

I now send you Rankine's paper[3] which I have looked through with great interest. It was forwarded to me to Largs where I was during the holidays, and I kept it without opening it for two days until I had finished a paper[4] on the same subject with which I had been occupied for some time. However I found that there is very little direct connection between our ⟨two⟩ [?] views, although I believe it will be important to investigate the relations between them. I am sorry that I have not been able to enter on Rankine's paper so fully as I would have liked, and I cannot therefore say I have verified or even fully understood all his investigations. So far as I can see, they appear exceedingly good, and I believe almost entirely new. I do not know enough of Sylvester's 'umbral' notation to be able to judge whether the analytical treatment is perfectly satisfactory. The author refers to Sylvester's papers[5] in the Cambridge & Dublin M. J., and in the Phil. Trans^{ns}, which of course ought to be read by any one before taking up Rankine's paper. Still if half a dozen lines added would make it formally complete, and intelligible to a competent mathematician without the necessity of reading a new branch of algebra, I think ⟨it⟩ such an addition would make the paper more acceptable.

On the whole, so far as I can judge, I think the paper well worthy of being published in the Transactions.

I am disposed to differ in opinion from what is said in the second and third paragraphs of the paper (see first leaf) and I feel convinced that I can show a body composed of parts subject to the special conditions there alluded to, & yet as a whole possessing elastic properties represented by 21 independent coeff^{ts}. However I suppose nearly all writers on the subject have taken the same view as that to which I object, regarding the properties of solids composed of centres of force, and I do not suppose it will be considered desirable to make any limitation of it in the present case.

Believe me, Yours very truly
WILLIAM THOMSON

P.S. It is rather late now to return your good wishes for 'the season'. However I wish you a 'good new year' as we say in Scotland.[6]

1 Kelvin misdated the letter 1855. Kelvin first reported on Joule's paper (see note 2) in November 1855, and Rankine's paper (see note 3) is dated June 1855. Besides, January 12th was a Saturday in 1856.

2 See letter 133, note 2.

3 W. J. M. Rankine, 'On the axes of elasticity and crystalline forms', *Phil. Trans.* (1856), 261–85.

4 Kelvin (91).

5 J. J. Sylvester, 'On the principles of the calculus of forms', *Camb. Dub. Math. J.*, 7 (1852), 52–97, 179–217, and 'On a theory of the syzygetic relations of two rational integral functions, comprising an application to the theory of Sturm's functions, and that of the greatest algebraical common measure', *Phil. Trans.* (1853), 407–576. Rankine added an explanation (dated 24 February 1856) of Sylvester's mathematics to his paper cited in note 3 (pp. 284–5).

6 Stokes has written at the bottom of the page, 'Rankine by Thomson 1855', 1855 being the year of the paper's submission, not of Kelvin's report.

146 Kelvin to Stokes, 28 January 1856
Stokes Collection, K89

2 College, Glasgow
Jan 28, 1856

My dear Stokes

Can you tell me whether it is usual to have any experimental illustrations, or apparatus, to show in delivering a Bakerian lecture? If desirable I could easily bring one or two rough pieces of apparatus which have been actually used, and so facilitate oral explanations. Of course actual experiments would be impracticable on such a subject, except something very slight & limited to one or two points.

I should be much obliged by your letting me know if there is anything besides what is described in your long paper[1] on change of refrangibility &c, and a short one[2] describing a method of showing fluorescence, which would be useful in explaining and illustrating your discovery in my lectures. A simple reference to anything published will be sufficient, but perhaps you could give me some useful hints (ever so brief) which I could not find in your published papers. Also I would like to know how much of your investigations on absorption you have published.

Have you seen Clerk Maxwell's paper[3] in the Trans R.S.E. on colour as seen by the eye? Are you satisfied with the perfect accuracy of Newton's centre of gravity principle on w^{h} ⟨the⟩ all ⟨such⟩ theories & nomenclatures on the subject are founded? That is to say do you believe that the whites produced by various combinations, such as two homogeneous colours, three homogeneous colours, &c, are absolutely indistinguishable from one another & from solar white by the best eye? It will be a corollary from this that every colour (except of course the purples) is identical, in the sensation it produces, with some homogeneous colour mixed with white light. Are you at all satisfied with Young's idea of triplicity in the perceptive organ?

Yours very truly
William Thomson

P.S. Have you thought on electrical crystals at all? They are, as described, electrically dipolar, since always one end becomes vitr[eousl]y & the other resin[ousl]y elec^{d} when heated: also geometrically dipolar, since they consist of prisms presenting different kinds of crystallization at the two ends. (De la Rive on Elec^{y}, Vol II, of which the transl^{n} has recently appeared)[4]

Will not such crystals probably possess optically dipolar properties – for instance Faraday's rotation of the pl. of polarization? Is any other optical property probably or possibly dipolar? Perhaps propagation with diff^{t} veloc^{y} in the two directions parallel to the dipolar axis.

1 Stokes (42), the main part of which was published in the *Philosophical Transactions* in 1852, with a short addition appearing in the *Philosophical Transactions* for 1853.

2 Kelvin has added a footnote here reading, 'I do not remember the reference for this paper and w[d] be obliged by your telling me it'. Kelvin probably meant the 1853 addition to Stokes (42) but may have meant Stokes (49).

3 J. C. Maxwell, 'Experiments on colour, as perceived by the eye, with remarks on colour blindness', *Trans. Roy. Soc. Edinb.*, **21** (1857), 275–98. The paper was presented in 1855.

4 Arthur Auguste de La Rive, *A Treatise on Electricity, in Theory and Practice*, trans. C. V. Walker, 3 vols. (London, 1853–8). La Rive (1801–73) was professor of experimental physics at the Académie de Genève.

147 STOKES to KELVIN, 4 February 1856

Kelvin Collection, S387

69 Albert St Regent's Park
London Feb 4th 1856

MY DEAR THOMSON,

I have been so busy between the Smith's Prize, the Royal Society and two courses of lectures that I have delayed hitherto answering your letter.

I have only been present at one Bakerian lecture, namely Tyndall's.[1] He had a good deal of apparatus. It would certainly add interest to the lecture to have some of the actual apparatus.

I have published only 3 papers on fluorescence, the long one (Phil. Trans 1852)[2] the shorter one (Phil. Trans 1853 I think)[3] and an abstract of my lecture[4] at the Royal Institution (Phil. Mag for I think Jan–June 1853). I have nothing to add that is material to these papers.

If by my ⟨experiments⟩ invetigations on absorption you mean a notice of the absorption-bands in solutions of permanganate of potash, of proto-salts, and of per-salts of uranium, you will find them in the long paper. If you mean the beautiful exemplification of the relation between surface-colour and substance-colour which is afforded by crystals of permanganate of potash you will find it in the Phil. Mag. I should think about 1853.[5] If you mean an explanation of the brushes seen *with common light* about the optic axes of certain biaxal crystals, I have published nothing about that yet. It is about the first thing I must publish when I have time.[6]

I have not made any experiments on the mixture of colours, nor attended particularly to the subject.

I had not thought particularly about pyro-electric crystals. I knew they were geometrically dipolar, but I had not thought about dipolar optical properties. I don't expect the effects you mention would take place. They could not without violating the principle of reversion,[7] which holds good in any system where the forces acting depend only on the positions of the particles. It does not hold good in the case of a magnetized medium. It would be proper to look out for quasi-magnetic rotation in a dipolar uniaxal crystal, though I don't expect it would be found.

Yours very truly
G. G. STOKES.

1 J. Tyndall, 'On the nature of the force by which bodies are repelled from the poles of a magnet; to which is prefixed an account of some experiments on molecular influences', *Phil. Trans.* (1855), 1–52.
2 Stokes (42).
3 *Ibid.*
4 Stokes (49).
5 Stokes (54).
6 This optical phenomenon involving 'common light' should be distinguished from that of Haidinger's brushes, which involves polarized light and on which Stokes had already published. (Stokes (39).) For a discussion of Haidinger's brushes, see Thomas Preston, *The Theory of Light* (London and New York, 1890), article 244. So far as I know, Stokes did not publish on the first-mentioned phenomenon.
7 Stokes defined his principle of reversion in his (26), in *MPP.*, II, 90.

148 KELVIN to STOKES, 9 February 185[6][1]
Stokes Collection, K77

Thornliebank
Feb 9, 1855 [*sic*]

MY DEAR STOKES

I enclose a note on the thermal effect of high pressure steam escaping through a porous plug, and I shall be obliged by your communicating it to the R.S. on the first convenient occasion, on the part of Joule & myself.[2] It is not meant for the Transactions, but we would be glad to have it published in the Proceedings.

Many thanks for your information about optics.

I am very busy trying to get out some more experimental results for my Bakerian Lecture,[3] besides writing out what I have, which, at my rate of composition, is a most formidable task.

Yours very truly
WILLIAM THOMSON

1 Kelvin misdated the letter. The two papers mentioned in notes 2 and 3, for example, were communicated to the Royal Society, respectively, on 21 and 28 February 1856.
2 Kelvin and Joule (6).
3 Kelvin (92).

149 KELVIN to STOKES, 19 March 1856
Stokes Collection, K103

2 College, Glasgow
March 19, 1856

MY DEAR STOKES

You will have begun to think my paper[1] is not going to make its appearance, but I hope it will really reach you very soon now. Unless you hear from me to the contrary you may rely on having it (I trust complete) before the next meeting after tomorrow. I felt very averse to do anything in the way of work that I could possibly delay when I got home from London, and so allowed the week during w[h] I sh[d] have finished it to pass with scarcely anything done. Then when I began I found the final preparation of the Tables a much more serious business than I expected, and I have had a great deal of trouble with it. However it is nearly done now & I have two days off lectures at the end of this week so that I hope to get the matter out of hands in a few days.

I have got a new exp[l] result which I shall try to get written out & sent either with or soon after the paper. It is on the electric conductivity of plat[m] wire under stress & proves by a differential experiment that the conductivity of the substance is diminished by longitudinal tension.

Will you hand the enclosed corrections on my abstract to D[r] Sharpey? If I notice any more I shall send them before the appearance of the next Number.

Yours always truly
WILLIAM THOMSON

1 Kelvin (92).

150 STOKES to KELVIN, 25 April 1856
Kelvin Collection, S388

Pembroke Coll. Cambridge[1]
April 25, 1856

MY DEAR THOMSON,

I promised D[r] Sharpey to write to you to ask you for an abstract of your paper on elasticity.[2]

I looked over it hastily yesterday in Town preparatory to reading it or rather small portions of it, which was done last night. I noticed one or two points at the end about which I will take the opportunity of making remarks.

(1). Sir J. Herschel did not discover the plagiedral faces of quartz, which were known before, but only the connexion between their direction of arrangement and the direction of rotation of the plane of pol[n].

(2). It is not 'in being separated from the salts' that right and left tartaric acid are obtained. It was on attempting to form the double racemate of soda & ammonia that Pasteur obtained instead a mixture of right & left double ⟨pa⟩ tartrates. These were separated by picking out, recrystallized, & the acids thence separately obtained. When mixed they reproduce racemic acid. Except the salt above mentioned & that in which the AmO is replaced by KO the salts of racemic acid examined by Pasteur crystallized as true racemates.

(3). Darker told me some time ago the history of the right and left oil of turpentine. The oil used in Britain comes from Canada & is obtained from trees of the pine tribe. This is left-handed (i.e. contrary to sugar) if I recollect right. The right-handed oil is obtained in France from, he said, a species of laurel.

(4) Your conclusion as to the necessary heterogeneousness of strain (I am not certain whether you said strain or medium, but I assume the former) to explain rotatory pol^n quite accords with my investigation in which the effect is made to depend on diff. cos. of the displacements of the 3^d not 2nd order (I speak of the terms appearing in the final equations of motion). The axis of a twisted wire (cylindrical) for instance would have the strain I require.

Yours very truly
G. G. Stokes

1 Kelvin wrote the following on the envelope, probably in 1903 (see letter 98, note 1, and letter 99, note 15):

Stokes Ap 25, 1855 [*sic*]
[Pasteur, racemates, rotation (chiral[ity])]

In the second line, the outside brackets are Kelvin's.

2 Kelvin (91).

151 Kelvin to Stokes, 16 May 1856
Stokes Collection, K90

2 College, Glasgow,
May 16, 1856

My dear Stokes

I now return Mr Phillips' paper[1] on Aurora, and I regret very much having kept it so long. I have been excessively occupied ever since I received it, first with winding up the session here, & then at Manchester, or else I sh^d have sent you it much sooner.

You will see by the accompanying memoranda, that I think there is a good deal objectionable in the paper. Besides I must add that I do not feel sure about the 'fire ball lightening' being a real phenomenon. I do not know enough of what has been reported by observers regarding it to have any decided opinion. But I am quite sure that the author of the paper has a wrong idea of shooting stars. ⟨I do⟩ It is I believe certain that they are in general, if not always solid meteors; and I have no doubt but that they become incandescent by friction in the atmosphere (Joule's explanation, & I dare say much older.) It would be rash to pronounce a shooting star *not* an electrical phenomenon, but I believe it to be as correct to pronounce the firing of a carriage wheel, as to pronounce a shooting star an electrical phenomenon. Indeed I think Berzelius & Joule ⟨was write⟩ were right in calling the heat of chemical combination, electrical; & I believe the heat of friction is so also. That a shooting star is electrical in what I gather is the author's sense, I believe to be not the case.

I am sorry that after all I have only been able very hastily to read the paper. It contains some curious views, ⟨&⟩ especially that of the formation of the auroral rays. Whether or not this view is correct, it is I think worth recording. There could be no harm in reading the paper at a meeting of the Royal Society I think, as far as regards countenancing heresy. Still it may be doubted whether ultimately it would be accepted for the Transactions. In saying this I am by no means saying I think it ought not to be published in the Transactions. If that comes to be a question, some of the meteorologists would have to say whether ⟨his⟩ the reported observations & descriptions he quotes are confirmed, and if they are, some little modification in the physical discussion might make it a very creditable paper.

I am sorry to be able to give only so undecided a report.

We leave Leith tomorrow for Rotterdam.

Would you oblige me by asking at the R.S. when you are there next, if a parcel containing scientific packets for Paris, Rome, ⟨&⟩ Stockholm, & St Petersburg reached, & whether the packets have been despatched to their several destinations.

You received I hope my letter & packet at Cambridge (⟨some⟩ two short articles[2] for the Proceedings & an addition to my paper[3] on the Electrodynamic Qualities of Metals).

Yours always truly
WILLIAM THOMSON

p 2. Striking distance very different from length of voltaic arc. The latter may be half an inch long when the striking distance between electrodes gradually approximated, before current commences, is insensible. The striking distance may be 6 inches or more (R[oyal] I[nstitution] plate glass elecl mach.) ⟨and⟩ in experimental arrangements, & I dare say several miles betw. thunder clouds.

pp 12 & 13 very good, accg [to] a hasty perusal.

foot of p 17 &c. I would like to know how to distinguish between the descent of negative electricity and the ascent of positive electricity!

p 19. foot. does anybody know that the tails of comets are formed of gaseous matter or is there the slightest probability that they are so?

pp 21 . . . 23 ⟨Commences with a true statement, followin⟩ [?] very good.

327 . . . 329 Very curious experiments (? phosphorescence;) but are the differences observed sufficiently accounted for by the positive & negative charges of the phial? This I think is certain, that if the two ends of the discharging train are symmetrical there will be no difference betw. the effects of positively and negatively charged phials. I have not been able to give suff^{t} attention to the description of his experiments to understand the circumstances.

Art 333 I think is wrong, as the author seems to be [*sic*] believe that positive electricity is a thing poured out into the thread when a positively charged phial is discharged, & negative another thing poured out from a neg^{vely} charged phial &c.

1 Reuben Phillips, 'On the Aurora', *Proc. Roy. Soc.*, **8** (1856–7), 214–15. *The Royal Society Catalogue of Scientific Papers* lists some thirty papers by Phillips on various physical topics, almost all being published in the *Chemist* or the *Philosophical Magazine*.

2 Kelvin (93) and (94), read before the Royal Society on 22 May and 12 June 1856, respectively.

3 Kelvin (92). The addition is article 143, marked 'received, May 10, 1856' in *MPP*, II, 293.

152 KELVIN to STOKES, 17 September 1856

Stokes Collection, K91

2 College, Glasgow
Sep 17, 1856

DEAR STOKES

I have added a few arbitrary examples regarding specification of strains, to the first Part of my paper on Elasticity, and I have inserted, with the date, Sep 16, 1856, an Example showing the Principal Strains of a body cubically isotropic, with those of a body perfectly isotropic as a particular case, in the Second Part.[1] I suppose there can be no objection to this being done on the proof, as a date is given to the only addition that can possibly be construed into an advance in science (although I had really worked it out three months before the paper was communicated, & only omitted it for want of time when I was sending in the paper.) I have not thought it worth while to give the date of insertion of the merely arbitrary examples in the 1st Part. I think it right to inform you as to the additions so that if any ⟨farther⟩ formal sanction is required or withheld the printers may act according to it.

I write in haste being up for a few hours to attend the meeting of Mechanical Engineers.[2]

We returned from Germany a fortnight ago, & cross to Arran tomorrow, to remain till November. My address is Invercloy Arran by Ardrossan, & I shall be very glad to have intelligence scientific or non-scientific of you.

Yours truly
WILLIAM THOMSON

1 Kelvin (91). For the solution to the problem of dating the additions, see the next letter and Kelvin's *MPP*, III, 84.

2 The Institution of Mechanical Engineers held its 'special general meeting' in Glasgow on 17 September 1856. (See the *Proceedings of the Institution of Mechanical Engineers* (1856), 125.)

153 STOKES to KELVIN, 24 September 1856
Kelvin Collection, S389

Aughnacloy Ireland[1]
Sept 24th 1856

MY DEAR THOMSON,

I had long been thinking of writing to you to Germany relative to the Bakerian Lecture,[2] but I thought it was no use as you could do nothing till your return. But first for your question.

I take for granted that the Council would allow the additions you speak of being made to your paper[3] on elastic solids; the only question is what to do about dating. It strikes me that the simplest plan would be after 'Read such a date' to put an asterisk & add a foot note describing in a general way the ⟨alterations⟩ additions which have been made to the paper since it was read, much as you have done in your letter.

The Bakerian lecture you know was not in the hands of the referees till long after it was delivered, and was not reported on till about the time you went to Germany. (Our report is dated June 2nd.) The question of printing could not of course be decided on till the next meeting of the Council after that. As one of the referees had made some suggestions about condensation relating to the first part I was directed to confer with you before the paper was printed. I think the best plan will be to send you the reports and let you judge for yourself whether any or what alterations are needed. I assume the leave of the referees to communicate the reports. Please to return them to me when you have done with them.

I will direct Mr Weld to send you the M.S. of the Bakerian lecture. Please to preserve it in the original state and write any fresh matter on new sheets making directions as to omissions or insertions in the old M.S. in red ink. Also should you think fit to make any new drawings please send the old as well. In short, I want to retain the old M.S. in its present shape to put in the Archives.

I hope your wife was benefited this year by her sojourn at Kreutznach, and that you yourself have laid in a stock of health for your winter's work. You say you will be glad to hear any intelligence scientific or non-scientific of me. I am afraid I have not much to tell you in the scientific line; my principal intelligence must belong to the non-scientific head which is that I am engaged to be married to Miss Robinson daughter of Dr Robinson.[4] The engagement is not yet of a fortnight's standing and the time is not yet fixed but I think it likely it may be in July. I must now make the neighbourhood of London my head quarters instead of Cambridge.

In case you want to write I will tell you that I remain here till Saturday or Monday; then Observatory Armagh till Wednesday then Cambridge from whence any letter will be forwarded to wherever I take lodgings for the winter in London.

Give my kind regards to your wife & believe me

Yours very truly
G. G. STOKES

1 Kelvin wrote the following on the envelope in 1903 (see letter 98, note 1, and letter 99, note 15):

Stokes Sep 24, 1856
Two enclosures (referees' reports on my Electrodynamical Properties of Metals) belong to R.S.
Larmor keep them
K[elvin] Dec 1, 1903

Just below Kelvin's notes, Larmor has written, 'sent to Roy Soc J.L.'

2 Kelvin (92). No reports on (92) are in the Royal Society's archive of referees' reports.

3 Kelvin (91). See the previous letter, note 1.

4 Mary Susanna Stokes (d. 1899) was the daughter of Thomas Romney Robinson (1792–1882), head of the Armagh Observatory from 1823 to 1882. A large correspondence from Stokes to Robinson survives in the Stokes Collection. The letters date mainly from the 1870s, and many concern Robinson's work on anemometers. Much of the correspondence, except that concerning anemometers, is published in Stokes's *Memoir*, 1, 324–81. Stokes's *Memoir* (1, 50–76) also contains a selection of personal letters from Stokes to his wife, the publication of which apparently caused some controversy. Surviving in the Stokes Collection are letters from Francis Darwin (Charles Darwin's son) and Larmor to Stokes's daughter, Isabella, regarding the advisability of publishing the letters from Stokes and his wife.

154 KELVIN to STOKES, 9 October 1856

Stokes Collection, K92

Invercloy, Arran
by Ardrossan, Oct 9 1856

MY DEAR STOKES

I missed the only day by which I could have written, to reach you before your departure from Ireland, after I received your letter of the 24th, or else I should not have delayed so long to send you my congratulations on the intelligence which it contains. It was the best answer you could give to my question for 'non-scientific intelligence,' and I wish you all the happiness, most sincerely, which it promises.

I am much obliged to you for the reports on my paper,[1] which I shall return you, with the paper itself, very soon, next week I hope at the latest. I shall attend carefully to all the suggestions, and add explanations or otherwise endeavour to do away with the objection to all the passages marked as wanting in clearness. I am going to Glasgow tomorrow & hope to get hold of Wheatstone's description of his method for resistances. It will depend chiefly on what I see in it & in Becquerel's, what I can do to alter my Part IV. according to the suggestions of the first referee.[2] I think I shall be able to do very little if anything to satisfy his somewhat sweeping requirements regarding Part I., and I can only wonder what he means as susceptible of abridgement in Part V., on which I labored much to give the necessary explanations of details in the simplest & shortest way I could. However I shall consider very carefully all suggestions, and shall follow as many of them as I can agree to as improvements, or perhaps even as not injuries, to the paper.

I shall do what you recommend regarding the added examples in my paper on Elasticity,[3] when I get the revise, which has not yet ⟨been sent to⟩ reached me. The printers have my address here & I suppose it must soon come.

I wish as soon as possible to send a short article[4] for the Proceedings, on the economy of submarine telegraphing to great distances, which I would like to take date and to be published in the Proceedings without more delay than must be. If I do not hear from you to the contrary I shall address it to Dr Sharpey, R.S. London, as soon as I have it ready, which I hope will be within a fortnight.

Mrs Thomson sends her kind regards, & her best wishes for your happiness, & I remain,

Yours very truly
WILLIAM THOMSON

1 Kelvin (92).
2 For Wheatstone and Becquerel, see *ibid.* in *MPP*, II, 298–301.
3 Kelvin (91).
4 Kelvin (95).

155 KELVIN to STOKES, 18 October 1856
Stokes Collection, K93

Invercloy,
Saturday Oct 18, 1856

MY DEAR STOKES

I now enclose (in a separate packet) the first part of my paper,[1] with explanatory notes (much shorter than they appear) added in consequence of the remarks of the referees as to clearness. The mails are so rare & irregular that I cannot despatch before Monday, but this & the packet ought to reach you on Tuesday. I shall shortly send the remainder of the paper, and, from what you say, I hope there will be no inconvenience in the farther delay, as what I now send can be put into the printer's hands at once. I have asked them for proofs in instalments of from one to two sheets; and a first instalment would as soon as I receive it, impel me at once to send off an additional instalment of M.SS., should I not before that time have despatched the whole. I am very sorry my copy for diagrams (w^h I wish to be chiefly wood cuts, in the pages) is so bad, & open to Prof [W. H.?] Miller's objection.[2] I find myself quite unable to make drawings with my own hand, and have much difficulty in getting them at all tolerably executed by others. I have asked the printers to let me see the working drawings (such as I believe are always made, no matter how good the copy is) before letting the cuts be executed.

I received the revise of my paper on Elasticity[3] at the end of last week in Glasgow, & it ought before this time to be in the printer's hands for press. I ⟨must⟩ was much obliged for your corrections as to Pasteur & Herschel. I do not know how I came to make the mistake regarding the racemic acid, as when you wrote to me in May about it,[4] I remembered that I knew the truth when we were speaking about it here this time last year.

Any proofs or letters for me posted in London on or after Monday Oct 25, had better be addressed to Glasgow.

Yours very truly
WILLIAM THOMSON

P.S. The notes & corrections on the MSS are written in the nearest approximation to red ink to be had in this island, which at all events can never be mistaken for black.

P.S. I enclose a Table belonging to the paper, accidentally omitted. Also directions for the printers w^h might rank as a letter, and w^h I shall be obliged by your handing to them.

1 Kelvin (92).
2 Although no reports survive for Kelvin (92), W. H. Miller did report on Kelvin and Joule (3)

(RR.2.131), whereas there are no surviving reports by W. A. Miller on any of Kelvin's papers.

3 Kelvin (91).

4 Letter 150. See also letter 132.

156 KELVIN to STOKES, 24 October [1856][1]

Stokes Collection, K366

Invercloy[2]
Friday Oct 24

MY DEAR STOKES

The plan of dividing the long diagram[3] will answer perfectly. The written heading was only intended in case it sh[d] be given as a woodcut, in the page, and may be omitted altogether if it is to be engraved, as I suppose it must be.

I write by return of post, a rare occurrence here now.

Yours truly
W. THOMSON

1 The letter naturally follows the previous two letters, also written from Invercloy. Also, in 1856, October 24th was a Friday.

2 Kelvin has written at the top of the letter, 'In the greatest haste ever since your letter came yesterday'.

3 For Kelvin (92).

157 KELVIN to STOKES, 1 November 1856

Stokes Collection, K94

2 College, Glasgow
Nov 1, 1856

MY DEAR STOKES

I only returned from Arran last night, and found R. Phillips' paper,[1] with your note accompanying it. If you will allow my letter[2] of last spring to stand for a report, I shall be very glad to escape the business of going into it again, and I therefore return you the paper by post along with this. General Sabine will probably know at once whether the author is right in his statements regarding various remarkable meteorological phenomena which I remember being struck with.

You ask me if my wife has benefitted by Kreuznach – a question which I cannot answer with much certainty. We spent six weeks at that place & then, after a short interval six weeks at Sc[h]walbach. I cannot say whether Kreuznach or Schwalbach or Time has been the means, (I hope the last, as it keeps going on, which the baths do not) but she is decidedly better than in spring. She was much better at one time, but fell back from walking too much which has been very strictly prohibited since, and she is still almost entirely confined to the sofa. I left her yesterday at Largs.

Believe me
Yours always truly
WILLIAM THOMSON

1 See letter 151, note 1.
2 Letter 151.

158 KELVIN to STOKES, 1 December 1856
Stokes Collection, K95

13 Ashley Place[1]
Monday evening Dec1/56

DEAR STOKES

I enclose the payment I accidentally omitted to bring with me along with the other parts of my paper[2] to give you today. You have it now all except Parts IV & V – of which the former will be very short, not more I think than 6 or 8 printed pages – & the last not long. I shall take care that the very last words are in your hands before the printers have to wait.

Will you give the same directions as before about the wood cuts viz. that the working drawings be sent to me? I got those of the first part, & I think the result, with some slight corrections I had to make, will be quite satisfactory.

Yours very truly
WILLIAM THOMSON

1 This was Edward Sabine's address.
2 Kelvin (92).

159 STOKES to KELVIN, 20 January 1857
Kelvin Collection, S390

69 Albert St Regent's Park[1]
London ⟨W.C.⟩ N.W. Jan 20/57

MY DEAR THOMSON,

You are a terrible fellow and I must write you a scolding. The vol of the Phil. Trans. ought to have been out by the 30th of Novr and here's your paper[2] won't be ready for a month yet. It is now received but a lot of woodcuts came in late and woodcutting takes time. I was waiting to complete Part 2 with your paper and then Miller's,[3] wh is printed at Government expense so as to have formed a distinct part according to the wish of the Council. Much dissatisfaction was expressed at the non-appearance of the part for so long a time. At last we (the officers) decided that it would be better to bring out part 2 and lump your paper with Miller's as part 3. And you have been keeping back Miller's paper too, w^h could not be struck off because the proper paging was not known. Verily I must make a rule not to send a paper to the press till it is complete. You know these things are laid on the Secretary, and the worst of it is that you are taking away the character of my wife to be; for people will be saying that she has been keeping me from my work, and lo it is all your fault.

Hoping you will be more punctual for the future I remain

Yours most sincerely
G. G. STOKES

1 Kelvin has written on the envelope, 'Stokes Jan 20 1857. Bakerian Lecture severely (not too severely) blamed for delay'.
2 Kelvin (92).
3 W. H. Miller, 'On the construction of the new imperial standard pound, and its copies of platinum, and on the comparison of the imperial standard pound with the kilogramme des archives', *Phil. Trans.* (1856), 753–946.

160 KELVIN to STOKES, 29 January 1857
Stokes Collection, K96

2 College, Glasgow
Jan 29, 1857

MY DEAR STOKES

I am exceedingly sorry my paper[1] has given you so much trouble and annoyance with reference to the publication of the Transactions. If I had known it would be so I sh^d certainly have petitioned to have it ⟨placed⟩ kept back till the publication of the first Part for 1853. [*sic*]

I have been doing my best ever since the end of September when I received the manuscript for revision, to get it ready for the press & brought through the press ⟨as⟩ with as little delay as possible. I have made great efforts not to keep the printers waiting as to proofs and, although I know I have sometimes failed, I think in general the delays were only such as are inevitable in bringing so long a paper, with numerous wood cuts, numerical tables, &c. through all the work that has to be done upon it. As to the numerical tables, I am afraid the printers must have been rather provoked with me; but I could not get all the plans completed for their arrangement, in manuscript (although I spent a great deal of time a year ago, & earlier trying to do so) & I had to alter & cut & carve a good deal after I saw them in type. I made a great mistake in not sending all the drawings up at once (which I could have done just as well as not) instead of sending them up with the instalments of M.SS. If I had done so, probably a good deal of loss of time w^d have been avoided. As to the printers they never complained (tho' perhaps they ought to have done) and I thought I was getting through the proofs on the whole as fast as they could manage them.

On the day your scolding arrived I sent off the very last set of *proofs*, and as there were not many corrections to be made I thought before this I sh^d have been able to write & say I had received & sent off the last revise for press. I have not yet received the revises however & I do not wish to delay longer writing to you to say how much I regret that I have occasioned all that annoyance & inconvenience.

On Saturday & Sunday I received India proofs of the last wood cuts & sent them off on Monday, so that there can be no more delay on their acc^t. At worst I think the very end will come within a fortnight (instead of within a month as you had been told) of the time you wrote.

Although I have been doing my best not to make delay, you will not of course scruple to throw the blame all on me when any accusations such as you tell me of are made.

Believe me
Yours very truly
WILLIAM THOMSON

1 Kelvin (92).

161 KELVIN to STOKES, 23 May 1857
Stokes Collection, K97

The Atlantic Telegraph Company Limited,
22, Old Broad Street,
London, E.C. May 23, 1857

MY DEAR STOKES

I have been prevented from answering your letter (forwarded from Moffat to Glasgow) sooner by being summoned here suddenly on Telegraph business. I went to the R.S. on Thursday night hoping to find you there but learned there was no meeting. I had some idea of taking a run to Cambridge this afternoon, but must give up the idea.

My brother's paper[1] was intended for the Proceedings & from your letter I presume you left it with Dr Sharpey for that destination.

I shall send detailed acc^t of expenditure of Gov^t Grant allowance up to the end of 1856 when the Bakerian lecture[2] was published. It will scarcely correspond to a finished research but more nearly to as much as is included in the Bakerian Lecture.

I have been [in] great difficulties about hydrodynamics for some time. I thought I had made out that a mote in a perfect (non frictional) liquid, would if set into in any state of motion, end with its centre of gravity at rest and all the energy of the given motion transformed into energy of rotation of the mote & of corresponding motion of the surrounding liquid. I am shaken however from this comfortable theory, and I fear it comes to nothing. Take a rectangular slip of paper & let it fall from a height, with its length nearly horizontal, & you will see a curious illustration. (Clerk Maxwell showed me this a long time ago.) I thought first, and am now again nearly convinced, that the generation of rotatory motion here is inexplicable without taking into account the viscosity of air. If the theory I supposed I had made out had held, the same kind of phenomenon would take place in a perfect liquid.

I have been thinking too a good deal about the resultant pressures on ⟨bodies⟩ solids rotating near one another in a perfect liquid. I can see clearly resultant repulsions in some cases and attractions in others, but I cannot yet make out an average ⟨resultant⟩ repulsion of each from its neighbours when there are a great many scattered through the liquid, each rotating, and all in different directions and about non parallel axes. If I could, I think I should have a very good medium for a mechanical illustration of light. If there was a preponderance of rotation in one direction about parallel axes, waves of transverse vibrations ⟨about⟩ in planes perpendicular to these axes would have Faraday's optical property of heavy glass, unless the time of rotation of a mote is infinitely small compared with the time of vibration of a wave. If ⟨each⟩ there is a preponderance of motes of a right handed spiral form, or of any right handed asymmetry, & if the number of motes in a wave length is not infinitely great, the heliçoidal rotatory property would exist. By the bye, I wish you would publish a short memorandum of what you have told me on this subject: a copy of or extract from a letter you wrote to me several years ago, for instance. I have it & could look it out for you if you care. However I suppose you will think very little of optics for some time now. Is *the* time fixed & when is it?[3] I expect to be in Manchester from the beginning to near the end of next week. If you have occasion to write address care of J. P. Joule Esq Oakfield Moss Side, Manchester.

Yours very truly
WILLIAM THOMSON

1 J. Thomson, 'On the plasticity of ice, as manifested in glaciers', *Proc. Roy. Soc.*, **8** (1856–7), 455–8.
2 Kelvin (92).
3 The date for Stokes's wedding in 1857.

162 KELVIN to STOKES, 17 June 1857
Stokes Collection, K98

Largs, June 17, 1857

MY DEAR STOKES

I hope you have duly received a communication for tomorrow evening's meeting entitled 'On the Electric Conductivity of Commercial Copper'.[1] I despatched it from Arran on Saturday, to be transcribed, with directions that the copy for you should be forwarded at the latest by yesterday afternoon's post from Glasgow.

By this post I send you a communication 'On the Electrodynamic Qualities of Metals:– Effects of Magnetization on the Electric Conductivity of Nickel & Iron'[2] which I hope will reach you in time to be 'read in' tomorrow evening.

If considered suitable, I w[d] like to have each of these communications published in the Proceedings. When I get a more extensive set of results to make a paper suitable for the transactions, I should be able to describe the whole more clearly and concisely, and make the continued treatment of the subject ⟨more⟩ equally complete and in better form & less compass, than I could by giving each additional set of results immediately in the form of a continuation of the 'Bakerian Lecture'.[3]

I hope sometime to have a good deal of conversation with you on hydrodynamics. I think the instability you speak of cannot exist in a perfect (i.e. incompressible & non frictional) liquid, but that it is a necessary consequence of fluid friction.

As to the mote theory I see enough now to be sure that one mote fired off in a perfect liquid of infinite extent, or limited by fixed boundaries, could never get its motion of translation stopped & the energy wholly converted into rotation, but I feel almost certain that if there be an infinite number or a large number of motes, the tendency of all motion among them & the liquid will be to rotation. This too I think is certain, that a single mote rotating with extreme rapidity about its centre, in a perfect liquid will experience a highly intens⟨ive⟩e repulsive action when brought near either fixed boundaries or other solids rotating or not. An infinite number of such motes all rotating with great angular velocities will repel one another & keep up the kind of stability & relative stiffness, required for luminiferous vibrations. If there be a preponderance of axes set in one direction, undulations among ⟨one⟩ [?] ⟨group⟩ the system would have Faraday's rotatory property, when the planes of the waves are perp[r] to this direction. When you last wrote to me on this subject & objected to a dynamic condition as accounting for Faraday's property of heavy glass *because a steel magnet* can produce & maintain it did you not overlook the true nature of heat which even in a solid seems to consist of motions. It ⟨seems⟩ appears to me that these motions only want *setting* in that axial fashion, to give rise to Faraday's property: & farther that Faraday's property *cannot possibly be explained without* some such dynamical conditions being admitted (Proc. R.S. for about last June contains what I cannot see but as an unanswerable argument to this effect.)[4]

I hope you will as soon as possible publish what you have on the heliçoidal rotatory property.

Yours always truly
WILLIAM THOMSON

1 Kelvin (96).
2 Kelvin (95A), which was received on 18 June and 'read' at the meeting of 15 June 1857.
3 Kelvin (92).
4 Kelvin (94).

163 KELVIN to STOKES, 7 November 1857
Stokes Collection, K100

2 College, Glasgow
Nov 7, 1857

DEAR STOKES

I should be glad to have an opportunity of proposing Whitehouse[1] as a candidate to be FRS but I do not know what forms have first to be gone through. I presume getting a ⟨paper⟩ form, & having it as well covered with good signatures as possible. If however the list of candidates is pretty well filled up for the next occasion, I would rather not propose him for another year. I sh[d] feel extremely obliged by your telling me if you consider the opening good at present, & think him likely to be elected. Any such information w[d] of course be entirely private & nonofficial. Will you give me a line if you can say anything as to his chances, & if you think the⟨m⟩ prospect fair, let me have a blank form. I do not know whether you are the right Secretary to apply to as to this last. If not perhaps you will kindly forward the request for me to the proper quarter.

Yours very truly
WILLIAM THOMSON

P.S. I hope D[r] Robinson has seen something of what he wanted to see, even if he did not go to Devonport.

1 Edward Orange Wildman Whitehouse (b. 1815) was electrician to the Atlantic Telegraph Co., of which Kelvin was a director. For the sharp differences between them regarding submarine cables, see Thompson, I, 325–96. Whitehouse did not become a fellow of the Royal Society.

164 KELVIN to STOKES, 20 December 1857
Stokes Collection, K101

2 College, Glasgow
Dec 20, 1857

MY DEAR STOKES

A good deal of routing out of old acc^{ts} was necessary before I could tell you how much I paid Silbermann for his painting of the three prismatic spectra.[1] Or else I sh^{d} have answered your letter sooner. I now find that the amount was ⟨£⟩fr. 180.

What is aesculine? Not having known before of its existence I was not prepared to appreciate properly your having got it crystallized.[2]

I have changed my mind greatly since my freshman's year when I thought it so much more satisfactory to have to do with electricity, than with hydrodynamics, which only first seemed at all attractive when I learned how you had fulfilled such solutions as Fourier's by your boxes of water.[3] Now I think hydrodynamics is to be the root of all physical science, and is at present second to none in the beauty of its mathematics.

You first told me that if the bounding surface of a perfect liquid originally at rest, be moved in any way (including of course change of shape) and be brought again to rest in its original position, or in any changed shape & position, the whole liquid will come to rest at the same time.

(interrupted, & sent off not to lose post – I shall continue tomorrow)

Yours truly
W. THOMSON

1 See letter 84.
2 See Stokes (65).
3 See Stokes (13) and John Herivel, 'The influence of Fourier on British mathematics', *Centaurus*, **17** (1972), 44.

165 KELVIN to STOKES, 23 December 1857
Stokes Collection, K102

2 College, Glasgow
Dec 23, 1857

[continued from Dec 21][1]

MY DEAR STOKES

That principle, in the hydrodynamics of a 'perfect liquid', which I first learned from you, is something that I have always valued as one of the great things of science, simple as it is, and I now see more than ever its importance. One conclusion from it is that instability, or a tendency to run to eddies, or any kind of dissipation of energy, is impossible in a perfect liquid (a fluid with neither viscosity nor compressibility.) It is a particular case of a more general theorem – that the solution of the equation of continuity, subject to the dynamic laws (which lead to $udx+vdy+wdz$ a complete diff[l], as the condition necessary & suff[t] for their fulfilment) is determinate when a mass of ⟨fluid⟩ liquid originally at rest has its boundary changed in any way. This theorem is easily proved mathematically, by showing that ⟨when⟩ $\int\int\int(u^2+v^2+w^2)\,dx\,dy\,dz$ is least possible consistently with the cond[n] at the surface, when $udx+vdy+wdz$ ⟨must be⟩ is a complete diff[l]. ⟨and that this is the sole cond[n].⟩

I have been trying to make out whether a slight degree of compressibility can possibly give rise to instability & eddies, but I cannot as yet fully convince myself that I am right in thinking not. Have you ever considered this question? If the true answer is negative, then viscosity must be the cause of eddies.

A perfect liquid with solids each free to move, distributed through it, would, as I thought last May, & still think, have the known properties of a physical liquid, if the number of the solids is suff[y] great, & their dimens[ns] suff[y] small. In trying to work out a definite mechanical theory of such a case. I have had to attack the problem of the free undisturbed motion of a single solid of any shape, surrounded by a perfect liquid, which at first seemed somewhat hopeless, but which I now find is quite as interesting, and not more than twice as complicated as the problem of the motion of a solid 'in vacuo', undisturbed by external forces.

Twenty one coeff[ts] ⟨being⟩ $[u^2], [v^2]\,[w^2], [wv,]\ldots[\omega^2], [\zeta^2], [\sigma^2], [\zeta\sigma],\ldots$ $[u\omega]\,[u\zeta]\ldots$, being determined by transcendental analysis (for an ellipsoid those of them that don't vanish will no doubt be algeb[y] expressible in terms of the transcendents involved in the expressions for the attr[n] of an ellipsoid) the *vis viva* of the motion of a solid & liquid surrounding it, when the ⟨fo⟩ motion of the former is (u, v, w) comp[ts] of velocity parallel to axes fixed in space and (ω, ζ, σ) compon[ts] of angular velocity round the same, will be a homogeneous quadratic function $[u^2]u^2+[v^2]v^2+\ldots 2[vw]vw+\ldots 2[\omega u]\omega u$ $+\ldots+\ldots$. By the consideration of this function, the equations of motion with or without external application of force, may be easily worked out. Thus, if an impulsive force, $(X, Y, Z,)$ & an impulsive couple (L, M, N) act on the body primitively at rest, the motion generated will be determined by the linear equ[ns]

$$X=[u^2]u+[uv]v+\ldots+[u\sigma]\sigma$$
$$Y=\ldots$$
$$L=[\omega u]u+[\omega v]v+\ldots+[\omega^2]\omega+\ldots$$

Then by considering that the coeff[ts] are altered with ref[ce] to the fixed axes, with the motion of the body, & expressing them in terms of coeff^ts for axes fixed in the body, & geom^l elements specifying the position of the body, the equations of motion under the action of continuous force may be investigated.

By choice of axes in the body (3 angular elem^s, & three elements to fix the origin of coordin^s) six of the coeff^ts may be made to disappear from the quadratic function, which may thus be reduced to the form

$$[u^2]u^2+[v^2]\langle[v^2]\rangle v^2+[w^2]w^2+[\omega^2]\omega^2+[\zeta^2]\zeta^2+[\sigma^2]\sigma^2+2[vw]vw+\ldots$$
$$+2[\zeta\sigma]\zeta\sigma+\ldots+2[\omega u]\omega u+2[\zeta v]\zeta v+2[\sigma w]\sigma w.$$

If there be nothing left handed or ⟨left⟩ right handed on the whole, in the body, the last three terms will disappear. But if the body be composed of fragments of screws rigidly connected, we may have the last three terms, with equal coeff^ts, and the six coeff^ts $[vw]$ &c, $[\zeta\sigma]$ &c may vanish. The mutual action of motes having this quality would, if the number of them within a cubic wavelength be not infinite, give the rotation of the plane of polarization which syrup, oil of turpentine, &c present in nature.

Some of the simplest applications of the theory are very interesting: for instance to the ⟨mote⟩ [?] case of a circular disc or oblate spheroid, moving ⟨solely in a m⟩ in a perfect [liquid] with no motion of the solid perpendicular to a certain meridian plane. What I am most anxious to make out however is the mutual action of motes, separated by a perfect liquid. A molecular theory on this foundation would include the generation of heat in liquids, by stirring, or in solids by electric conduction: Magnetic force & Faraday's rotation of the plane of polarization under magnetic influence, as most elementary deductions. It might include a great deal more, such as the elasticity of gases & solids, chemical affinity, thermoelectricity, &c &c.

As to Faraday's magneto-optic experiment, I think my argument that it must depend on a peculiar state of motion induced by magnetic influence (Proceedings R. S. June or July 1856)[2] is unanswerable. Have you considered it? I can now give a much more apposite mechanical illustration than that of the rotating Blackburn's pendulum.[3] Suppose there to be an oblate or prolate spheroid rotating about a diameter, or an ellipsoid rotating about any one of its principal axes, in a perfect liquid, and suppose[4] its centre to have a position of stable equil^m, under the action of any kind of external influence. The problem of finding the inf^y small vibr^ns following a disturbance in a plane perp^r to the axis of rotation has, if the angular velocity of rotation be kept uniform, the same solution as that which I gave for the rotating pendulum, & the constants may be determined (I feel confident) in terms of the ordinary elliptic transcendents used for expressing the attr^n of an ellipsoid.

Rankine gave, I believe first, the idea that double refr^n may be due to the difference of an equivalent for inertia, according to the line of motion, of a non spherical solid vibrating in a fluid. To include the explanation of Faraday's discovery all that is necessary is to suppose the vibrating bodies to have rotations, independently of the undulatory movement, & these rotations to be set with their axes on the whole along the lines of propag^{n} of the light.

It seems like old times for me to be writing you so long a letter, and I am afraid you will be less disposed to be so bored. Your redress simply be not to read it.

With best wishes for a 'merry Christmas' of which there can be no doubt now, I remain

Yours always truly
WILLIAM THOMSON

1 These and all the brackets through the paragraph ending *&c present in nature* are Kelvin's.
2 Kelvin (94).
3 See Thompson, I, 52.
4 Kelvin has written above the line and deleted, '⟨it to be⟩ [?]'.

166 STOKES to KELVIN, 12 February 1858
Kelvin Collection, S391

69 Albert Street Regent's P^{k}
London N.W. Feb 12, 1858

MY DEAR THOMSON,

I have been so very busy of late that your letter has remained for a long time unanswered. I now set to to answer it, though I have still got plenty of work before me. And first for Dec. 21.[1]

Aesculine is what has been hitherto considered *the* substance to which the fluorescence of horse chestnut bark solution is due, but I have discovered that the bark contains a second fluorescent principle, also crystallizable, which I call Paviine.[2] I have found a new and easy as well as cheap mode of obtaining crystallized aesculine. I mean to try the method in the preparation of analogous substances, but I have no leisure at present for such things. In fact I must in great measure bid good bye to all original investigations.

Without having a decided opinion either way I have always inclined to the belief that the the [*sic*] motion of a perfect incompressible liquid, primitively at rest, about a solid w^{h} continually progressed, was unstable. I was well aware (and I have mentioned it at art 4. of my paper On some cases of fluid motion Camb. Phil. Trans. Vol 8 Part 1)[3] that the motion (under the usual assumptions as to continuity) was determinate, and the same as would be produced impulsively if the solid and fluid were ⟨at first⟩ at rest in the actual positions, and the solid then suddenly got its actual motion. But this is on the *assumption* of continuity, and I have always been rather inclined to believe that surfaces of discontinuity would be formed in the fluid, i.e. surfaces in passing across which the velocity resolved in a direction tangential to the surface would alter abruptly. On this supposition the solid would experience a resistance of the ordinary kind which is known to exist, and the work of the resistance, if the solid were moved through any space and then stopped, would be measured of course by the vis viva of the solid's wake. This motion, i.e. that of the wake, would go on indefinitely and non-molecularly. Fresh surfaces of discontinuity might be produced however. If now friction, ever so slight, started into existence, the finite change of velocity in passing infinitely little across a surface would be destroyed, eddies would be formed, and the motion would gradually pass from hydrodynamical to molecular (heat).

In short if μ be the index of friction, and the work of the resistance as the solid moves through a given large space be denoted by $f(\mu)$, I am disposed to think

(1) That $f(\mu)$ does not vanish with μ

(2) That the unique motion w^h^ would be determined mathematically, whether for small values of μ or for $\mu = 0$, on the *assumption* of continuity is not the actual motion, the latter being of the unstable kind.

(3) That the actual motion is of the stable kind if μ be *great* enough, and the solid's motion be not changed too abruptly, nor its velocity made too great.

I need hardly say that if $\mu \neq 0$ the discontinuity involved in instability cannot be such as to allow an abrupt change of vel^y^ in passing infinitely little across a surface.

I believe Rankine was the first to publish the idea that double refraction might be due to different inertia in different directions. The idea occurred to myself a long time ago. When engaged in the investigations which led to my paper On some cases of Fluid motion w^h^ was read in 1843, I could not fail to be struck with the analogy between the expression

$$M_1 = M \cos^2 \alpha + M' \cos^2 \beta + M'' \cos^2 \gamma$$

(art 5) and Fresnel's equation

$$v^2 = a^2 \cos^2 \alpha + b^2 \cos^2 \beta + c^2 \cos^2 \gamma$$

and I was very much tempted to account for double refraction by considering the ether as a continuous medium bathing a system of nonisotropic solids. Since we have only to deal with relative motions in considering the resultant pressure of the fluid on the solid, it signifies not whether the solid move to and fro, the fluid being otherwise at rest, or the fluid move to and fro past the solid. We have thus a different inertia in different directions. But in following out this view I saw that it would not lead to the expression for determining the wave velocities

$$v^2 = a^2 \cos^2\alpha + b^2 \cos^2\beta + c^2 \cos^2\gamma$$

but to

$$\frac{1}{v^2} = \frac{\cos^2\alpha}{a^2} + \frac{\cos^2\beta}{b^2} + \frac{\cos^2\gamma}{c^2}$$

The wave surface for Iceland spar would be not the sphere and spheroid of Huygens but the sphere and a surface gen^{d} by the revolution of an oval having merely the same principal axes as the ellipse. And on calculating the difference between the results of the two theories at an angle of 45° to the axis I found it so large that I did not feel myself justified in publishing expressions which were in contradiction to those already received as expressing the facts of observation without first carefully scrutinizing the measures of Woollaston [i.e. Wollaston] and Malus, especially the latter, to make out whether the difference could fairly be attributed to errors of observation. Or I should rather say lay within the fair unavoidable errors of observation. I have long had in view a mode of experimental investigation in w^{h} the errors of observation in the case of Iceland spar would not amount to more than perhaps the $\frac{1}{100}$th part of the discrepancy between the two theories but I know not when I shall have leisure to carry it out.

In speculating a good while ago (in fact ⟨shortly⟩ no great time after Faraday's discovery) as to the cause of magnetic rotation I naturally tried rotations of the luminiferous ether as suggested by Ampère's theory but found that the proper law for the rotation as a function of λ would not thus come out. The law shows that the effect depends on diff. cos of the *third* order in the eq^{ns} of motion, and I investigated what *must* be the form of the terms. In fact I arrived at the equations of motion but I seemed farther than ever from a mechanical theory. It was different with syrop of sugar &c. for w^{h} the eq^{ns} lead most naturally to a plausible mechanical theory. I certainly am by no means clear that magnetic rotation must be due to motions going on independently of luminiferous vibrations.

Your paper[4] on ice is down for next Thursday.

Yours very truly
G. G. Stokes

P.S. I have investigated the properties of the wave-surface (w^{h} is of the 16th degree) corresponding to the $\frac{1}{v^2} = \frac{\cos^2\alpha}{a^2}$ &c. theory. For this surface as well as for Fresnel's the curves of plane contact are circles. I have also a theory of double absorption to publish if I live to have leisure.

1 Letter 164.

2 See Stokes (65).
3 Stokes (13).
4 Kelvin (105).

167 STOKES to KELVIN, 13 February 1858
Kelvin Collection, S392

69 Albert St Regent's Park
London N.W. Feb 13, 1858

MY DEAR THOMSON,

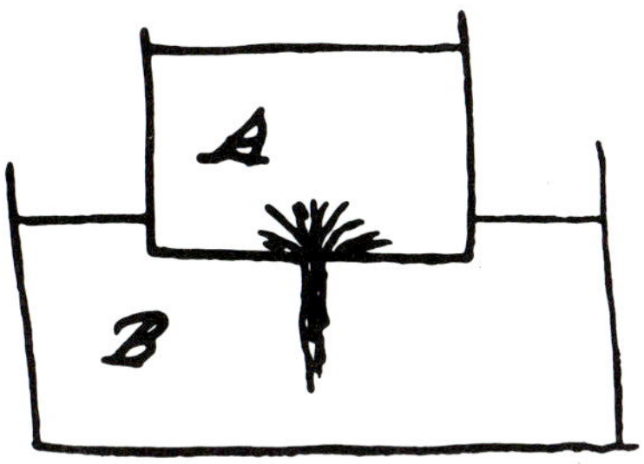

I may as well mention a case of fluid motion which leads me to think that the motion of a perfect and incompressible fluid about a continually progressing solid is unstable. Let fluid flow from a vessel *A* to a vessel *B* through a small orifice. If you like you may suppose the two masses of fluid infinite and separated by an infinite plane with a small orifice, or orifice not necessarily small, but suppose it small to fix ideas. The motion on the two sides of this plane is very different. In the mass *A from* which the fluid flows the motion is insensible except near the orifice, towards which the fluid flows from all sides. In the mass *B* the issuing fluid instead of spreading out tends to keep to a stream of its own. This follows readily from known principles, as I have shown in my paper On the Steady Motion of incompressible fluids Camb. Phil. Trans. Vol 7.[1] I am in the habit of showing it in my lectures by colouring the upper fluid. Now unquestionably the solution obtained from the eq^{ns} of motion (supposing the motion to begin from rest, or ⟨to be⟩ being of the same character i.e. ⟨to have⟩ having $udx + [vdy + wdz]$ an exact differential, to be steady) would be symmetrical with respect to the orifice, and the fluid when it got through would spread out. The function ϕ of course would answer to temperature in an analogous problem of heat. Now why does an actual fluid not spread out but keep to a stream of its own? Not I believe in consequence of friction, though friction does exist and modify the motion, but because the spreading-out motion w^{h} would result from the equations on the assumption of continuity is unstable. But when a sphere is moved onwards in a fluid (or the fluid flows past it which comes to the same) the motion behind the broadest part (in a dir^n $\perp$ to the motion) of the sphere w^h would follow from the equations is of the spreading out kind and is unstable.

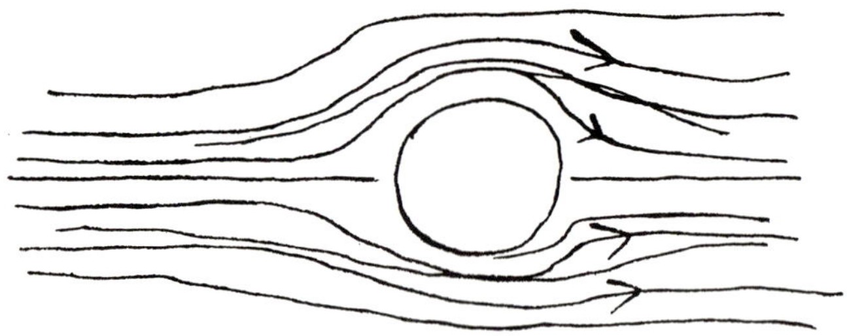

Problems relating to the motion of a perfect incompressible fluid of the kind for w^h $udx + [vdy + wdz]$ is an exact differential differ from the corresponding problems in ⟨tem⟩ heat in this respect that ϕ w^h answers to the temperature represents nothing physical, but only its diff. cos., whereas p which does represent something physical has no analogue in heat. Thus the problem of the two infinite masses of fluid, supposing for simplicity that no forces (like gravity) act and that the pressure in the two masses at a distance from the orifice has two given values, is not the same as that of the flow of heat in two infinite masses separated by a plane w^h except at one place is impervious to heat. In the former case if we suppose the fluid to be initially at rest (we may suppose that at first the pressure in the two masses is the same and that then suddenly a pressure is let on to the first mass w^h at a distance from the orifice has a given value w^h remains constant) the solution I feel sure would be that the velocity would go on continually being accelerated, the solution, that is, obtained on the *assumption* of continuity.

In what I say of instability I have in view only a fluid moving in mass and treated as continuous. The motion of a medium penetrating intramolecular spaces might be very different.

Yours very truly
G. G. STOKES

1 Stokes (1).

168 STOKES to KELVIN, [1 April 1858]
Kelvin Collection, S393

[The letter is missing, only the envelope surviving. According to the postmarks, the letter was posted in London on 1 April and arrived in Glasgow on 2 April 1858. See letter 173, note 1.]

169 Stokes to Kelvin, 5 October 1858
Kelvin Collection, S394

The Athenaeum
Oct 5/58

My dear Thomson,

I called on Darker this evening, and not finding him at home I have left the mirror with his son and I have just written to him on the subject. When the mirror reached me it had the form of two planes inclined at an obtuse angle perhaps 170° or 165° but each a good plane as far as the naked eye enabled me to judge. I was uncertain whether this was a defect or from design, thinking it might be to save the loss of an observation when the magnet swung so far as to throw the image out of the field of view, but putting everything together I supposed it was a break which took place at the file. Also the nature and use of the little piece of rat tail file was to me a mystery. To stick on to the magnet by mere attraction? I did not dare to suggest any doubts to Darker lest he should put off the making: if you can get him to do it by the first intention that's your best chance. So I boldly ordered 6 mirrors of about the size and weight of the pattern and with the little piece of rat tail file, taking for granted they were to be plane and not composed of two planes.

It is a great pity to see the cable in its present state after apparently so successful a laying down. Still the thing has been done and even if this should be utterly lost the matter will not I presume rest there.[1]

I did not go to Leeds this meeting.[2] On the morning of the 27th my wife was safely delivered of a fine boy.[3] She is going on very well but I am afraid her complete recovery will be slow.

Yours very truly
G. G. Stokes

1 For the state of the Atlantic cable in October 1858, see Thompson, I, 365–82.
2 The British Association met at Leeds in 1858.
3 Arthur Romney Stokes (1858–1916), who took his B.A. at Cambridge in 1881 with no particular distinction and went on to become assistant master at Shrewsbury School from 1889 to 1916.

170 Stokes to Kelvin, 12 February 1859
Glasgow University Library, Kelvin Papers, S81

Lensfield Cottage Cambridge
Feb 12th 1859

My dear Thomson,

At the last meeting of the Govt Grant Committee it was recommended, and the Council afterwards sanctioned the recommendation, to assign to you out of the grant the £100 you applied for. I was directed to point out that it was somewhat irregular to apply the grant to the defraying of past expenses but it was decided that you might in the present case so apply a part of it as you had mentioned.

I have another iron in the fire now: I have just been appointed an additional secretary of the Cambridge University Commission.[1] Their powers expire at the end of the year and they have a great deal of work before them. I hope the Commission may turn out beneficial to the University.

Yours very truly
G. G. STOKES

1 See D. A. Winstanley, *Early Victorian Cambridge* (Cambridge, 1955), 287–8 and 314–72, respectively, for a description of the powers and an account of the activities of the Statutory Commission, which began work in 1856.

171 KELVIN to STOKES, 1 March 1859
Glasgow University Library, Kelvin Papers, S81

2 College, Glasgow
⟨Feb⟩ March 1, 1859

MY DEAR STOKES

I am sorry that my application to the Government Grant Committee was not in proper order and I beg to express my thanks ⟨to the Committee⟩ for the manner in which it has been received.

I may explain ⟨that⟩, with reference to the experiments on conductivity, that I was led to continue them as belonging strictly to the subject for which I had applied to the Committee in May 185 [*sic*] in consequence of the extraordinary results indicated by an examination of specimens for the Atlantic Telegraph Company ⟨and⟩ but that I deferred making any farther application for assistance out of the Government Grant until I saw my way to prosecute this and other strictly scientific investigations in a more continued manner than my engagements during the past year permitted.

With reference to the old standing accounts due to Mr Robert Bryson of Edinburgh,[1] including the amount due for a valuable balance, which I have now had in use ⟨in gasious experiments of some of which the results have been already communicated to the Royal Society,⟩ for ⟨gases⟩ [?] several years I have ⟨applied hitherto⟩, after repeated efforts, entirely failed to get a statement of it. About a fortnight before I wrote to you last I intimated to Mr Bryson that unless he sent in his account immediately I could no longer hold myself accountable [?], but I have had no reply. Until I can see him, I do not know what more I can do, as writing seems to have no effect, but I am anxious to get the matter settled, and I shall take the first opportunity of being in Edinburgh to have it finally disposed of.

Yours &c
WT

1 Robert Bryson and Sons, chronometer, watch and clock makers in ordinary to the Queen.

172 Stokes to Kelvin, 28 April 1859

Kelvin Collection, S395

Lensfield Cottage Cambridge
April 28th 1859

My dear Thomson,

I am very much obliged to you for kindly undertaking to be ready for the 12th with Earnshaw's paper.[1]

No candidates for medals have yet been named. The 12th is the first day for naming them. I was not aware that Tyndall was going to be named. I shall bear in mind what you say.[2] I have not myself given more than a very cursory attention to the subject of glaciers in general, and to Tyndall's doings in particular.

I was very much struck with Forbes's mode of accounting for the re-mending of broken ice. It is a theory on which my mind can rest with perfect complacency.

My ordinary address will now be Lensfield Cottage Cambridge. I have taken a lease of this house for 3 years, renewable for three years more. My lectures will soon be beginning now, so that though I am appointed an additional Secretary of the ⟨Royal Society⟩ Cambridge University Commission[3] I shall be more here than in London.

Yours very truly
G. G. Stokes

1 Samuel Earnshaw, 'On the mathematical theory of sound', *Phil. Trans.* (1860), 133–48. Earnshaw (1805–88) was senior wrangler in 1831 and coached students for the mathematical tripos until 1847 when he became chaplain on Queen Mary's Foundation in the church and parish of Sheffield.

2 The next letter is an extensive account of Kelvin's views. Letter 172 appears to be a response to an earlier letter from Kelvin which has not survived. As it turned out, neither Forbes nor Tyndall received a Royal Medal. Forbes had previously won a Royal Medal in 1843, but Tyndall never received one.

3 See letter 170, note 1.

173 KELVIN to STOKES, 29 April 1859

Glasgow University Library, Kelvin Papers, S82

Partially printed in J. S. Rowlinson, 'The theory of glaciers', *Notes and Records of the Royal Society of London*, *26* (1971), 195–6.

[Copy][1]
Birkenshaw Cottage
Thornliebank, Glasgow
April 29, 1859

MY DEAR STOKES

Since I last wrote to you[2] I have been informed that friends of Professor Forbes intend bringing matters to an issue by reproposing that one of the medals should be conferred on him for his researches on glaciers: it having been proposed by Sir C. Lyell in 1846 [1848?] that the Royal ⟨Society⟩ medal should be given at that time for his papers[3] in the Phil Trans on the subject. Altho' I should feel some right, merely as a fellow of the R.S., to take the more invidious position of protesting, even formally to the Council, against conferring such a mark of approbation on Tyndall, for what he has done on the same subject,[4] because I believe that to do so would be to commit an act of injustice to others, and especially to Forbes, I feel that it is only to you as a private friend that I can unasked, offer any opinion on the positive side, when the proposition in favour of Forbes is under consideration. On this footing however and because I have already written to you regarding Forbes' position, I now say that I believe all unpre⟨d⟩judiced naturalists & men of science generally who know the circumstances would judge that the R.S. had done well if it is decided now to recognize his merits in this particular branch by awarding a medal for what he has done. It surely is a very great thing to have distinctly demonstrated by accurate measurement continued summer and winter for years, that the middle of a glacier moves faster than its sides and bottom, and that the whole mass accommodates itself to the bends and general form of the bed down which it moves. No one denies that he has done so, and after this is done, it is merely a matter of words whether 'the Viscous Theory' is to be accepted as a correct designation for the aggregate of truth which he has demonstrated. Even as regards the name 'Viscous' which has been found so hard of acceptance, Forbes has taken a very sure and philosophical position: and his illustrations with mortar, Plaster of Paris, and Stockholm pitch, shows that he leaves it an open ⟨to⟩ question for farther investigations as to the physical properties of ice, to discover whether it is that the glacier moves on the whole and on the large scale like a viscous mass because of yielding & mutual sliding among finite appreciable parts, or by a true molecular viscosity. If there is anything in Tyndall's 'fracture & regelation' as propounded with so much parade, it is merely to account for Forbes' theory in one particular way, and so to confirm it. But how inappreciably little there is in this of Tyndall's in comparison with my brother's physical theory[5] which first, on thermodynamic principles, establishes the lowering of the freezing point by pressure and the lowering of temperature consequent on the application of pressure, and shows exactly how much; and which both explains the freezing together of pieces of ice, to which Faraday had called attention[6] and shows that clear ice at 32° must have the plastic quality without fracture, you will I think admit even if you think Forbes' explanation of the consolidation satisfactory. Where, in all this, is Tyndall's theory & what is it? Some of the glacial writers whom Forbes quoted long ago refer to making a snowball, & to the consolidation of snow in a boy's slide (Bp. Rendu[7] I think, but you will find it all in Forbes' book[8] republished a few weeks ago from papers which were written & published long before Tyndall was heard of.) What more can be said as to the making of snow into compact ice in a glacier & the healing up of fractured ice below a cascade than 'witness a boy's slide?' or at least what more did Tyndall say in his 1856 paper?[9] Faraday's lecture on the subject ⟨which⟩ was to melting analogous to Black on boiling water. A well known old truth was brought up and a wonderful property of matter elicited from it. Tyndall did well to to [*sic*] refer to Faraday about that, but was it not a most impudent imposture to pretend that in doing so he had over thrown Forbes' viscous theory of glaciers? ⟨But⟩ This supposed destruction of an old theory and establishment of a new one, only lasted out the interval from the R[oyal] I[nstitution] lecture, & abstract in the Proc. R.S., to the appearance of his paper in the Phil. Trans.[10] when his case against the Viscous Theory had quite collapsed. In this paper Forbes was referred to on *that point fairly*, ⟨altho' a pretence against the viscous theory was still slightly kept up against the Viscous Theory⟩ in substance, (after for a year the public had been led to suppose that his theory was over thrown)[11] and Forbes' own viscous model brought forward as an illustration. Tyndall's experiment on the moulding of ice in the Brahmah's press was good & well worthy of being communicated to the R.S. If its author had had the common candour, honesty and good feeling to show the relation of what he had done and wished to enforce, to what Forbes had done, there would (as I think I have shown) be no one more ready than myself ⟨(and [?] as I think I have shown)⟩ nor, I believe, than Forbes (as he has shown in the ⟨republis⟩ preface to the reprint of his papers) to give him ample credit for all he ⟨was due⟩ could claim. But besides the nefarious attempt to place himself in the position of the first true expounder of the motion of a glacier there was in that paper[12] of Tyndall & Huxley's a *most improper* polemic on the veined structure. That this was admitted at all for publication was I am convinced an unintentional oversight on the part of the Council and its referees who (from what I have heard) I believe took considerable pains to remove unfair or objectionable attacks against Forbes from the paper before allowing it to ⟨be printed⟩ appear. Tyndall in this case neither for the first nor for the last time adopted his peculiar policy of raising a controversy on a false issue, taking the credit of a victory where the field had been evacuated before he appeared at all, and leaving the general impression that he had conquered the whole ground for himself.

With reference to the veined structure, I am quite ready to admit that Tyndall's pressure experiments are valuable illustrations and that in this, as in the question of viscosity, he deserves great credit for showing in the laboratory the same kind of actions as take place on a large scale in glaciers. But the seemingly incurable obliquity of the man comes out on this point too. After I have forced him to see that the infiltration theory was not held by Forbes for years before Tyndall came on the ⟨ground⟩ field, he again ⟨takes⟩ ⟨took⟩ takes up a controversy on a false issue. In his last R.I. lecture[13] he flatters a Mr Ball[14] by *attacking* and *disproving* his theory that the blue & white veins are the remains of stratification of the snow as it originally fell! Ball is delighted (I don't mean so much with the flattery, as with the genuine observations Tyndall describes) & writes to the Phil Mag,[15] admitting &^{c} &^{c} – at the same time saying that there is more to be done. So Tyndall is paraded as again the establisher of a truth: and what truth? – *That the veined structure is induced in the snow after it falls, and quite irrespectively of the stratification of its original deposit: and that it really originates in the dynamic action accompanying the motion of the glacier.* Now what is more palpable than that all this was thoroughly enforced by Forbes? It is truly a *fine observation* and one well worthy of credit, and of the admiration Ball expresses for it, that which Tyndall made on the ?Furrge glacier, where he saw the blue veins crossing the stratification of the fallen snow. But *in every case*, according to Forbes, the blue veins must cross the original stratification, and it is really a beautiful confirmation of this theory to find in one place, as T. did, the appearance of the stratif^{n} remaining unobliterated. Where, and what now is Tyndall's theory on the blue veins? If he knew the math^{l} theory of elasticity of

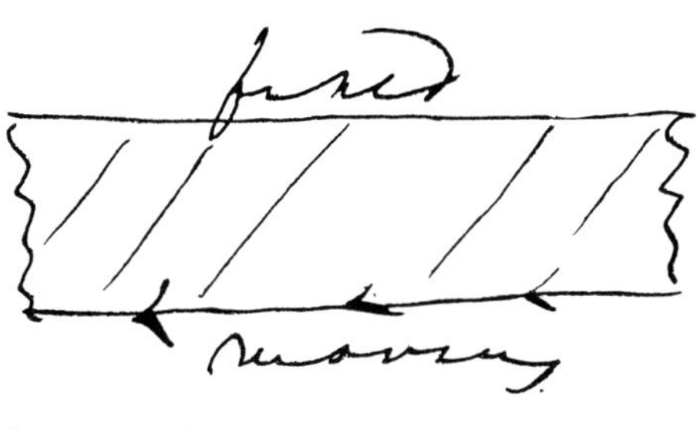

solids at all he would see (perhaps he does see) that the surfaces of greatest pressure would be as indicated in the diagram, in a piece of clear ice acted on as indicated, and that the blue veins would, if he is right, begin to appear in such planes, but that as the motion is continued they would get laid more and more

nearly into positions such as those shown in the second sketch. Tyndall has never distinctly pointed out the⟨se⟩ relations of difference of pressure in different directions, to differential motion (deformation) neither has Forbes I believe (nor in fact so far as I know has any one but Hopkins[16] in the glacial questions.) If both or either had done so and if both instead of only one had been candid, I believe there would have been no difference of opinion. Tyndall's stratif^{n} naturally turns through 45° or very nearly so till it lies as Forbes finds it. Forbes might have traced his surfaces of tearing backwards historically, and found them nearly at right angles to the lines of maximum pressure. I look upon neither Forbes' statement, nor Tyndall's, either one or other considered pointedly, as complete, and neither as more correct than the other but both partly right. Tyndall quite overlooks that differential motion is a necessary consequence of difference of pressure in different directions[17] that a large amount of differential motion must actually take place before the

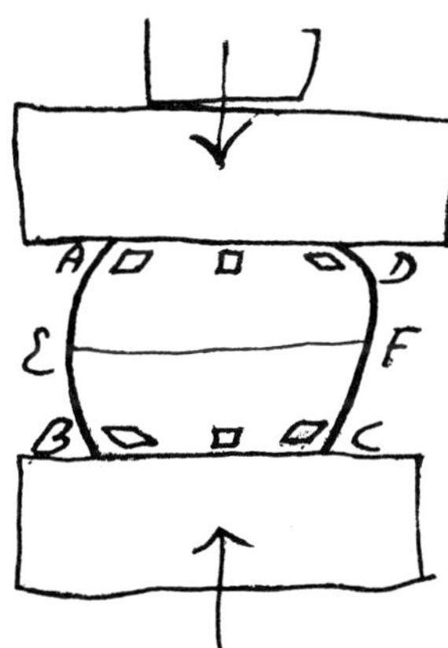

veined structure is induced in snow. In a glacier this is experienced by the varying mass in the same manner as it is experienced by the parts of block of ice in a press, in the positions *A*, *B*, *C*, *D*. According to both Forbes & Tyndall there ought to be lamination in these positions. But Forbes' view would not apply at all to lamination in the plane *EF*. That there would be lamination in this plane would, I believe from my mechanical theory, be found to be the case, and I suppose we are to understand that Tyndall has actually observed such. I believe therefore that Tyndall's statement & experiment, when fully interpreted and worked out, and my thermodynamic explanation of the phenomena, give a physical theory which comprehends, and to some extend [*sic*] rectifies Forbes' theory of the blue veins. But that Tyndall has gone so far as to distinguish between what is right and what is not completely satisfactory in Forbes' view, is certainly not the case, nor is it even true that he has given any correct or full definition of what I am crediting him by calling his own statement, by which I mean what a person who understands the math^{l} theory of elasticity would take up as the only interpretation of his statement. Tyndall's 'Theory' as stated by himself, does not get farther than to say that the blue veins are caused by pressure and are perp^{r} to the directions of pressure. (About 1841 Forbes first described them as having the appearance of being perp^{r} to the directions of greatest pressure.) Forbes ten years ago had published as his final theory that they are caused by the *kneading* effects of pressure and of long continued differential motion, and had constantly spoken of the *bruising effect* of the rocky sides, obstacles, &^{c} also had spoken of a ⟨whole⟩ vein as a ⟨'⟩broad 'bruise.' Does Tyndall say it is not a bruise but is an effect of pressure? No – he (& Huxley unfortunately associated with him) proves that the blue vein is not a frozen infiltration of water (which Forbes had never maintained, but had only thrown out as a conjecture which he was glad to abandon) and next, Tyndall (in his last R.I. lecture)[18] says that the only hypotheses worthy of attention are the stratific^{n} hypoth^{s} and the pressure hypoth^{s}: he disp⟨roves⟩oses of the former by observation (correctly) and walks the course as the victorious establisher of the latter. Senarmont,[19] Faraday, and Weber[20] on the several different grounds on which Tyndall has approached them, have all been treated similarly by him. His last move in glacial theory is to appropriate the liquefaction by pressure theory. Of course he 'referred' to my brother & myself but referred in such a way that the reporters in the Saturday Review[21] & the Athenaeum[22] (and if they, how many not, if all the hearers) entirely failed to perceive that this advance (which they treated as most important) in the physical theory was ours not his. In the same way he 'referred' to Faraday as having experimentally proved difference of magnetic inductive capacity in different directions in a crystal, yet scarcely anyone who did not know this before hand could read his paper without the impression that the discovery was his & that he had found Faraday at fault with ref^{ce} to it – in fact that Faraday had failed to establish it but he, T. succeeded. (See Tyndall's own paper. Also Matteucci & my observations in reply Com[p]tes Rendus March or thereabouts 1854 ⟨or th⟩ [?])[23]

As to the glacial theory particularly, I consider however his agression against Forbes as infinitely of more importance than his agression on my brother & myself, tho' nothing could well be more flagrant than the latter, because in the former he tries to put himself into the position which Forbes achieved by the best part of a life of work, carried out with consummate skill, and unremitting perseverance.

I do hope that whatever may be the decision as to the medals, which after all is but a feeble way of asserting the right in such a matter, the Council of the R.S. will set its face against future injustice, & take some way of repairing what I cannot but feel to have been unjust in what has already been done, or omitted or both, by it, in regard to Forbes' great work.

Believe me

Yours very truly
WILLIAM THOMSON

P.S. The viscous theory is of course, as I need scarcely ⟨say to⟩ remind you, quite independent of my special explanation of the precise modus operandi by which the blue veins grow, & rests on the broad foundation of the facts established by Forbes with his theodolites & measurements. It is not a physical hypothesis ⟨not⟩ at all – & there is ample room for physical explanations & investigations on this as on every other theory of natural history.

1 Kelvin's brackets. This is Kelvin's copy of the original. It is probably a copy he sent to Archibald Smith, for on 4 May 1859 he wrote Smith (Kelvin Collection, S171) a long letter on the whole subject and asked him 'to return the copy of my letter to Stokes'.

The previous year Kelvin had published two brief articles on the same topic as this letter in the *Proceedings of the Royal Society* (Kelvin (104) and (105)). They were received on 23 January and 3 April 1858, respectively. Although they were in the form of letters to Stokes, they are not reproduced here.

2 See the previous letter, note 2.

3 J. D. Forbes, 'Illustrations of the viscous theory of glacier-motion: – Part I. Containing experiments on the flow of plastic bodies, and observations on the phenomena of lava streams', *Phil. Trans.* (1846), 143–56; 'Illustrations of the viscous theory of glacier-motion: – Part II. An attempt to establish by observation the plasticity of glacier-ice', *ibid.*, 157–76; and 'Illustrations of the viscous theory of glacier-motion: – Part III. On the motion of glaciers of the second order. On the annual motion of glaciers, and on the influence of seasons. Summary of the evidence adduced in favour of the theory', *ibid.*, 177–210.

4 Between 1857 and 1859, Tyndall published several papers on glaciers, the earliest being a lecture delivered on 23 January 1857, 'Observations on glaciers', *Proc. Roy. Inst.*, **2** (1854–8), 320–7.

5 See J. Thomson, 'Theoretical considerations on the effect of pressure in lowering the freezing point of water', *Trans. Roy. Soc. Edinb.*, **16** (1849), 575–80, and 'On the plasticity of ice, as manifested in glaciers', *Proc. Roy. Soc.*, **8** (1856–7), 455–8. See also his 'On recent theories and experiments regarding ice at or near its melting point', *Proc. Roy. Soc.*, **10** (1859–60), 152–60.

6 Faraday's lecture on the subject at the Royal Institution on 7 June 1850 was reported in *The Athenaeum* (1850), 640.

7 L'Abbé Rendu, 'Théorie des glaciers de la Savoie', *Mémoires de la Société Académique de*

Savoie, **10** (1841), 1–158, and 'Théorie sur les glaciers en général', *Bulletin de la Société Géologique de France*, **1** (1843–4), 631–6. Louis Rendu (1789–1858) became bishop of Annecy in Savoy in 1843.

8 J. D. Forbes, *Occasional Papers on the Theory of Glaciers* (Edinburgh, 1859).

9 So far as I have found, Tyndall published none in 1856 on glaciers, so perhaps Kelvin meant the paper cited in note 4.

10 The sequence noted by Kelvin can be seen in the following papers by Tyndall: 'Observations on glaciers', *Proc. Roy. Inst.*, **2** (1854–58), 320–7; 'Observations on the Mer de Glace', *Proc. Roy. Soc.*, **9** (1857–9), 245–7, and 'On the physical phenomena of glaciers', *ibid.*, 668–70; and 'On the physical phenomena of glaciers. Part 1. Observations on the Mer-de-Glace', *Phil. Trans.* (1859), 261–307.

11 Kelvin has added a footnote here reading, 'For instance the Abbé Moigno headed an article "M. Tyndall, Glaciers. Théorie de M. Forbes renversé", or exactly to that effect'. See the review of Tyndall's work in 'Théorie des phénomènes de structure observés sur les glaciers', *Cosmos*, **10** (1857), 246–52, which states 'Nous nous rallions complétement à cette bonne et belle synthèse de M. Tyndall . . . .' (P. 252.)

12 J. Tyndall and T. H. Huxley, 'On the structure and motion of glaciers', *Phil. Trans.* (1857), 327–46. Thomas Henry Huxley (1825–95) was, of course, associated with Tyndall in more than glaciers – agnosticism and the politics of science, for example. Also like Tyndall, he did scientific battle with Kelvin, in his case, over the age of the earth. Again like Tyndall, despite his differences with Kelvin, Huxley got along without animosity with Stokes, with whom he was fellow secretary of the Royal Society from 1872 to 1881.

13 J. Tyndall, 'On the veined structure of glaciers', *Proc. Roy. Inst.*, **3** (1858–62), 72–8, delivered 4 March 1859.

14 John Ball, 'On the structure of glaciers', *Phil. Mag.*, **14** (1857), 481–504, is the article criticized by Tyndall. Ball (1818–89) also published a few articles on botany and geology and became a fellow of the Royal Society in 1868.

15 J. Ball, 'On the veined structure of glaciers', *Phil. Mag.*, **17** (April 1859), 263–8.

16 W. Hopkins, 'On the motion of glaciers', *Trans. Camb. Phil. Soc.*, **8** (1849), 50–74, 159–69. See also his several papers in the *Philosophical Magazine* for 1845 and his much later 'On the theory of the motion of glaciers', *Phil. Trans.* (1862), 677–745.

17 Kelvin has placed an asterisk in the text here indicating he intended to add a footnote, but there is no footnote.

18 See note 13.

19 Tyndall cited Sénarmont's work on the influence of crystalline structure upon the transmission of heat in his 'On molecular influences. – Part 1. Transmission of heat through organic structures', *Phil. Trans.* (1853), 217–31. For Sénarmont's papers, see letter 74, note 2.

20 For the Tyndall–Faraday–Weber–Kelvin controversy, see Tyndall's 'On the nature of the force by which bodies are repelled from the poles of a magnet; to which is prefixed an account of some experiments on molecular influences', *Phil. Trans.* (1855), 1–51, and also the numerous contributions by the four in the two volumes of the *Philosophical Magazine* for 1855. Wilhelm Eduard Weber (1804–91) was professor of physics at Göttingen for most of his career with an interval at Leipzig from 1843 to 1849.

21 The anonymous article, 'Structure of glaciers', in *The Saturday Review of Politics, Literature, Science, and Art* (19 March 1859), 333–5, summarized and praised Tyndall's lecture at the Royal Institution on 4 March 1859 (see note 13).

22 The anonymous review of Forbes's book (see note 8) in *The Athenaeum* (9 April 1859), 478–80, stated, for example: 'Even though Prof. Forbes should survive his own theory, even though he should be an unwilling attendant at its funeral obsequies, yet, should it be

consigned to that vast vault where so many lifeless theories repose in inglorious obscurity, he may take his stand upon the sepulchral stone, and exclaim, "*Non omnis moriar*"'.

23 Kelvin (70), dated 22 March 1854, discusses Tyndall in commenting on three memoirs published by Matteucci in *Comptes Rendus*, **36** (1853), 740–4. Carlo Matteucci (1811–68) was professor of physics at the University of Pisa.

174 KELVIN to STOKES, 11 May 1859

Stokes Collection, K104

Birkenshaw Cottage
Thornliebank, Glasgow
May 11, 1859.

MY DEAR STOKES

Although I have not yet had time to enter so fully on the very difficult subject of Mr Earnshaw's paper[1] as to allow me to perceive thoroughly the bearing⟨s⟩ of every part of it on the theory of sound, I have now seen enough to conclude very decidedly that it should not be published in the Transactions in its present form. In the first place, it is a very unsatisfactory beginning that the author 'found reason for believing that the equation $\frac{dy}{dt}=F\left(\frac{dy}{dx}\right)$ must always be satisfied'. As no indication of the author's reason is given, the reader is left to investigate for himself the meaning of this condition and to find whether it is or is not an arbitrary restriction. As the whole treatment is founded on it, a feeling of suspicion is raised at the beginning as to the general character of the processes and results, which is not relieved by any explanations contained in the paper. Unless therefore the reader can work out the whole theory for himself independently, he can scarcely derive much light on the subject from the paper as it stands. At the same time he can scarcely read it without receiving some striking suggestions, and perceiving that something important lies under the not very intelligible, and the very frequently erroneous statements and formulae which it contains. But on the other hand, on looking to Poisson's paper 'Journal de l'Ecole Polytechnique Vol VII',[2] he finds that an apparently much less incomplete solution of the equation is given there: and that exactly the same solution as that which Mr Earnshaw ⟨deems⟩ considers a novel solution if 'an equation hitherto deemed altogether intractable & incapable of integration' is given by the Astronomer Royal, as having been suggested by Professor DeMorgan (Phil. Mag. Jan–June 1849 p 402).[3] It does appear that Mr Earnshaw has made something more of this ⟨equ⟩ solution than the Astronomer Royal thought ⟨it⟩ [?] to be deducible from it: but yet in a form which so far as I have yet been able to examine it, appears too vague and incomplete to be at all satisfactory. I think something really satisfactory may be made out from his equations

$$\zeta\zeta' = \zeta_0^2; \quad w = \sqrt{\mu\epsilon^{-\frac{U}{\sqrt{\mu}}}} \quad \langle\text{wh}\rangle\ [?] \quad ww' = \mu$$

(w & w' being what he estimates as velocities of transmission of densities $\zeta\ \sigma\ \zeta'$) but I think I am not doing the author injustice when I say that this is not done unless the reader does it for himself, with the paper as it stands. I am anxious especially to see the relation between these formulae and yours expressing 'a difficulty in the theory of Sound' Phil Mag July to Dec 1848, p 349 &c especially pp. 354, 355.[4] I think I see my way to completely explain this difficulty[5] and understand [?] all your formulae, but I have not yet been able to make out the bearings of Earnshaw's formulae (some of which are identical with yours).

Again, I have to remark that the most striking part of Mr Earnshaw's paper is anticipated, and in my opinion more reasonably treated, in your own article, and that of the Astronomer Royal (published in 1848 & 1849 respectively) – that namely which has reference to the formation of an 'aerial bore,' and that anything in the same subject) [*sic*] to be now eligible for publication, should throw more light on it, which I do not find is done ⟨by the author⟩ in the present paper.

Lastly, in the present paper, there are many statements and suggestions which are founded on very mistaken physical notions: (such as the whole treatment of the question when will & when will not the development of heat affect aerial motions, which you have so thoroughly dealt with Phil. Mag. . . .) there are most unsound positions from results of formulae applicable (if at all) only to a non viscous fluid in a uniform straight perfectly rigid tube, to draw conclusions as to phenomena 'in aere aperto' (as for instance the doubling of the sound of a cannon & the greater velocy of the sound of a cannon than of a gentler sound): ⟨and⟩ there are adduced very doubtful experimental confirmations of results (for instance of the two last –) which are so much at variance with general observation that they cannot be accepted without a very thorough sifting of the evidence on which they are given: and lastly, ⟨there is⟩ perhaps even more than the [this?] usual amount of fallacy is introduced into the consideration of the supposed velocity of ⟨of⟩ air rushing into a vacuum, and the (in my opinion) certainly false supposition that a ball moving through the air at a sufficy great velocy leaves a perfect vacuum behind it.

Without such a remodelling of the paper as to take acct of what has been done by previous writers in the same direction, to correct many fallacious and erroneous ⟨parts which⟩ processes contained in it as it stands, and to show some more definite & complete definition of what the author's investigation really proves, I think it ought not to be admitted to the Transactions.

I remain

Yours very truly
WILLIAM THOMSON

Professor Stokes

1 See letter 172, note 1.

2 S. D. Poisson, 'Mémoire sur la théorie du son', *Journal de l'École Polytechnique*, 7 (1808), cahier 14, 319–92.

3 G. B. Airy, 'On a difficulty in the problem of sound', *Phil. Mag.*, **34** (1849), 401–5. As Airy indicated on page 402, De Morgan's suggestion was made privately to him. Augustus De Morgan (1806–71) was fourth wrangler in 1827 and professor of mathematics at University College, London, from 1828 to 1831 and 1836 to 1866.

4 Stokes (22).

5 Kelvin has added a footnote here reading, 'See P.S. to my letter of this date addressed to you Cambridge'. This letter is missing.

175 KELVIN to STOKES, 30 July 1859

Stokes Collection, K105

Invercloy, Isle of Arran,
July 30, 1859

MY DEAR STOKES

I am afraid I forgot to write to you about ⟨6⟩ 8 weeks ago, at the request of Dr Nichol,[1] to ask if you can give any reference for your extensions of the spectrum, ⟨or⟩ especially to any published drawing. He wishes to introduce the best information into a new edition of his Cyclopedia[2] w^h he is just bringing out. If you cannot give references to published articles or drawings, any ⟨assistance⟩ written statement or sketch you feel disposed to give for the purpose w^d be thankfully rec^d.

I am chiefly occupied with atmospheric electricity, and am laboriously preparing at a very slow rate, a paper[3] for the R.S. on the subject. I have got a new apparatus set up here, &, after waiting a fortnight for the electrometer, have at last got it in good order. I find an effect indicated this morning by as much as 190° of torsion, under a serene sky – the ordinary earth-negative effect of serene weather. Dr Nichol has had one of my app^s made for the observatory, & I expect will have it in action today. My old app^s is still in action at the College, & observed by my assist.

My last app^s with electrometer costs £17.10 & is adapted for a permanent observatory.

I hope the British Assoc^n will have something done to promote electric observation in various latitudes & localities, & by travellers. Will you attend at Aberdeen?

Yours always truly
W. THOMSON

1 John Pringle Nichol (1804–59), as professor of astronomy at Glasgow from 1836 until his death, was a friend of the Thomson family and an early influence on Kelvin. (See Kelvin (375P).)

2 J. P. Nichol, ed., *A Cyclopaedia of the Physical Sciences* (London, 1857), 2nd edn, revised and enlarged (London and Glasgow, 1860). Kelvin's contributions to the *Cyclopaedia* were (100A), (121A), (121B), and (121C).

3 Kelvin (111) on atmospheric electricity was presented at the 1859 meeting of the British Association, and Kelvin published several other papers in 1859 and 1860 on atmospheric electricity, but none in the publications of the Royal Society.

176 KELVIN to STOKES, 6 October 1859
Stokes Collection, K106

Invercloy
Isle of Arran
by Ardrossan
Oct 6, 1859

MY DEAR STOKES

Why have you not (if you have not) become a candidate for the Professorship of Astronomy, vacant by the death of Dr Nichol which took place during the meeting of the Association? I do not know why it did not occur to me sooner, but I have this morning had a letter from Arch Smith suggesting it.[1] It is not a very tempting situation in point of emolument – £300 a year and a house (very good, & delightfully situated.) If the smallness of the income is not a barrier, I think the situation would suit you in every other respect. Conducting the observatory (and making ⟨physical⟩ electrical observations, physical & optical experiments ad libitum) would be the natural and proper occupation. A short course of lectures in the University every session would be desirable. If students could be got to come, I think you would find it a satisfactory kind of lecturing.

Cayley is thinking on being a candidate, but it will probably be considered, & I believe justly, that he is not physical enough. I say this to you in confidence, knowing that you will not misunderstand what I say of so good a friend of us both. I think Cayley ought to be provided for by the country – mathematician laureate would be his right post – but there is no doubt but that popular or physical lines of science are not in his way, & that in a situation where either may be required, he might not be well placed. Now I wish you most seriously to consider whether you will not apply for, and therefore I trust take, the Glasgow Professorship, & resign your Cambridge Prof^{p} in favour of Cayley? I am assuming of course that both as regards income, and scientific occupation the Glasgow situation is at least equally eligible for yourself.

Not knowing your present address, I send this to Armagh as giving the best chance of finding you soon. I intended, had time before the post permitted, to have taken the liberty of writing to Dr Robinson, to enlist if possible his influence in favour of my proposition. I have however instead marked that this is to be read by him if you are not with him when it comes, and I trust he will excuse my doing so, and will kindly consider our Glasgow interest in the question, as well as the importance to science of getting you out of London & Cambridge, those great Juggernauts under which so much potential energy for original investigation is crushed. There is now no difficulty as to tests,[2] the recent act having thrown open the Professorships to all denominations. You should *immediately* give in your application to the Home secretary, if you decide on being [a] candidate.

Believe me
Yours always truly
W THOMSON

P.S. I have had a letter from Helmhol[t]z,[3] asking about your investigation of hydrodynamical equations with fluid friction taken into acc^{t}. I can only give him the reference to the Camb. Trans. for the first,[4] and I forget whether your second[5] with numerical results for water and air, is Camb. or London. Will you let me know *date* & ref^{ce} for each? If you have any spare copies of either or both, they could not possibly be better bestowed than on Helmhol[t]z. He writes to me that he has investigated the equations, & I find his results w^{h} he gives in his letter, are identical with yours (so far as I recollect). Some one had told him that you had done the thing, but he could not find your paper anywhere. His address is Prof. Helmhol[t]z Heidelberg.

1 The letter has not survived. Archibald Smith (1813–72) was senior wrangler in 1836 and was called to the bar in 1841. Like Kelvin, he was from Glasgow and attended Glasgow University before entering Cambridge. He was one of the founders of the *Cambridge Mathematical Journal*, became a fellow of the Royal Society in 1856, and received a Royal Medal in 1865. See Kelvin (198A).

2 See letters 34 to 43, concerning Stokes's application for the Glasgow professorship of mathematics in 1849.

3 H. von Helmholtz to Kelvin, 30 August 1859, Kelvin Collection, H65.

4 Stokes (11).

5 Stokes (47).

177 KELVIN to STOKES, 6 October 1859
Stokes Collection, K107

Invercloy, Isle of Arran
by Ardrossan,
Oct 6, 1859

DEAR STOKES

I have sent a letter[1] to Dr Robinson for you as the best chance of finding you quickly but as I wish to lose no time, I send a line also to your optical residence, to try & persuade you to apply for the Glasgow Professorship of Astronomy vacated by the death of Dr Nichol about three weeks ago. The income is only £300 a year. There is an excellent official house (in the observatory, delightfully situated[)]. Assuming that this is at least equal to your Cambridge position *materially*, physically I am sure you will find it in every way eligible. Conducting & directing Astronomical observation, physical & optical exp^s, & scientific research generally, are the primary duties. A short course of lectures in the University every session would also be desirable. Cayley has some idea of being a candidate but, if you would make up your mind to take the situation, w^h I suppose would imply (altho' possibly not) resigning your Cambridge professorship I think this last would be a much better thing for Cayley, who is so much more devoted to mathematics, than to any kind of physical observation. I think Cayley ought to be provided for by the country in some highly honorable scientific position, but I doubt whether the work of an observatory would suit him. I say this to you in confidence, knowing you will not misunderstand it, or think it inconsistent with the friendship I feel for Cayley.

I do hope you will decide to become a candidate for the situation. There are now no tests you are aware, according to the recent act, & only a declaration, which Christians of all denominations may cheerfully make.

I think it would be a great thing for science to get you to such a place as the Glasgow observatory; & I would feel it an immense advantage to myself to have you so near.

I have written to Dr Robinson or yourself, at Armagh hoping to enlist him in favour of my proposition. Do if possible accede to it, & *immediately* make your application to the Home Secretary. I think to apply for & to take the situation ought in your case to be synonymous.

Yours always truly
W THOMSON

1 The previous letter.

178 KELVIN to STOKES, 13 December 1859
Stokes Collection, K108

2 College, Glasgow
Dec 13, 1859

MY DEAR STOKES

I enclose an application which I make as directed by the British Association, to the Gov^{t} Grant Committee, for funds to construct apparatus for observing atmospheric electricity in the manner which I explained at the Aberdeen meeting.[1] I shall be obliged by your laying it before the Committee at its first meeting. I shall have an independent application to make on my own acc^{t}, for funds to enable me to prosecute certain experimental researches in my laboratory; but I delay in the mean time to allow me [to] make up my accounts, & also, to write out a statement of some results obtained with the assistance of the sum allowed last year.

Yours very truly
WILLIAM THOMSON

P.S. Have you seen Le Verrier's researches[2] on Mercury's motion according to w^{h} he is led to conclude the existence of sensible influence from small bodies circulating inside his orbit? This I said was to be looked for, a long time ago, (Addition No IV. to my paper 'on the Mechanical Energies of the Solar System'.)[3] It is a most important confirmation of my theory of the Sun's Heat.

1 Kelvin (111).
2 U. J. J. Le Verrier, 'Sur la théorie de Mercure et sur le mouvement du périhélie de cette planète', *Comptes Rendus*, **49** (1859), 379–83.
3 Kelvin (73) in *MPP*, II, 23–5. See also Kelvin (118).

179 KELVIN to STOKES, 21 December 1859
Stokes Collection, K109

2 College, Glasgow
Dec 2⟨0⟩1, 1859

MY DEAR STOKES

Having heard that Dr Matthiessen is to communicate some of his experiments on electric conductivity tomorrow evening to the Royal Society,[1] I send you the accompanying Table of results, which altho' obtained nearly all, as they stand, two years ago, I have kept back until I should have an opportunity of going through them again to correct an error which had somehow occurred with reference to absolute measures and which affected them all in the same ratio. It is only a few days since I have completed this (3^{d}) series of determinations, & the results which I now send cannot be far wrong.

I have partially prepared a description of the methods followed, with drawings of apparatus, & I hope very soon to be able to communicate it: but as the results are of more importance than the methods, of which the principles are perfectly well understood by most experimenters, the publication of the Table in the Proceedings need not I think be delayed.

Excuse extreme haste in preparing the abstract. If it will not do I shall rewrite it for the Proceedings.[2]

Hofmann's report[3] is necessarily delayed, as I cannot find it to send it by this post.

Yours very truly
W Thomson

1 A. Matthiessen, 'On the electric conducting power of alloys', *Phil. Trans.* (1860), 161–76. Augustus Matthiessen (1831–70) took his Ph.D. at Giessen in 1853 and was lecturer in chemistry at St Mary's Hospital Medical School from 1862 to 1868.
2 Kelvin (113).
3 A. W. Hofmann's report to Kelvin, dated 10 March 1858, is included in *ibid.* in *MPP*, II, 118–21. August Wilhelm von Hofmann (1818–92) was professor at the Royal College of Chemistry.

180 Kelvin to Stokes, 25 January 1860
Stokes Collection, K110

2 College, Glasgow
Jan 25, 1860

My dear Stokes

I send by this post the paper[1] on the electrostatic forces required to produce sparks in air ⟨to⟩ which I promised in my last letter to you. I hope it and its predecessor[2] will not be considered too long for the Proceedings. I have still a great deal to do on both subjects, and hope some time to have a more complete paper, suitable for the Transactions. The results I now send might however I think be published in the mean time. I shall have to try the effects of curvature of the opposed surfaces with a view if possible to explaining the strange conclusion that ⟨less⟩ more electrostatic [stress] is supported by ⟨air boun⟩ a thin stratum of air than by a thick one: also to test the effects of changes of temperature, pressure, & moisture, on the strength of the insulating power; and I should like to have made some progress in such farther investigations before preparing a paper for the Transactions.[3]

I am adding a 10th and 11th page to show a summary of results in a Table, and I shall have to change the diagram of curves somewhat to make it fit copy for the engravers; because the results of the different series turn out to agree *too well* to be distinctly represented by different curves exactly in the way set forth in the text. This I shall do without delay & send to you, so that if it is desired to put the copy into printer hands at once, you may reckon on it complete next week. I have also to send you Hofmann's report on analysis,[4] for my previous article on electric conductivity.

I send the present communication because altho' not quite complete, it may perhaps be admitted for tomorrow evenings meeting, & at all events it will show the Grant Committee that I am getting some decided results.

I enclose, also for the Grant Comm[ee], a letter from M[r] Sandeman[5] expressing his willingness to make electric observations. I have assurance to the same effect from Prof. Everett.[6]

Yours very truly
W THOMSON

1 Kelvin (115).
2 Kelvin (113) or (114).
3 Kelvin did not publish such a paper.
4 See the previous letter, note 3.
5 Patrick Sandeman to Kelvin, 19 October 1859, Stokes Collection, S191. Sandeman wrote the letter from Glasgow but was on his way to Demerara, where he proposed to make the observations.
6 Joseph David Everett (1831–1904) studied with Kelvin at Glasgow in the 1850s and was professor of natural philosophy at Queen's College, Belfast, from 1867 to 1897. From 1859 to 1864 he was professor at King's College in Windsor, Nova Scotia.

181 KELVIN to STOKES, 7 February 1860
Stokes Collection, K111

2 College, Glasgow,
Feb 7, 1860

MY DEAR STOKES

I am much obliged to you for your letter informing me of the resolutions adopted by the Council regarding my applications on the part of the British Association, and for assistance to my own experimental investigations, out of the Government fund.

If the latter is to cause any difficulty such as you tell me has been felt by the Council regarding it, it must be simply withdrawn, as nothing could induce me to accept of any assistance unless the council is perfectly satisfied with the plan I have followed, and propose to follow still, in my experimental investigations.

You tell me it was objected to my application 'that it seemed rather for carrying on the general experimental work of my laboratory, than for some specific object involving considerable outlay'. – This is a most erroneous impression, and is quite inconsistent with the terms of my application, in which are proposed several specific objects of research involving, some of them, very considerable expense; and constituting on the whole a course of experimental investigation naturally following that of previous years, and not to be effectively carried on for one year more, without an expenditure much exceeding the sum for which I have applied. The '*general* experimental work of my laboratory' includes not only endeavours to investigate new truth, for which alone, as the terms of my application clearly show, I have asked assistance from the Royal Society. It involves as the primary and essential work the preparation of illustrations for my lectures during the winter six months. I have besides instituted a system of experimental exercise for laboratory pupils in which I am induced to persever; devoting a great deal of time to it, and a larger expenditure out of my private resources than I feel to be altogether consistent with other claims. During the winter session I have about 20 such laboratory pupils, (all volunteers and paying no laboratory fee;) and I generally find several among them efficient for original investigation, to whom I give ⟨some of the⟩ work under some of the heads specified in my applications to the Grant Committee. I keep my laboratory open during nearly the whole of the summer six months, exclusively for the prosecution of our[?] final investigations. My permanent assistant, whom I engage during the whole year at my own private expense, is in the summer months solely occupied in assisting me in such investigations, and in carrying them on according to my directions when I am not on the spot. I have also repeatedly employed one of my laboratory volunteers who had proved to be efficient, to act as a second assistant during the summer months, and under the sanction of the Committee, made him a small payment (in each case £20 for a period of five months,) out of the sums granted to me.

I enter upon these particulars to make it perfectly clear that my application is not, 'for carrying on the general experimental work of my laboratory', but for enabling me to do some thing more in original investigation than I could accomplish without such assistance.

I remain
Yours very truly
WILLIAM THOMSON

182 KELVIN to STOKES, 8 February 1860
Stokes Collection, K112

2 College, Glasgow
Feb 8, 1860

MY DEAR STOKES

By this post I send a paper on the insulating power of gutta percha which, at my suggestion, Mr F. Jenkin has prepared for the Royal Society.[1] I shall write to him for an abstract, for the Proceedings, which I suppose should be sent to you direct. If you have any other direc^{ns} to give as to [the] abstract, or can say at what time you want it, I shall be obliged by a line.

I think you will see that my own paper[2] for the Proceedings, re Daniell's battery, is complete, in form. It was the spark paper[3] which wanted one curve. ⟨on⟩ Before I had your letter I was going to write to you that the curves may be withdrawn, so far as publication in the Proceedings is concerned. The tables represent the results suff^{y}, & it will perhaps be better to defer the publication of curves until I have pushed the investigation farther.

You have by this time received a copy of Hofmann's report.[4] ⟨The⟩ I had it from him a long time ago & had mislaid my copy. He told me at the time that he would like to see it before publication. In consequence of what you told me I did not trouble him by writing to him about it just now, but perhaps you will not object to let him see a proof which I have no doubt will allow him to make any change (which if any I believe was very slight, probably only verbal) he may wish to make.

I hope to send you both Earnshaw's[5] & Donkin's[6] paper before next week's meeting. If I do not hear from you to the contrary I shall address them to you at Burlington House. Let me know if you would wish to have Earnshaw's paper sooner, & if so I shall try to get it off on Sat^{y} or Monday.

Yours very truly
W THOMSON

1 F. Jenkin, 'On the insulating properties of gutta percha', *Proc. Roy. Soc.*, 10 (1859–60), 409–15. Henry Charles Fleeming Jenkin (1833–85) and Kelvin worked together a great deal, especially on underwater telegraph cables. Jenkin became professor of engineering at University College, London, in 1865 and at the University of Edinburgh in 1868. There are numerous letters from him to Kelvin in the Kelvin Collection and in the Kelvin Papers in Glasgow University Library.
2 Kelvin (114).
3 Kelvin (115).
4 See letter 179, note 3.
5 See letter 172, note 1.
6 W. F. Donkin, 'On the analytical theory of the attraction of solids bounded by surfaces of a class including the ellipsoid', *Phil. Trans.* (1860), 1–11.

183 KELVIN to STOKES, 15 February 1860
Stokes Collection, K113

2 College, Glasgow
Feb 15, 1860

MY DEAR STOKES

I return Earnshaw's paper[1] to you by this post.

Many of the objectionable passages are removed, but still I cannot feel satisfied with it. The 'appendix' containing professedly a demonstration of the fundamental assumption

$$\frac{dy}{dt}=F\left(\frac{dy}{dx}\right)$$

involves so far as I can see no other principle that [i.e., than] that of what is commonly called the 'equation of continuity', and which in this case is merely $\zeta=\frac{\zeta_0}{\frac{dy}{dx}}$, and so far as I can see nothing but this equ^n should be obtainable. I should be glad to hear what you find as to this.

The supposed relation $\frac{dy}{dt}=F\left(\frac{dy}{dx}\right)$ is certainly not, as Earnshaw seems to consider it, essential to motion of air in a tube. Take the common integral of what (he however will not admit to be) the approximate equ^n

$$\frac{d^2y}{dt^2}=\alpha^2\frac{d^2y}{dx^2} \quad \text{viz.} \quad y=f(x-\alpha t)+F(x+\alpha t)$$

Then

$$\frac{dy}{d\langle t\rangle x}=f'+F' \quad \frac{dy}{dt}=\alpha(f'-F')$$

and $\frac{dy}{dt}$ is decidedly not a fu^n of $\frac{dy}{dx}$. It is quite clear that rigorous solutions may exist not fulfilling the supposed law. For instance if two pistons be gradually moved towards one another, the velocity at the middle part of the air between them will always be zero, while the density there will gradually rise.

Surely it is taking the name of nature in vain to invoke her for ⟨mending⟩ strengthening the solution of a partial differential equ^n threatened with rupture (p. 9). If there were any latent or subtle property of matter concerned, or if some organic act were the subject the statement might be appropriate, but I think it is not satisfactory in a purely mathematical investigation. I have not yet managed to make myself so thoroughly acquainted with either the formulae of the paper, or those of Poisson &^{c} to which you referred me, to feel quite clear as to what exactly Earnshaw's ⟨pap⟩ formulae do express, or what relation they bear to the others. Earnshaws $\zeta_1\,\zeta_2=\zeta_0^2$ (p 5) & the interpretation he gives seem real & substantial but is not the same as that of yours we discussed last May? I would be unwilling to advise the rejection of Earnshaw's paper while not quite sure that it may not contain something of importance distinct from what previous writers have done. I do not think however that it could at all appear as it stands. It must be made much more distinct what is really expressed by the solutions, & how far they may be extended to express consequences of arbitrary initial conditions. There are besides still some very decided error[s], besides what I have referred to above & which I cannot but think is a fundamental error. For instance (p 13, No 25) ' – general property of motion through a tube – that a gas cannot be conveyed faster than, $\sqrt{\quad}$', suppose a portion of gas were enclosed betw. two pistons & these kept at const. dist. and pushed along at more than that rate; would the gas betw. them refuse to keep between them? Again (same page) ' – may be nothing more than a hole through a part[itio]n &^{c}'. The motion through a thin part^{n} is of course not in straight lines, & is not subj^{t} to the same set of equ^{ns} at all. There are many other unsatisfactory expressions &^{c} through the paper. I really feel at a loss what to re[c]ommend about it. I must whether with ref^{ce} to ⟨a⟩ the question of its appearance in the Trans^{ns} or not, learn the subject thoroughly; and if I could do any more with ref^{ce} to your decision, I sh^{d} be glad to have the paper again from you with your remarks.

Donkin's paper[2] will be sent by this evening's post, & should reach you a few hours after this ⟨comes to⟩ is delivered tomorrow. I may not be able to get a report written out, but if not I shall send one with very little delay.

Yours very truly
W THOMSON

1 See letter 172, note 1.
2 See the previous letter, note 6.

184 KELVIN to STOKES, 15 February 1860
Stokes Collection, K114

2 College
Glasgow 15th Feby 60

MY DEAR STOKES

I send Donkin's paper[1] with my report[2] upon it, which will be followed in a few days, by the extension of investigation referred to; and probably some additional remarks which I am obliged for the present to delay. I have a heavy mass of proof sheets of evidence[3] for the Government Telegraph Commission on hands which I must despatch first, but I hope to be able to let you have the proposed supplemental statement on or before Thursday of next week when I presume you will again be in London.

Yours very truly
W THOMSON

1 See letter 182, note 6.
2 Kelvin's report on Donkin's paper is in the Royal Society, RR.4.68.
3 See Kelvin's evidence, given 17 December 1859, in the '*Report of the Joint Committee*

Appointed by the Lords of the Committee of Privy Council for Trade and the Atlantic Telegraph Company to Inquire into the Construction of Submarine Telegraph Cables; together with minutes of evidence and appendix', Parliamentary Papers (1860), vol. LXII, pp. 744–61.

185 KELVIN to STOKES, 14 March 1860
Stokes Collection, K115

2 College, Glasgow
March 14, 1860

MY DEAR STOKES

I am much obliged to you for your letter regarding my paper on the conductivity of alloys.[1] I do not know whether Matthiessen had the paper itself in his hands; but I think if he had, he would have seen that I had guarded against the supposition that I believed their stated composition could be relied on. Indeed I have pointed out that the electrotype copper from which the two sets were prepared gave considerably different results; and I have all along been fully alive to the kind of doubt w^h Matthiessen expresses to you.[2] If I have not suffic^y stated this in the paper (which was very hurriedly finished to have it in time if possible to be read in the same evening with Matthiessen's) I shall take care to make such an additional note as shall not leave it possible for any mistaken impression to be produced.

I may mention that in a previous communication[3] to the R.S. I had proved that mechanical differences cannot (as have sometimes been supposed) be the cause of the diff^ces in conductivity, and that I therefore am ready to conclude that there is a chemical difference whenever I find a considerable diff^ce in conductivity.

The unsatisfactory character of my trials with alloys (w^h the title of the paper indicates) I spoke of in reply to the Gov^t Telegraph Committee last August or September, & I stated that to arrive at trustworthy results the alloys must be prepared under the supervision of a scientific chemist who would be responsible for their purity. I had been asked to continue the investigation for the Committee, & I said that I would be willing to do so if such guaranteed alloys were sent to me; but at the same time I mentioned Matthiessen as probably able to take up the subject with the requisite chemical guarantee.

Later, I was called to give evidence in London, & I expressed the same in substance so far as my own investigations were concerned, & in reply to Wheatstone asking me could not reduction of oxyde by the lead &c, give rise to the improvement, I replied that very possibly it might.

Perhaps a short extract from that evidence, which is now printed, may be not not [sic] *mal a propos* as an addition to my paper.[4]

From your letter I am relying on having a proof of the spark paper[5] sent me by Dr Sharpey; but to make sure I shall write & ask him for a proof of this one also.

Yours very truly
WILLIAM THOMSON

1 Kelvin (113).
2 Although no report by Matthiessen on Kelvin (113) survives in the Royal Society, there is a later letter from him giving 'my reasons for assuming that his [Kelvin's] experiment[s] are not quite right: (that is to say if they are the same as I have already seen viz where he states that some alloys of copper conduct better than pure copper)'. (Matthiessen to Stokes, 6 September 1860, Stokes Collection, M393.)
3 Kelvin (96).
4 See letter 184, note 3. Wheatstone was a member of the committee on submarine telegraphs. See his questions 2464 and 2465 and Kelvin's answers to them.
5 Kelvin (115).

186 KELVIN to STOKES, 28 March 1860
Stokes Collection, K116

2 College, Glasgow
March 2⟨1⟩8, 1860

MY DEAR STOKES

I have had great difficulty in making up my mind what to advise about Earnshaw's paper.[1] Since there is something decided in the way of solution in it I am reluctant to advise its rejection, & yet as it stands it would inevitably cause confusion, without I fear giving much enlightenment in return. If your advice as to the omission were followed the objections would no doubt be removed if not wholly, at least so as not to preponderate so much as they do over good qualities. Now however as you tell me he seems not disposed to do what you recommend I feel in a greater difficulty than ever.

I was much obliged to you for your article in the relation of solutions;[2] and have been only prevented by too much pressure of occupations from writing to you regarding it & returning Earnshaw's paper before. I shall make a great effort to go through the whole once more immediately and write to you again, with the least possible delay. I have several days holidays here beginning Wednesday (this day week) & I could then get my mind applied. Let me know however by return of post if you would wish Earnshaw's paper sooner than say the end of next week; & if so I shall send you it instantly.

Believe me

Yours very truly
W THOMSON

1 See letter 172, note 1.
2 Apparently, Stokes (67).

187 KELVIN to STOKES, 30 April 1860
Stokes Collection, K117

Birkenshaw
April 30, 1860

MY DEAR STOKES

Earnshaw's paper[1] is lying in my house in College, and I have not it by me therefore to refer to at present, but I shall send it to you by same post with this in the afternoon.

I am sorry to have given you the trouble of again writing to me about it, and I wish you to know that I have not been neglecting it and that my delay has been entirely from a difficulty in making up my mind. I have had the subject continually before me, & yet do not feel perfectly satisfied as to the true state of the case. I have however little or no doubt left but that the paper will not do for the Royal Society. Besides the objections on which you have written to the author which he certainly does not meet in a satisfactory way in his replies;[2] it appears to me that §17 is fatal. In that article it is admitted that the equations are not applicable to what the author calls 'discontinuity' (an abrupt – how abrupt? – a ⟨transition⟩ variation of density. ⟨)⟩ If they are applicable at all they must be applicable to *very* sudden and in the limit, by proper interpretation to infinitely sudden variations.) Supposing ⟨this⟩ ⟨however the matter is⟩ this to be a true limitation of equations which are true until the limit is reached, he says the limit cannot be reached because it would be disagr[e]eable to nature who therefore commences sending offshoots from the wave which violate the equations *before* the limit is reached. I wish I had the paper by me to read that & consider it once more; but I believe I am not doing ⟨the paper⟩ injustice when I say that it makes the equations be both satisfied & violated in one & the same condition of the fluid – satisfied as throughout they are supposed to be, rigorously – violated because the conclusion is abhorrent to – nature – i.e. the author's residual common sense. If I could not understand anything of the mathematics of the subject I would simply infer that there was a fundamental error, which the author has been forced to discover, because of one of the conclusions to which his theory leads.

There is another conclusion not less abhorrent to nature than the supposed but not admitted 'bore' – the absolute vacuum left behind a piston moved at 5000 feet per second. (See second part in which changes of temp^{re} are taken into acc^{t}.) The impossibility of this result does not seem to strike the author, otherwise he would have to begin breaking down his theory on that side too, in time to avoid it.

I still think it will be found that the fallacy of the whole lies in the unproved assumption stated on the first page. (The appendix is as I think you will admit ⟨)⟩ no proof at all.) No doubt the equation he finds is a solution but does it really fulfil the conditions of air in a pipe originally at rest, & set into motion by the arbitrary motion of a piston? If the author would give a single worked out case, for instance ⟨a⟩ the piston commencing with a uniformly accelerated motion, & continuing so forever, then it would be seen when ⟨&⟩ how and where his equations begin to fail if in reality they ever hold. As to myself I am not satisfied, because I have not yet done this. I think it probable that there will be found to be ⟨continual⟩ continuous violation⟨s⟩ of his fundamental principle by the continual reflection backwards from every portion of the wave, of effects which in the approximate theory would consist of a superimposed ⟨waves⟩ variation propagated backwards so that if, at any time after the piston commences moving, the fluid were suddenly to assume such a law of pressure as would make it fulfil the approximate equations rigorously, it would be found in such a state as to condensation & velocity as would make both terms of the solution

$$F(x-\alpha t)+f(x+\alpha t)$$

be required to express its subsequent motion.

Although I have not yet been able to work this out to my own satisfaction, it still appears to me that a paper which leaves such an investigation necessary for explaining mutually contradicting passages, is not suitable for publication in the Transactions. If it is published it will remain for the reader to find out what if anything is true of it, & for some other if not the same author, in a new communication, to give a self-consistent theory. The notice which I suppose has appeared in the Proceedings must, I should think contain all that the author ⟨can⟩ has deduced from his equations besides the result that 'the transmission velocity is superimposed on the wind velocity'. This conclusion which is after all the chief substance of the first part of the paper, & is that which leads to the *bore*, is still more obvious from Poisson's equation ⟨$u=F(x-t-f\alpha)$⟩ [?] which, if ⟨it⟩ I remember right, is $u=f(x-(\alpha+u)t)$. However this may be, Earnshaw's solution is as you have shown virtually the same as Poisson's. The want of essential novelty may therefore be fairly considered as a sufficient reason for not accepting Earnshaw's unless, which certainly appears to me to be not the case, it had some decided superiority in point of form or of clearly & correctly worked out conclusions.

I shall be in London for nearly a week before the 18th on wh day I am to lecture on Atmospheric Elecy to the Royal Institution.[3] I hope I shall have some opportunity of seeing you. Shall you be in town for an R.S. meeting on the 17th?

Yours very truly
WILLIAM THOMSON

1 See letter 172, note 1.
2 See Earnshaw to Stokes, 5, 7, and 20 March 1860, Stokes Collection, RS162, E8, and E9. He

began the second letter: 'I have examined the articles of my paper, to which you refer, again; & am utterly at a loss to conceive what objections there can be to them'.

3 Kelvin (108), delivered on 18 May 1860.

188 KELVIN to STOKES, 7 May 1860
Stokes Collection, K118

2 College, Glasgow
May 7, 1860

MY DEAR STOKES

I shall be very glad indeed to be relieved from the difficulty about Earnshaw's paper[1] by the appointment of a 3[d] referee, should the council consent to this course. I certainly would wish to be of the same opinion with you on the subject, but if, after all, you are still satisfied with the validity of the main investigation I ⟨certainly⟩ should not wish to stop the publication of the paper even altho' I could see quite clearly through my present difficulties.

I have read your letter carefully, and I believe it answers my objection to the result as to a finite velocity for air following a piston, which therefore may be valid as a rigorous consequence of the ⟨equation⟩ law $\zeta \propto p^{1/k}$ supposed rigorous.

As to the novelty of Earnshaw's solution, I perhaps did not sufficiently distinguish its applicability to the law $p \propto \zeta^k$. This seems to follow so obviously from the plan by w[h] he finds the solution for the hypothesis $p \propto \zeta$, that I was in the habit of considering the equation as solved for one hypothesis when solved for the other. Perhaps taking the question up in Poisson's way (with which I am unacquainted) would not make this so obvious and it may be that I have taken advantage of Earnshaw's method against himself. I have been very anxious to avoid underestimating what he has done, & from what you say I have perhaps not succeeded.

I always recur however to a feeling of dissatisfaction with the main investigation as applicable in the manner supposed, and from what I wrote to you before, I did not think this would surprise you. I cannot feel quite convinced yet as to the reality of the working up to a bore. I do not feel absolutely convinced that this is impossible, but I am less nearly convinced by Earnshaw's investigation that it is real. Is there not 'a screw loose' as to the initial condition of the whole mass of the fluid. This must be such that, with the prescribed motion of the piston, the *particular* solution shall apply to the subsequent motion. How is it known that with the fluid ⟨will be⟩ everywhere at rest ⟨if⟩ & the piston commenc⟨es⟩ing ⟨at⟩ from rest,[2] the particular solution will be applicable? May not some particular initial distribution of density or of velocity be required to ensure that the one sided propagation shall go on after the piston commences moving. Earnshaw & you & I have all virtually the same idea it appears, as to reflecting away from the bore – backwards. But does not this reflection, as I suggested in my last letter, commence at the very beginning of the motion and from every part of the air? I think it probable that it really does so, unless some very ⟨special⟩ peculiar (? finite & possible) initial distribution of density & velocity be given. With reference to this you say 'In this case it seems evident and indeed I think Earnshaw's reasoning suffices to show, that until $t = 0$ the fluid will be at rest.' Of course my objection falls, (& the bore ⟨must⟩ becomes inevitable I suppose) if this is the case, but is it made out?

Supposing Earnshaw's investigation & application of it to be really valid, it would be an inconceivably great addition to the merit of his paper to show the thorough working out of any one chosen case. Do you think this could not be done? We should then see exactly how the crisis is reached, & what the finite motion of the fluid immediately preceding it would be.

Yours very truly
WILLIAM THOMSON

P.S. As to the *novelty* of Earnshaw's solution, you will see that I referred to it as insufficient *only if the main application was wrong*. Supposing the main application to be unsatisfactory, a very decided novelty in the form of [a] solution might of course still be a sufficient merit for a place in the Transactions.

1 See letter 172, note 1. The Royal Society has reports only from G. B. Airy and Kelvin (RR.4.70, 71).

2 It appears that, before Kelvin's modifications, this passage read, 'How is it known that the fluid will be everywhere at rest if the piston commences at rest'.

189 KELVIN to STOKES, 8 May 1860
Stokes Collection, K119

Birkenshaw Cottage
Thornliebank
Glasgow
May ⟨7⟩8, 1860

DEAR STOKES

I enclose a very short notice[1] for the 'Proceedings,' of a continuation of the experiments on thermodynamic⟨s⟩ properties of gases which Joule and I commenced together some years ago, and which we hope to be able now to push farther.

I intend to be in London all next week. Before & after, till the end of June, my present address as above will be the best to find me without delay.

I hope to see you in London.

Yours very truly
WILLIAM THOMSON

1 Kelvin and Joule (7).

190 KELVIN to STOKES, 20 June 1860
Royal Society of London, RR.4.71

Oakfield[1]
Moss Side
Manchester
June 20, 1860

MY DEAR STOKES

By this post I send our paper[2] of w^h I wrote to you.

On the whole I think if called on to vote, it would be against the publication of Earnshaw's paper.[3] I should (as I could not otherwise be called on to vote) be in that case fortified by *one* of the referees. You may judge that if the two others are in favour of the paper I shall not protest.

I still think there must be a reflection at every point ⟨of⟩ where there is unequal density from point to point, & that this will destroy Earnshaw's solution so far as applied to express consequences of a single piston.

His solution for the true physical property (altering temperature, according to pressure) will surely express the superposition of transmission velocity according to the altered temperature, on wind velocity: & he in this respect quite analogous to Poisson's.

On speaking to Rankine I found *the* idea he had taken from Earnshaw's paper, as represented by abstracts, was superposition of transm^n vel. on wind vel.: & he thought it good. This however is of course fully expressed in Poisson's solution.

Yours very truly
W THOMSON

1 This is Joule's address. (See letter 161.)
2 Kelvin and Joule (8).
3 See letter 172, note 1.

191 Kelvin to Stokes, 2 July 1860
Stokes Collection, K120

Manchester July 2, 1860

My dear Stokes

I should be much obliged by your putting me down as a subscriber to the Hopkins testimonial fund for £10.10, which I believe is the amount subscribed by yourself and others. I understand that you are on the committee, and therefore write to you on the subject. If not will you kindly forward this to Liveing[1] or some other member of [the] committee? I had a circular, with a note appended by Liveing, sometime ago, but had to leave it at the moment intending to bring it with me here and answer it. Somehow it has fallen aside, &, after writing to have it looked for in Glasgow without succeeding in getting it, I am at a loss how to address for Liveing, and am not even sure that I saw your name on the committee. It is on this account that my reply is so late & that I now trouble you with it.

I hope you have had a good meeting at Oxford.[2] I have been very busy here with Joule, & we have got through a number of good experiments on Carbonic acid, Oxygen, Nitrogen, & common air. I leave for Arran tomorrow, & my address during the summer will be

Invercloy
Isle of Arran
by Ardrossan

Yours very truly
William Thomson

1 George Downing Liveing (1827–1924) was eleventh wrangler in 1850 and placed in the first class of the natural sciences tripos in 1851. He was professor of chemistry at Cambridge from 1861 to 1908.

2 The British Association met at Oxford in 1860.

192 Kelvin to Stokes, 9 November 1860
Stokes Collection, K121

2 College
Glasgow
Nov 9, 1860

My dear Stokes

I have read through Hopkins' paper[1] with much interest and only deferred giving in my report because I was not sure that Dulong & Petit's formula, with a term for convection, is as applicable as is supposed in the paper. It appeared to me at first, that the convection in Hopkins' calorimeter must be much less than in the metal globe employed by Dulong & Petit; & that very probably its effect might be insensible in comparison with that of radiation. Then I saw that the results of the exp^s do not in fact agree very closely with D. & P.'s formula: and I have kept the paper to test the agreement of the term for pure radiation, with the results. This will require a considerable amount of calculation, but I hope to have it done in a few days. In the mean time I shall make the requisite extracts & send you the paper at once, as you may desire to put it into some other person's hands: unless I hear from you to the contrary. If it would cause you no inconvenience whatever for me to keep the paper a fortnight from this time I should be much obliged by a single line to say so as this would save me the trouble of making extracts.

I did not see Bowditch's paper,[2] but was requested by the author to allow it to be communicated in my name. From what you say I should think the publication of it in the Proceedings must be satisfactory. Mr Bowditch I believe intends some time becoming a candidate for admission to the R.S. as a fellow.

I remain
Yours truly
W THOMSON

1 W. Hopkins, 'On the construction of a new calorimeter for determining the radiating powers of surfaces, and its application to the surfaces of various mineral substances', *Phil. Trans.* (1860), 379–408. Kelvin's report on Hopkins's paper is in the Royal Society, RR.4.136.

2 W. R. Bowditch, 'On coal gas', *Proc. Roy. Soc.*, 11 (1860–2), 25–40. William Renwick Bowditch (d. 1884) took his B.A. at Cambridge in 1843 and was vicar of St Andrew's, Wakefield, Yorkshire, from 1845 to 1884. He did not become a fellow of the Royal Society.

193 KELVIN to STOKES, 17 December 1860
Royal Society of London, RR.4.137

2 College, Glasgow[1]
Dec 17, 1860

MY DEAR STOKES

I think Hopkins' paper might be published as it is, with a caution (if my remarks are assented to) as to the doubt that exists how much of the effect is due to radiation proper & how much to aerial convection & conduction. Perhaps also ⟨some modification of⟩ a change in, or omission of the statement which occurs somewhere early in the paper, to the effect that the agreement with Dulong & Petit's formula is so good as to confirm the law which this formula represents, ought to be made.

I believe that with Hopkins' actual apparatus excellent absolute results might be obtained by experimenting on polished silver & on lampblack, separately (as I stated in my report.) If this opinion is considered correct, some indication that such experiments might or may be made, & that those already made are published in the meantime provisionally, might perhaps be of use to the reader.

I am sorry to hear of 'the accident' you refer to, of w^h I had not learned before.[2] I hope it is not of a serious character, & I sh^d be much obliged by a line from you informing me.

Yours very truly
WILLIAM THOMSON

1 Stokes has written at the top of the letter, 'In answer to a request for an explicit recommendation as the report only *implied* a $recom^n$'. For references to Hopkins's paper and Kelvin's report, see the previous letter, note 1.
2 The accident involved Hopkins. (See letter 196.)

194 KELVIN to STOKES, 16 January 1861
Stokes Collection, K122

Auchinean Largs,
Jan 16, 1861

MY DEAR STOKES

I am much obliged to you for your letter of the 11th which has been forwarded to me here. If all is well in another year I hope to have some experimenting in hand, & if so I sh^d probably apply to the Royal Society for assistance. At present however I am stranded here on my back, in a rigorously horizontal ⟨cond⟩ position, in consequence of a fall w^h I met with in 'curling' on the ice, on the Saturday before Christmas, by which the neck of my left thigh bone was fractured within the joint. The fall, so far as whole vis viva is concerned, was by no means severe, but it was so nicely adjusted that as nearly as may be the whole momentum was ⟨as nearly⟩ stopped by a single blow on the apex of the trochanter major, & hence the injury. I experienced not the slightest shock or strain either on my shoulder or leg. It is a favo[u]rite theory of the doctors that there is never 'bony union' after a fracture of this kind. The truth seems to be however that complete recovery is only a little less frequent than the fracture itself, in persons under 60. I wonder what Dr Sharpey would say as to this. Syme,[1] of Edinburgh, says I shall be as well as ever; but then he did not say he was sure it was a fracture, only that it must be treated as a fracture.

It will be 5 or 6 weeks yet before I shall be able to return to Glasgow, & go on with my lectures again. Even then it is not likely I shall be able for any experimental or other extra work. Just now I am feeling perfectly well & beginning to get good enough sleep at night. The constant uniformity of my position prevents me from getting any refreshment & makes any thing like head work to be avoided. Also gravity, not withstanding capillary attraction, prevents the use of ink. You will therefore excuse this ink roll.

Yours very truly
WILLIAM THOMSON

1 James Syme (1799–1870) was professor of clinical surgery at the University of Edinburgh from 1833 to 1869.

195 KELVIN to STOKES, 28 January 1861
Stokes Collection, K123

Auchinean Largs
Ayrshire
Jan 28, 1861

MY DEAR STOKES

Matthiessen with whom I have had a good deal of co⟨nversation⟩rrespondence & interchange of specimens for measurement as to conductivity has sent me a paper[1] on the subject, explaining how it was that certain lead–copper alloys ⟨sent⟩ prepared for me by Matthey & Johnson[2] conducted better than their electrotype copper. He proposes to send this to the R.S. for the Proceedings, to be communicated by me. If you receive it from him will you therefore let it appear as communicated by me? He proposed first that we should make a joint communication, but as I had only a very small share of the work (some determinations of conductivity) & as the explanation is his, we have agreed that the plan now proposed is better.

I am still in the same ⟨place⟩ position as when I wrote last, and having been getting on as well as could be expected – keeping in my good general health. At the end of next week the splint is to come off & then I shall be allowed to try to move.

Yours very truly
WILLIAM THOMSON

1 A. Matthiessen, 'On the electric conducting power of copper and its alloys', *Proc. Roy. Soc.*, 11 (1860–2), 126–30, *Phil. Trans.* (1862), 1–27.
2 Johnson and Matthey were assayers in London. See Kelvin (113).

196 KELVIN to STOKES, 14 February 1861
Stokes Collection, K124

Auchinean, Largs
Ayrshire Feb 14, 1861

DEAR STOKES

I have no objection to my report being put into Hopkins' hands.[1] If you see him will you express my sympathy for him about his accident? I hope he has quite recovered from its effects now.

I am in a fair way of recovery. The splint was removed 6 days ago & the doctor is satisfied with the position of the limb and with the degree of motion (which Joule's train of levers for showing the elongation of bars by magnetism, would render visible to the naked eye) of which it is capable. I am myself rather disappointed at the slowness of my progress but I am assured that quicker could not have been expected.

I quite agree with you about the observ^y.[2] I think it would be a great evil to saddle any of the math prof^ps with it. It is certainly much to be desired that the University should have an astronomical professorship, or directorship of the observ^y whether with or without the professorial title. I am afraid astron^l observ^ies are on the whole rather ineffective at present for want of some comprehensive system of arranging work. What is the use for instance of all the volumes of Cambridge & of Edinb^ro observ^ns? Possibly a great deal, but uninformed as I am I have doubts. If a division of useful work for the Cambridge Observ^y were arranged ⟨in connection⟩ by agreement with with [*sic*] the Astr. Royal,[3] and it could be shown that astron^l science would be promoted by it. I think its maintenance, or large assistance towards its maintenance would be a fair claim on gov^t as a national object. Without any such arrangement however I suppose it can only be regarded as an adjunct to the University chiefly useful for promoting the thorough teaching of astronomy. It will certainly be a pity if some enterprising astronomer does not get the Northumberland telescope, which I suppose is really available for discovery.

Yours very truly
W. THOMSON

P.S. I was stopped by my accident from writing to ask if you would care to have Liouville's Ed^n of Monge.[4] I got 2 copies by mistake (presented) & I w^d be glad to send you one if you care for it.

1 See letter 192, note 1.
2 This paragraph concerns the Cambridge Observatory, the directorship of which in 1861 became associated with the Lowndean professor, rather than the Plumian. In 1861, the Plumian professor, Challis, resigned the directorship which then went to J. C. Adams, who

had been Lowndean professor since 1859. When Adams died in 1892, Robert Stawell Ball (1840–1913) became both Lowndean professor and director of the observatory.

3 G. B. Airy.

4 Gaspard Monge, *Application de l'analyse à la géométrie*, 5th edn, ed. J. Liouville (Paris, 1850).

197 Kelvin to Stokes, 27 March 1861
Stokes Collection, K125

Auchinean, Largs
Ayrshire
March 27, 1861

My dear Stokes

I am much obliged for your letter, w^h I have rec^d here today, telling me that the Gov^t Grant Committee have granted Joule's & my $applic^n$.

I intend to send you Tyndall's paper by the end of the week – with report.[1] I am much interested in it.

I hope by this time, or very soon, you will receive Monge,[2] which I sent off for you last week via Maclehose[3] & Macmillan. As I got ⟨it⟩ both copies as a present I wished you to accept one on the same terms.

Helmhol[t]z must have meant *peine* instead of pain I think. My movements are certainly attended with much trouble, but little pain, none unless with some unusual strain. The bad leg is still so stiff in each chief joint that it is a great encumbrance, but with the other & crutches I get about a little. Since yesterday I have taken to living down stairs so as to get out of doors, which it is to be hoped will accelerate my progress. I have no prospect of getting back to my class this session.

Yours very truly
William Thomson

1 J. Tyndall, 'On the absorption and radiation of heat by gases and vapours, and on the physical connexion of radiation, absorption, and conduction', *Phil. Trans.* (1861), 1–36. Kelvin's report is in the Royal Society, RR.4.273. See also the next three letters.

2 See the previous letter, note 4.

3 James Maclehose, the Glasgow publisher.

198 Kelvin to Stokes, 17 April 1861
Stokes Collection, K126

Largs Ap 17, 1861

My dear Stokes

I enclose my report on Tyndall's paper[1] which I have delayed in the hope of seeing my way decidedly through the Franz difficulty.[2] I hope you will excuse its being written in pencil, as I am still unable to sit up freely enough to use pen & ink. I am sorry too to have to send it off ⟨without⟩ hurriedly to reach you [in] time for tomorrow's meeting, in accordance with your letter rec^{d} this morning. I hope however you will find it legible & intelligible.

I am getting on very slowly – [illegible word] out every day on crutches but able yet to do no good, (or not enough to compensate the trouble of having it in this way,) with the bad leg. The prospect of a complete recovery however seems good.

Yours very truly
W THOMSON

1 See the previous letter, note 1.
2 See Rudolf Franz, 'Ueber die Diathermanität einiger Gasarten und gefärbten Flüssigkeiten', Poggendorff's *Annalen der Physik und Chemie*, **94** (1855), 337–56. Franz (b. 1827) taught at the gymnasium of a monastery in Berlin.

199 KELVIN to STOKES, 19 April 1861
Royal Society of London, RR.4.272

Largs Ap 19, 1861

MY DEAR STOKES

In my report[1] which was necessarily somewhat hurriedly written to be in time for your meeting of Council I did not perhaps make suffic^{y} clear what I meant as desirable for Tyndall's columns' 'calculated results'.

What I suggest could only be carried out when the ratio of absorption to whole that would pass through the exhausted tube. If this ratio is denoted by x, and if A denote the whole, Tyndall's column & that which I propose to substitute for it would be as follows

T's 'calc^{d} results'	Proposed plan
Ax	$A\{1-(1-x)\}$
$2Ax$	$A\{1-(1-x)^2\}$
$3Ax$	$A\{1-(1-x)^3\}$
⋮	⋮

As long as x and n are so small that $A \times n\dfrac{n-1}{2}x^2$ is insensible the two will sensibly agree.

Would it be desireable [*sic*] to have a few atmospheric curves done by the mirror recording electrometer at Kew, (which Stewart[2] has now got into excellent working order,) to show at one of the R.S. Conversaziones? If you think so I shall ask Stewart to send some, or send some myself.

By the way what do you think of Kirch[h]of[f]'s '"new" proposition in the theory of heat' (Phil Mag April).[3] Is it not identical with Stewart's's? K.

probably did not know of Stewart's when he emitted it, but the Phil Mag should have known.

As to solar & stellar chemistry I have for 8 or 10 years quoted you as the inventor of its principles & mean to do so whenever I have an occasion to refer to it.

Yours very truly
WILLIAM THOMSON

1 See letter 197, note 1.
2 Balfour Stewart (1828–87) was director of Kew Observatory from 1859 to 1871 and professor of natural philosophy at Owens College, Manchester, from 1870 until his death. An extensive correspondence from him to Stokes survives in the Stokes Collection, and there are eleven letters dated 1861 from him to Kelvin in the Kelvin Collection.
3 G. R. Kirchhoff, 'On a new proposition in the theory of heat', *Phil. Mag.*, **21** (1861), 241–7. See Daniel M. Siegel, 'Balfour Stewart and Gustav Robert Kirchhoff: Two independent approaches to "Kirchhoff's radiation law"', *Isis*, **67** (1976), 565–600.

200 STOKES to KELVIN, 17 May 1861
Kelvin Collection, S396

Lensfield Cottage Cambridge[1]
17 May 1861

MY DEAR THOMSON,

The last soirée of the Royal Society took place on the 11th inst and I only got your letter today, therefore too late.

I did not write a report on Tyndall's paper[2] but merely recommended it, stating however my dissent from the author's view as to the relation between conduction and radiation, but saying that if the author wished it retained he had done such good work that he had, I thought, a right to keep it. There were other points of less importance about which I said I would write to the author, and have already done so.[3] The points which struck me were mainly the same that struck you.

As to Lorenz's paper[4] I have not yet (on account of my lectures) *fully* gone in to his mathematics but from the examination I have given it I think I shall be satisfied with it though I may not find it quite so complete as my own solution. But he is entirely mistaken in saying that I have neglected the reflected ray and that my solution is consequently incomplete. I have taken it into account just as much as he has. I shall probably put into the July N° of the Phil. Mag. something about it.[5] I think he can hardly have seen my paper or he would not have made the mistake. His experimental results completely confirm mine, and he seems to have arrived independently at the result I had previously obtained from my experts that the diffraction must be conceived to take place just *before* the light reaches the grating when the diffraction takes place at the common surface of two different media.

I have been busy at lectures and have not yet cons[d] Maxwell's papers.[6]

Yours very truly
G G STOKES

1 Kelvin has written on the envelope, 'Stokes May 17/61. His Report on Tyndall's paper on radiant heat. Reply to my question as to Lorenz & important remarks on Diffraction'.

2 See letter 197, note 1.

3 Stokes wrote to Tyndall:

> As to the connexion between radiation and conduction, I feel that it is not for me to dictate to, hardly even to advise, one who has done such capital work. I can only say that while the generality of the alleged physical connexion seems to me to need further examination, the reasoning by which it is attempted to account for the connexion seems to me altogether inconclusive; perhaps chiefly because it points to a state of things so different from what optical considerations would lead us to regard as possible. [Stokes to Tyndall, 16 May 1861, Royal Institution, Tyndall Papers.]

4 L. V. Lorenz, 'On the determination of the direction of the vibrations of polarized light by means of diffraction', *Phil. Mag.*, **21** (May 1861), 321–31.

5 He did not do so.

6 J. C. Maxwell, 'On physical lines of force. Part I. – The theory of molecular vortices applied to magnetic phenomena', *Phil. Mag.*, **21** (March 1861), 161–75, and 'On physical lines of force. Part II. – The theory of molecular vortices applied to electric currents', *Phil. Mag.*, **21** (April 1861), 281–91 and (May 1861), 338–48.

201 KELVIN to STOKES, 25 May 1861
Stokes Collection, K127

Auchinean, Largs
May 25, 1861

MY DEAR STOKES

I omitted to say when I wrote to you about my brother's paper on ice,[1] that he wishes 50 copies from the printers at his own expense. (I presume it to be still the rule that private copies of articles in the Proceedings are not given by the R.S.) He would also be glad to see a proof, although he does not consider it necessary. His address is

Prof James Thomson
Donegal Square
Belfast.

Perhaps I should have written to this effect to D[r] Sharpey. If so will you oblige me by forwarding this letter to him?

I hope in a few days to have a short paper completed for the Proceedings on the measurement of Electrical Resistance.[2] I shall address it to you either to London, posting on Tuesd. or Wed[y] next or to Cambridge later, unless in the mean time I hear it should go direct to D[r] Sharpey. I hope either he or you as the case may be will excuse the form in which it will appear partly ink and partly pencil as in my present more than half disabled condition I could not get a fair copy ready without much delay and labour which if you will kindly overlook the informality of the copy I propose to send would be needless. I have taken care to make it as good for the printers as a fair copy could be, & it will be much freer from mistakes than ⟨it could possibly be if I were to give it out to anyone to copy it⟩ a fair copy made by anyone else could be.

My progress is very slow, but in other respects satisfactory. I walk about very lamely on crutches but there seems every reason to hope that in time [my leg] will be as useful, and it now is as long, as ever.

Yours very truly
W. THOMSON

1 J. Thomson, 'Note on Prof. Faraday's recent experiments on "regelation"', *Proc. Roy. Soc.*, 11 (1860–2), 198–204.
2 Kelvin (125).

202 KELVIN to STOKES, 6 June 1861
Stokes Collection, K128

Kilmichael, Brodick
June 6, 1861

MY DEAR STOKES

By book post yesterday I sent you a short article, by Mr J. J. Murphy, of Belfast, for the Proceedings if approved of.[1] It seems to me to give a correct explanation of a very remarkable phenomenon, but I do not know enough of what has been written on the subject to form an independent judgment as to the novelty of the views contained in it. I enclose a letter I had regarding it from my brother, who has been thinking a good deal about meteorology lately, and who will soon I hope have a paper[2] ready which I have been urging him to write, for the Royal Society, giving a mechanical explanation of the prevalence of wind from the south, in middle latitudes.

On Monday I sent by book post, addressed Cambridge, my own article[3] on the Measurement of Electric Resistance, regarding which I wrote to you before.

My address will be as above, during the summer, and I should be much obliged by your giving it for anything (Numbers of the Proceedings, &c) which may have to be sent to me by post from the R.S.

Have you seen something in Crelle's Journal, of I believe rather more than a year old, on the integration of the equations of sound?[4] Helmhol[t]z told me about it, & as far as I could judge it seemed to be analogous to Earnshaw's solution but less restricted. I have not seen the article however and Helmhol[t]z only told me ⟨what⟩ of ⟨the⟩ its general character.

Yours very truly
WILLIAM THOMSON

1 J. J. Murphy, 'On great fluctuations of temperature in the arctic winter', *Proc. Roy. Soc.*, 11 (1860–2), 309–12. Joseph John Murphy (b. 1827) lived in Dunmury, Co. Antrim.

2 He did not send one to the Royal Society, but see J. Thomson, 'On the grand currents of atmospheric circulation', *Brit. Assoc. Rep.* (1857), part 2, pp. 38–9.

3 Kelvin (125).

4 H. Helmholtz, 'Theorie der Luftschwingungen in Röhren mit offenen Enden', Crelle's *Journal für die reine und angewandte Mathematik*, 57 (1860), 1–72.

203 KELVIN to STOKES, 7 August 1861
Stokes Collection, K129

Edinburgh
Aug 7/61

MY DEAR STOKES

I have not yet received a proof of the article I sent some time ago 'On the Measurement of Resistance'.[1] I am not quite sure whether I asked one to be sent to me. If not, & if not too late, I should be obliged by your directing the printers to send me a proof addressed

Kilmichael
Brodick
by Ardrossan.

Also, 50 copies of the article when ready to be sent for me to care of M^r^ D Macfarlane[2]

Natural Philosophy Labor^y^
College, Glasgow.

This latter order I can give the printers myself *if I am to have a proof*, & in this case do not trouble yourself about it.

Is any thing decided about Stewart's paper![3]

I am considerably bored by a paper[4] of Sir W. S. Harris' which you have sent me. It is so bad, like all he has done – that it would not be creditable to England & the R.S., except that (as in Challis' case & the C[ambridge] P[hilosophical] S[ociety]) the full amount of discredit has already been achieved, & Sir W. is too well known to excite any new sensation by such characteristics. At the same time there are curious & so far as I know novel results of long & varied observation which are worth publishing. However I shall write more distinctly if I can when I return the paper which I hope will be soon.

Yours truly
W THOMSON

I return to Arran this week, & leave Edinburgh today where we have come to get Syme's opinion about my leg. He says it was really broken (which he did not think when he first saw it) but is remarkably well mended & that I shall be quite free from lameness. My progress has been very tedious so far & I suppose must be so yet. I still require crutches.

1 Kelvin (125).
2 Donald McFarlane was Kelvin's assistant. There are nice discussions of him in A. Gray, *Lord Kelvin: An Account of His Scientific Life and Work* (London, 1908).
3 B. Stewart, 'On internal radiation in uniaxal crystals', *Proc. Roy. Soc.*, 11 (1860–2), 193–7. Kelvin's referee's report on the paper is in the Royal Society, RR.4.258.
4 W. Snow Harris, 'On some new phenomena of residuary charge, and the law of exploding distance of electrical accumulation on coated glass', *Proc. Roy. Soc.*, 11 (1860–2), 247–57. The next letter is Kelvin's referee's report on the paper.

204 KELVIN to STOKES, 25 October 1861
Royal Society of London, RR.5.96

Kilmichael, Brodick,[1]
by Ardrossan
Oct 25/61

MY DEAR STOKES

Your letter only reached me yesterday, the day of the Council meeting which you intimated. I am sorry to have, through spending all my limited working time almost ever since I received it on some work of which I hope soon to be able to send results to the Royal Society,[2] been led to delay so long sending you back Sir W. S. Harris' paper.[3] The question of its being accepted for publication seems to me to stand thus. The author's conclusions, which are in my opinion nearly all erroneous, are published in the Proceedings in the Abstract of the paper. But it is rather his *facts* than his *conclusions* that may be regarded as valuable, and they are wanted to allow those who read the abstract to judge for themselves as to the conclusions. But to have the facts explained & experiments described, with sufficient fulness, the paper in full is refused. As I have no doubt but that many of the experimental facts are of a novel and instructive character, although to be explained otherwise than as the author explains them, and as the general character of Sir W. S. Harris' reasoning and conclusions has long been made known through the Philosophical Transactions, and has formed the subject of published discussions, I think that the paper now in question should be published.

I must look for a reference to Riess[4] which I believe will show a previous publication of one of the rather remarkable results of Sir W. Harris' present communication, before I return you his paper, as I think it may be right that the author should have his attention called to it, but I hope by next post, that is three days hence, to be able to despatch the paper, & any final remarks I may have to make.

Yours very truly
W THOMSON

1 Stokes has written at the top of the page, '1861 Harris by Thomson'.
2 Probably Kelvin (130). See the next letter, note 4.
3 See the previous letter, note 4.
4 Peter Theophil Riess (1804–83) was a professor at Berlin and published a number of papers on electricity, none of which are mentioned in Harris's paper.

205 KELVIN to STOKES, 29 October 1861
Stokes Collection, K130

address 2 College Glasgow
after this
Oct 29/61

MY DEAR STOKES

Enclosed with this[1] I send ⟨the⟩ a copy of the paper by Riess in full,[2] to w[h] I promised a reference. It unmistakeably anticipates a large part of Sir W. S. Harris' results,[3] & even if Riess' explanation is not adopted by the former his result ought certainly to be referred to. Although this anticipation takes away a good deal of the novelty of Sir W. S. Harris' paper, I do not see sufficient reason to change the advice I gave.

Do you know if Poisson or any one else has solved the problem of finding the whole effect produced through a uniform elastic sphere by any arbitrarily given stresses applied to its surface or by any arbitrarily given displacements of all the points of its surface. I have worked out a full solution of this problem for isotropic material, & two independent coeff[ts] (rigidity & compressibility.) But I find a hollow spherical shell too hard a nut to crack.

I find that the earth as a whole must be far more rigid than glass otherwise the solid would yield so much to tidal influence of sun & moon as to leave no sensible tides of water relative to solid land.

I find that Laplace's theory of tides requires radical amendment for the compressibility of water. You will see this by considering the velocity of a long wave of water calculated on the supposition of incompressibility:

because it comes out greater than that of sound in water if the depth exceeds some thing not many miles or even less (I cannot repeat the calc^{n} before post.) This of course is absurd & therefore compressibility must be taken into account. Have you ever thought of Laplace's vanishing of the diurnal tide as to elevation, in a sea of uniform depth? It is merely a matter of rotation. The earth solid rotates about one axis, & the water, at same ang^{r} vel., about an axis nearly coinciding with it, in the case of motion which constitutes that particular solution of Laplace's. The *force* part of the problem is not satisfactorily explained.

I hope to send you a paper for the R.S. soon on the rigidity of the earth:[4] but I shall be able for little or no extra work after next week when I begin my lectures.

I only know Laplace on the tides as yet through Airy.[5]

Yours truly
W THOMSON

1 Kelvin has written above the line, 'The paper goes by same (book) post'.
2 See the previous letter, note 4.
3 See letter 203, note 4.
4 Kelvin (130).
5 G. B. Airy, 'Tides and waves', in Edward Smedley, Hugh James Rose, and Henry John Rose, eds., *Encyclopaedia Metropolitana*, 20 vols. (London, 1845), v, 65*–135*. (Asterisks are in the original.)

206 STOKES to KELVIN, 4 November 1861
Kelvin Collection, S397

Lensfield Cottage Cambridge[1]
4 Novr 1861

MY DEAR THOMSON,

Sir W. S. Harris's paper has safely arrived, with your report,[2] or rather the completion of your former report. If the paper were cut down to about a fifth of its bulk, the reserved part consisting of a description of the experiments and a slight notice of the theoretical views which led him to undertake them (for whether the views be right or wrong they serve to show the drift of the exper^{ts} as the author conducted them) there would I think be no doubt of the eligibility of the paper for publication, due recognition of Riess[3] being inserted. Whether the merit of this part will carry through a paper of its bulk remains to be seen.

Not many years ago it was proposed as a prize subject by the French Academy to determine the equilibrium of an isotropic elastic solid not small in one of its dimensions (a rectangular parallelopiped or a sphere for example) subject to the action of given forces, whether acting all over the mass or at the surface only I am not sure, nor do I imagine it would make much difference in the difficu[l]ty of the solution. I don't think anything satisfactory was sent in in answer to this invitation, and I feel pretty sure that the solution of the problem has not hitherto been effected. I take for granted I am right in supposing that you take the forces acting all over the surface perfectly arbitrary in magnitude and direction except of course that they must satisfy the 6 conditions of eq^{m} of a rigid body. The solution of the problem will be an important step in the application of mathematics to elastic solids.

I should not have supposed a priori that if you had solved the problem for a sphere you would have been unable to effect the solution in the case of a spherical shell, though it would seem very likely that the solution in the latter case would be far more complicated, regarding merely the length of the expressions, than in the former.

Your result regarding the action of the tide-producing forces on the earth considered as an elastic solid is very remarkable, and not at all what I should have expected.[4] As regards the physical result brought to light by confronting the solution (assumed to be correct) with observation, it seems to me that it is more likely to be that the rigidity of a solid is greatly increased when the solid is subjected to an enormous hydrostatic pressure than that the materials of which the earth are composed are under like conditions vastly more rigid than glass.

I have never thought of the problem of a long wave taking account of the compressibility of the liquid. The consideration you mentioned seems to show that compressibility will have a sensible though probably small (assuming $3\frac{1}{2}$ miles as an average depth) influence on the phenomena of the tides. If however the cubical compressibility of water notably diminishes when the water is subjected to a pressure of many atmospheres the results will not be much affected by compressibility. It is a long time since I have attended to the tides.

Believe me
Yours very truly
G. G. Stokes

1 Kelvin has written on the envelope: 'Stokes Nov 4/61. on Sir W. S. Harris' last paper – Elastic Tides; spherical shell &c. Veloc. of waves in compress. liqu.'

2 See letter 203, note 4.

3 See letter 204, note 4.

4 See Kelvin (130).

207 KELVIN to STOKES, 7 November 1861
Stokes Collection, K131

Thornliebank
Glasgow, Nov 7/61

MY DEAR STOKES

I am much obliged to you for your letters. I am glad you think the elastic sphere a problem which it is worth while to work at.[1] My solution is general, as you describe. For the case of a solid sphere, with ⟨uni⟩ [?] arbitrarily given superficial displacements, it is very simple – being this.

$$n\nabla^2\alpha + m\frac{d\delta}{dx} = 0$$

$$n\nabla^2\beta + m\frac{d\delta}{dy} = 0$$

$$n\nabla^2\gamma + m\frac{d\delta}{dz} = 0$$

where α, β, γ are the displacements of any p[t] xyz,

$$\delta = \frac{d\alpha}{dx} + \frac{d\beta}{dy} + \frac{d\gamma}{dz} \quad \text{for brev}^{\text{y}}$$

$$\& \quad \nabla^2 = \frac{d^2}{dx^2} + \frac{d^2}{dy^2} + \frac{d^2}{dz^2}$$

Let also $r = (x^2 + y^2 + z^2)^{\frac{1}{2}}$

The above are the general equ[ns] w[h] I know from your paper. Then ⟨solution⟩ for the following superficial condition,

$$\alpha = \phi_i, \quad \beta = \psi_i, \quad \gamma = \chi_i; \quad \text{when } r = a$$

where ϕ_i &c are rational integral homog[s] functions of xyz of degree i ⟨each⟩ fulfilling the equ[ns]

$$\nabla^2\phi = 0, \quad \&c$$

(⟨So that⟩ They are in fact Laplace's functions, of order i); the solution is

$$\alpha = \phi_i + I(a^2 - r^2)\frac{d\langle\phi\rangle\psi}{dx}$$

$$\beta = \psi_i + I(a^2 - r^2)\frac{d\psi}{dy}$$

$$\gamma = \psi_i[sic] + I(a^2 - r^2)\frac{d\psi}{dz}$$

where $$\psi=\frac{d\phi_i}{dx}+\frac{d\psi_i}{dy}+\frac{d\chi_i}{dz},$$

and $$I=\frac{m}{2[(i-1)m-(2i+1)n]}$$ errors excepted[2]

You will readily see that the equations & superficial conditions, are verified by the solution so expressed.

When the force at each point of the surface is given a solution of a similar character holds, but it requires rather more work.

The great rigidity in the interior I supposed to be explained by the great pressure. Indeed it seems almost quite certain that, if it is solid at all, the great pressure must make it much more rigid than the same matter under ordinary pressure. The coefficient of ⟨compressibility⟩ resistance to compression must be enormously increased, for certain.

Yours very truly
W THOMSON

P.S. My theory of the elastic tides is even more decisive than Hopkins' argument against the prevalent geological theory of internal fluidity.[3]

1 See Kelvin (133).
2 Kelvin has added a footnote here reading, 'I have this *booked* right, but work it out hastily just now as above'. Kelvin's brackets.
3 In his (130), Kelvin cited W. Hopkins, 'On the phenomena of precession and nutation, assuming the fluidity of the interior of the earth', *Phil. Trans.* (1839), 381–424; 'On precession and nutation, assuming the interior of the earth to be fluid and heterogeneous', *Phil. Trans.* (1840), 193–208; and 'On the thickness and constitution of the earth's crust', *Phil. Trans.* (1842), 43–56.

208 KELVIN to STOKES, 27 November 1861
Stokes Collection, K132

Thornliebank, Glasgow
Nov 27, 1861

MY DEAR STOKES

I am sorry that I shall not be able to be present on the occasion of the dinner to Hopkins, in consequence of living at so great a distance. I should have been much pleased to be present, and I hope you will mention to Mr Hopkins my regret at being absent, and include me with others if you are making any statement of the kind from the chair.[1]

If there is any mention, in connection with the proceedings on that occasion, to be made of Hopkins' scientific investigations, his demonstration of the earth's solidity ought not to be omitted. It always *appeared* to me to be a very important speculation, and to contain a perfectly sound argument against the common geological theory of a thin crust & all fluid within. The only way to answer it would be by *viscosity*. It is possibly just possible that viscosity might in a fluid globe enclosed in a thin shell give rise to the same *precession* as exists in reality; because in the course of a 26,000 years' period, the average axis of rotation might be prevented from lagging much behind the actual axis of rotation of the crust. But surely no amount of viscosity conceivable as probably real in the circumstances could prevent ⟨a⟩ large monthly & yearly nutations from being manifested by the solid crust, which is not the case in reality, or could produce the close agreement which observation shows between the actual 18½ years nutation & that which theory gives on the hypothesis of solidity. These considerations incline me to give full weight to Hopkins' argument, & I have written a slight statement to that effect as an introduction to my projected paper on the rigidity of the earth.[2] I should be very glad to know if you have thought on the subject, & if you agree with me.

I find a great deal wanting as to 'Laplace's functions' for practical use, & any one who would work at it could be well rewarded.

As to B. Stewart's paper,[3] I quite agree with you that he does not bring out satisfactorily the dynamical principle of his argument. That piece of mathematics I was not sure of might have been right, only I did not verify it the way Stewart gave it, & I did find a different result working out a particular case for myself on what I thought to be his principles. I saw him a few days ago here, & I found that he was disposed to send the paper for publication in the Proceedings but would try to improve it & make it clearer.

I have at last got some experiments of what I think are a satisfactory kind, to determine thermal conductivities. I find marble conducts 11 or 12 times as well as the number given for it in Rankine's 'Prime Movers'[4] as derived from Despretz,[5] would allow. I knew before hand that copper (in Calvert & Johnstone's [*sic*] experiments)[6] conducted more than three times as much heat as the number for copper in the same table would give. It is strange that on so important & obvious a practical subject, the available information hitherto published should be so worthless.

Yours very truly
WILLIAM THOMSON

1 The dinner probably took place in connection with the Cambridge Philosophical Society of which Stokes was president during 1861. Unfortunately, the society's *Proceedings* skip from March 1861 to October 1863.

2 Kelvin (130).

3 See letter 203, note 3.

4 W. J. M. Rankine, *A Manual of the Steam Engine and Other Prime Movers* (London and Glasgow, 1859).

5 César Mansuète Despretz (1792–1863) published a number of articles on heat and electricity, mainly in the *Comptes Rendus*. He was professor of physics at the Sorbonne.

6 See R. Johnson and F. Crace-Calvert, 'On the relative power of metals and alloys to conduct heat. – Part I', *Phil. Trans.* (1858), 349–68, which gives conductibility figures. Richard Johnson (1810–81) owned a wire factory in Bradford. Frederick Crace-Calvert (1819–73) took his Ph.D. at Giessen and was lecturer at the Royal School of Medicine and Surgery in London.

209 STOKES to KELVIN, 22 February 1862

Glasgow University Library, Kelvin Papers, S83

Lensfield Cottage Cambridge[1]
22 Feb[y] 1862

MY DEAR THOMSON,

I don't think I spoke of *doubtful experiments*, but of doubtful uses of the application of theory to experiment: at least that is what I speak of at p [94].[2] It was not the ⟨actual⟩ accuracy of the experiments I felt doubtful of but the applicability of the theory, partly because I was apprehensive of the effect of eddies, partly because I did not I believe feel quite sure of the proper condition to assume at the surface of a solid; and if there *was* any finite sliding it seemed that its effect must be very much greater in the flow of a liquid through a capillary tube than in the case of a pendulum swinging in an unlimited mass of fluid surrounding it. I hardly believed in the possibility of a finite sliding, but my scepticism has been shaken by an experiment by Helmholtz & Piotrowski (Sitzumsberichte [*sic*] of the Vienna Acad. for 1860 or 61 I think)[3] in which it was found that the diminution of the arc of oscillation (slow bifilar oscillations) in the case of a stoppered bottle completely filled with water and oscillating about its axis of figure was sensibly lessened by chemically silvering the inside. Girard[4] too found that the flow of water through a capillary copper tube was greater than through a ⟨b⟩ [?] glass one of equal diameter; but this experiment if it stood alone might be referred to an errour [*sic*] in the measurement of the diameter. If the sliding be made out it may be connected with an imperfect wetting of the surface by the liquid. Water is rather indisposed to wet silver. According to Helmholtz & Piotrowski there is however a sliding even in the case of watery fluids against glass.

I did not know of Poiseuille's experiments when I wrote my paper but only of Girard's. I see the value of μ' for water which Helmholtz has calculated from ⟨Poiss⟩ Poiseuille's experiments, that which he has calculated from Piotrowski's experiments on the decrement of the osc[illatio]ns of a hollow sphere filled with water, and the value I calculated from Coulomb's and Bessel's experiments are not very different.[5]

I have not made up my mind as to the most probable of the slightly (or rather slightly) different values of μ' (water). In the calculation of Coulomb's the calculation was only approximate, each annulus being deemed to behave like a portion of an infinite plane moving with the same velocity. Still the value got from Coulomb's disks is backed by Bessel's pendulums. Perhaps eddies operated sensibly in both.

I believe the ratio of μ' for different gases to μ for air may be got from Graham's exper^{ts} on the transpiration of gases in two papers in the Phil. Trans.[6] But a strange result of his experiments is apparently that μ and not μ' is independent of the density, quite in accordance with Maxwell's speculations.[7] This follows from Graham's law that the vel^{y} of discharge from a reservoir into a vacuum through a long capillary tube (estimated at the density in the reservoir) $\propto \frac{\text{pressure}}{\text{length}}$. Also I found that the times in which the ⟨guage⟩ [*sic*] mercury fell from 28^{in} to 24, 24 to 20 &c. as given by experiment followed very well the laws given by theory on supposing μ constant (as to a change of ϱ) but not on supposing μ' constant.

I could not get absolute results from Graham's exper^{ts}, the tubes not being regular enough, but a rough comparison for the tube *H* gave assuming $\sqrt{\mu'} = .116$ about 5/3 of the experimental discharge, which is pretty close as there were probably several constrictions in the tube (which was a patched one) not taken into account.

Yours very truly
G. G. STOKES

1 Kelvin has written at the top of the letter, 'W.T. to P.G.T.' [i.e. P. G. Tait] I had yesterday afternoon & this morning driving in convinced myself (out of Helmhol[t]z' paper w^{h} I had just found) that H. is right as you see Stokes admits. [one illegibly deleted word] You need not return this as long as it can be of any use to you. I shall send you mine, particularly as to Helmhol[t]z' slipping'.

2 Stokes (47). Stokes's brackets, which are also around the page numbers in the original published version of the paper.

3 H. Helmholtz and G. Piotrowski, 'Ueber Reibung tropfbarer Flüssigkeiten', *Sitzungsberichte der Mathematisch-naturwissenschaftlichen Classe der Kaiserlichen Akademie der Wissenschaften*, 40 (1860), 607–58. Gustav Piotrowski (1833–84) was professor of physiology at Cracow.

4 P. S. Girard, 'Mémoire sur le mouvement des fluides dans les tubes capillaires, et l'influence de la température sur ce mouvement', *Mémoires de la Classe des Sciences mathématiques et physiques de l'Institut* (1813–15), 249–380. Pierre Simon Girard (1765–1836) was director of buildings and canals at Ourcq from 1802 to 1820.

5 Jean Léon Marie Poiseuille (1799–1869) was a medical practitioner in Paris and published a number of papers on the circulation of blood, on circulation in capillaries, and on the movement of liquids through glass tubes of small diameters. For Coulomb and Bessel, see C. A. Coulomb, 'Expériences destinées à déterminer la cohérence des fluides et les lois de leur résistance dans les mouvements très lents', *Mémoires de l'Institut national des sciences et arts – sciences mathématiques et physique*, 3 (1800), 246–305, and F. W. Bessel, 'Untersuchungen über die Länge des einfachen Sekundenpendels', *Abhandlungen der Königlichen*

Akademie der Wissenschaften zu Berlin (1826), 1–256. Both Coulomb's and Bessel's papers are cited in Stokes (47). (See Stokes's *MPP*, III, 92–107.)

6 T. Graham, 'On the motion of gases', *Phil. Trans.* (1846), 573–632, and (1849), 349–92. Thomas Graham (1805–69) studied chemistry at the University of Glasgow and held professorships in chemistry at Anderson's College, Glasgow, and University College, London, before succeeding John Herschel as master of the mint in 1854.

7 J. C. Maxwell, 'Illustrations of the dynamical theory of gases. Part 1. On the motions and collisions of perfectly elastic spheres', *Phil. Mag.*, **19** (1860), 19–32, and 'Illustrations of the dynamical theory of gases. Part 2. On the process of diffusion of two or more kinds of moving particles among one another', *Phil. Mag.*, **20** (1860), 21–37.

210 STOKES to KELVIN, 25 February 1862

Kelvin Collection, S398

Lensfield Cottage Cambridge[1]
25 Feb[y] 1862

MY DEAR THOMSON,

I find there is to be a meeting of the Gov[t] Grant Committee on Thursday, so if you should have recovered the £5 for breakage of apparatus please let me know by a line addressed to Burlington House if time permits. What the Committee recommend will have to come before the Council, so the thing won't be finally settled before the 3[d] Thursday in March or thereabouts.

The result I mentioned as to the constancy of μ follows readily from Graham's laws[2] if we ⟨ma⟩ [?] may assume that eddies &c may be neglected. The ground of this assumption is that Graham found the velocity of 'effusion' (that with which a gas was discharged thro' a hole in a thin plate into a highly exhausted receiver) varied for different gases as $\frac{1}{\sqrt{\varrho}}$, while in 'transpiration' (discharge through a capillary tube of sufficient resistance) the velocities of different gases were as certain numbers independent of the particular capillary employed i.e. if $v_O v_H \ldots$ are the velocities for oxygen hydrogen &c

$$\frac{v_O}{k_O}=\frac{v_H}{k_H}=\frac{v_N}{k_N} \quad \&c$$

where the ratios of $k_O : k_H : k_N$ &c are independent of the tube employed

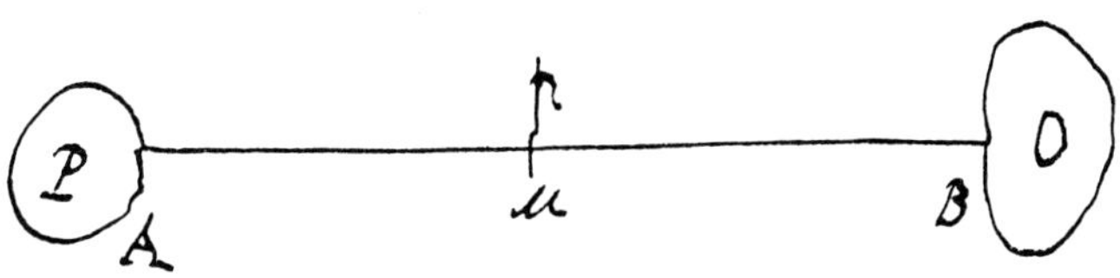

Let AB be a long uniform capillary, given except as to length, and let a given gas be discharged through it from a reservoir at pressure P into a vacuum. If V be the vel[y] at A, then according to Graham

$$V = k\frac{p}{l} \tag{1}$$

where $l = AB$ and k is of the form

a quantity depending on the gas
× a quan^{y} depending on the bore of the tube

Let $MB = x$, and let p, v be the pressure and velocity (mean vel^{y}) at M. Then by Graham's laws applied to MB regarded as a complete tube fed by a reservoir at pressure p

$$v = k\frac{p}{x} \tag{2}$$

and by the condition of continuity

$$pv = PV \tag{3}$$

the temperature being supposed constant so that $p \propto \varrho$.

The resistance for the element dx of the tube is proportional to $\frac{dp}{dx}$. Assuming the resistance to be some function of the velocity and density or what comes to the same the vel^{y} and pressure we have

$$\frac{dp}{dx} = \phi(v, p) \tag{4}$$

Elim^{g} v from 2 and 3 we have

$$PVx = kp^2$$

$$\therefore \quad PV = 2kp\frac{dp}{dx} = 2kp\phi(v, p)$$

and now eliminating PV from this equation by means of (3) we have

$$\phi(v, p) = \frac{v}{2k}$$

or varies as the first power of the velocity and is independent of the density.

Now if we may assume that the motion was truly linear without finite slipping at the surface the sliding must have been proportional to v, and since the resistance $\propto \mu \times$ the sliding we have $\mu v \propto v$ or μ is independent of ϱ.

However it may well be doubted whether with such high velocities there was not a continual formation of eddies within the tube. Graham himself found that the ratios of transpiration velocities for the different gases became somewhat abnormal from 'excessive resistance' when the difference of pressure at the two ends of a tube was but small. Whenever (if ever!) I undertake my pendulum experiments I propose to include in them exper^{ts} on the discharge of gases through pipes.

There is one result of Graham's which I cannot at all explain on the supposition that the motion was linear. In the case of a long (only moderately long it is true) capillary tube which was decidedly conical he found a discharge 3 times as rapid when the narrow end was high-pressure as when the tube was reversed.

Yours very truly
G. G. STOKES

1 Kelvin has written on the envelope, 'Stokes Feb 25/62. Graham's exp^{ts} on Transpiration of air &c'.

2 See the previous letter, note 6.

211 STOKES to KELVIN, 7 March 1862
Kelvin Collection, S399

The Royal Society,[1]
Burlington House, W.
7 March 1862

MY DEAR THOMSON,

At the meeting of the Gov^{t} Grant Comm^{ee} yesterday your application for £50 for contact elec^{y} &c. and the *first* of your three applications for elec^{l} instruments amounting to £14–10 were granted, subject to confirmation by the Council.

The second (relative to Nova Scotia)[2] was postponed on the ground that it would be a bad precedent to grant money for observations to a person who merely wished to have the observations made. It was thought that the application should come from, or at least be accompanied by an application coming from, the person who was going to make the observations, or else the institution at which they were to be made.

The 3^{d} (relative to Melbourne) was postponed for a still stronger reason, or rather considered as *withdrawn*, for Genl Sabine was informed by [W. P.] Wilson in a letter from Melbourne direct that Neumayer[3] was returning or had returned to Germany, and I don't know who takes his place.

When Graham's paper[4] came before the R.S. I was much struck with his observation relative to oil of vitriol compared with the same somewhat diluted, and exactly the same ideas occurred to me as those which you have expressed in your letter.

I doubt the feasability [*sic*] of your suggestion relative to a suspended disc. Of course the whole mass of fluid would be set in rotation, so that there would be friction on the upper as well as the under surface of the disc. Besides I have a sort of feeling that the least defect of figure of the vessel from perfect uniformity round the axis of motion would tell comparatively largely on the result.

Yours very truly
G. G. STOKES

1 Kelvin has written on the envelope:
Stokes March 7/62
Decision of Gov^{t} Grant Comm^{ee} on applications
Exper^{t} on Viscosity
2 See the next letter.
3 Georg Balthasar Neumayer (1826–1909) was founder and director of the marine observatory in Hamburg from 1876 to 1903.
4 T. Graham, 'On liquid transpiration in relation to chemical composition', *Phil. Trans.* (1861), 373–86.

212 KELVIN to STOKES, 10 March 1862
Stokes Collection, K133

Natural Philosophy Labor^{y}
Glasgow College
March 10/62

MY DEAR STOKES

I have been in correspondence for some time with Prof Everett regarding electrical observations,[1] in fact ever since he went ⟨there⟩ to Nova Scotia. Before he went out he told me that he was anxious to make observations if possible and he learned the mode of doing so with the instruments I then had completed. Rather more than two years ago I wrote to him that I had improved plans and instruments, and asked if he would wish to have a set made for him to be his own property or that of the university or of some meteorological society in the place. He answered that he had no available funds, but that if I could send him instruments, out of the grant I then had from the Royal Society, he would observe with them, and he felt sure of being able to do so with advantage, and could get assistance in using them from skilled persons there.

Since that time we have had repeated letters from him asking for the instruments and saying he was ready to use them with advantage and anxious to have them without delay. He also did all he could on the spot to get the electrometer I sent out last summer repaired, but as the main jar was broken, he could not get it into action, and I have therefore written to him to send it home for renewal.

I should be obliged by your stating as much of the above particulars to the Committee at their next meeting, as you think proper, because I am anxious they should know that I have taken no step, either as to sending out instruments or applying to the Royal Society for assistance, without sufficient ground for conviction that good results were to be expected.

I shall write immediately to Everett asking him to make an application, direct to the Royal Society, addressing it to you. If this is not right, or if there is any special circumstance requiring attention as to the form of the application I should feel much obliged by a line from you informing me.

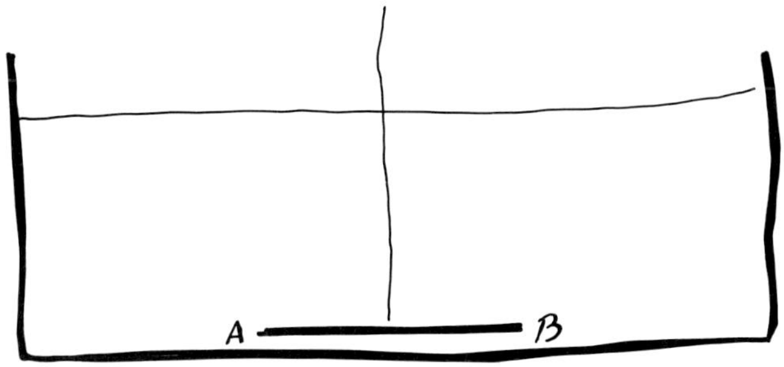

As to viscosity and slipping – I do not see how inequalities of figure could have any disturbing influence, on the plan I proposed, except such as could be easily eliminated. By using some such proportions as in the sketch, it would be secured that the main slipping and distortion would take place in the space *AB* between the discs. The rest of the fluid would revolve sensibly with the vessel. The motion would be slow that, centrifugal force would make scarcely any sensible disturbance of the surface (although this condition would not be essential) and that no slight motion of *AB*, swinging horizontally, would gather disturbances through instability of any kind.

It is very desirable some one should be induced to experiment on the subject. Will you not? I see no way more likely to bring results than the one I have suggested, but perhaps I am wrong.

Yours very truly
WILLIAM THOMSON

1 See the letters from Everett to Kelvin (26 December 1860 to 15 May 1861) in the Kelvin Collection, E97 to E100, in which he discusses his work on underground temperatures as well as volunteering to make electrical observations.

213 KELVIN to STOKES, 12 April 1862
Stokes Collection, K134

Glasgow College
April 12, 1862

MY DEAR STOKES

I send, by book post with this, a non-mathematical paper 'on the Rigidity of the Earth' for the Royal Society.[1] Two or three of the propositions asserted in it require mathematical proof; but in each such case I have referred to a coming mathematical paper.[2] I hope this will not be considered fatal to the present, as a communication for the Transactions. I thought it better to make the mathematical treatment altogether detached, because I know how much even the appearance of a few pages of symbols repels many readers who could quite enter into the general argument of such a paper.

The mathematical paper will be on the Distortion of ⟨an⟩ Elastic & Fluid Spheres (if I cannot get a better title) & will of course require a good deal of spherical harmonic functions ('Laplace's coefficients') I could send it in a few days if I do not wait to put the solution for the shell into shape. There should be a good deal too about the effects of rotation, but I think I may get it all into moderate compass.

I am to be in London about the middle of May & to remain a month or 6 weeks, as juror on philosophical instruments, & I can bring the paper with me if there is no object to be gained by my sending it to you sooner.

Yours very truly
WILLIAM THOMSON

1 Kelvin (130).
2 Kelvin (133).

214 KELVIN to STOKES, 14 April 1862
Stokes Collection, K135

Glasgow College, April 14/62

MY DEAR STOKES

I have been thinking that perhaps it will look like expecting too much, from the solution of the precessional problem for a spheroidal solid nucleus surrounded by a lighter liquid, to say what I have said in the sentence in which 'barest possibility occurs'. Will you therefore draw your pen through this sentence & any others in connection with it that you think had better be omitted?[1]

Have you ever remarked ⟨what⟩ how Airy, every time he refers to the equil^{m} Theory calls it contemptible except when he calls it 'miserable' §571?[2] I felt impelled to take some notice of this, from the great prominence Airy gives to it (the 'contemptibility' I mean,) but I hope the way I have done so is not objectionable. Perhaps it might be better to substitute justly, for 'reasonably' in the parenthesis which follows my allusion to this. If you think so will you make the change?

I hope 'kinetics' & *naturalist*, will not be objected to, fatally. I know that pigeon fanciers & butterfly & beetle collectors will be desperately offended at being classed with Newton & Faraday, but still I think propriety & convenience of language renders it necessary to disregard their feelings.

As for *kinetics*, I have Lushington[3] (who was not senior wrangler because he was senior classic) strong in favour of it, & of keeping kinematics as it is. Tait[4] agrees to use it in the book we are going to bring out; but I hope the Royal Society will not be considered an improper place for calling attention in the first instance to either conservative reforms, ⟨(such as naturalist,)⟩ or innovations, of the kind.

I thought, till just before I sent off the paper, that I had a good theoretical authority from Airy to give in favour of the anticipation that the ⟨semi da⟩ [?] fortnightly tide must on the whole follow nearly the equilibrium law, at good stations. I find nothing of the kind, very tangible. Will you therefore add the accompanying leaf to the § of my paper where I say the equil^{m} ⟨tid⟩ theory will probably be fulfilled by the fortnightly tide at well chosen stations. I add also an extract from Airy to show all that is known by observation on the subject. It may find its way somewhere, foot note or otherwise, into the paper. I am sorry to trouble you thus.

Yours very truly
W THOMSON

1 Kelvin (130).
2 G. B. Airy, 'Tides and waves', in Edward Smedley, Hugh James Rose, and Henry John Rose, eds., *Encyclopaedia Metropolitana*, 20 vols. (London, 1845), v, 390*. (Asterisk is in the original.)
3 Edmund Law Lushington (1811–93) took his B.A. from Cambridge in 1832 as senior classic and was professor of Greek at Glasgow from 1838 to 1875.
4 Peter Guthrie Tait (1831–1901) was senior wrangler in 1852 and professor of natural philosophy at Edinburgh from 1860 to 1901. Kelvin and Tait's *Treatise on Natural Philosophy* (Oxford, 1867) was an enormously influential textbook which went through several editions. It is 'the book' mentioned in this paragraph. An extensive correspondence between Tait and both Kelvin and Stokes survives in the Kelvin and Stokes Collections and the Kelvin Papers in the Glasgow University Library. The surviving letters are almost entirely *from* Tait, as apparently no significant collection of Tait manuscripts – which would have contained the other sides of the correspondences – has been preserved. There are, however, some seventy letters between Tait and Kelvin in the Edinburgh University Library.

215 KELVIN to STOKES, 19 April 1862
Stokes Collection, K136

Thornliebank, Glasgow
April 19, 1862

MY DEAR STOKES

I have a paper[1] (experimental) on Contact Electricity nearly completed, which I should be glad to have an opportunity of communicating before the end of the session. I hope to be able to present, and if necessary (& permitted) to answer objections that may be made in case of there being any discussion upon the subject. I suppose however that so late in the session there will be little time to spare for any single paper, and perhaps to get it 'read in' before the end of the session may be as much as I need expect. If I do not hear from you about it, I shall only aim to have the paper in your hands in time to 'read in' at the last meeting.

I hope also to have a short communication[2] on Conducting Powers of solids for heat, which may be admitted for the 'proceedings' as a temporary statement. I have obtained results so much in advance of those commonly published, (as regards quantity of heat conducted through) that I think it might be useful to publish them provisionally, without waiting for more accurate results which I hope to obtain by pursuing the investigation.

By today's post I send you an Abstract of the paper you have.[3]

Yours very truly
W THOMSON

1 I find no reference to such a paper. However, Kelvin had published a note on contact electricity earlier in 1862 – Kelvin (122A).
2 I find no reference to such a paper.
3 Kelvin (130), which was received on 14 April 1862.

216 KELVIN to STOKES, 19 April 1862
Stokes Collection, K137

Thornliebank,[1]
Glasgow
April 19, 1862

MY DEAR STOKES

There was one point connected with my paper[2] on the Rigidity of the Earth that has given me a good deal of trouble and anxiety. I found on looking into Hopkins' papers[3] that, so far as the mathematical problems he attacks are concerned, they are all wrong. This I think you will admit to be the case if you consider his fundamental proposition, which I first learned as quoted by Pratt (Figure of the Earth, recent edition)[4] & which is obviously wrong. It is this. A spheroidal homogeneous liquid mass, enclosed in a rigid shell[5] of which the outer and inner boundaries are similar ellipsoids, & of which the density is homogeneous & the same as that of the liquid, would experience the same precession and nutation as if it were rigid throughout. His proposition (3^d Memoir) about Iso thermal surfaces is also obviously wrong.

I have therefore felt great difficulty in referring to his investigation. It was necessary I should do so because he was the first to propose the argument, & I think his conclusion valid. Yet, having been one of his pupils, and having experienced the greatest possible benefit from his teaching, I would certainly shrink from the task of finding fault with his investigation: and, as it was necessary I should not commit myself to assenting to it, I therefore ⟨added⟩ introduced a statement in a foot note near the beginning, which you may perhaps have noticed. I do not know whether this will be considered a satisfactory solution of a very difficult problem, but I could find no better; and I should be glad to change it if anything better is suggested. I fear Hopkins, if he notices it, may not be pleased with the foot note, & I should be very sorry to be called on to justify it to him.

Yours very truly
WILLIAM THOMSON

1 Kelvin has written at the top of the letter, 'Private (second letter under one envelope)'. That is, this and the previous letter were apparently sent in the same envelope.
2 Kelvin (130).
3 See letter 207, note 3.
4 J. H. Pratt, *A Treatise on Attractions, Laplace's Functions, and the Figure of the Earth* (Cambridge and London, 1860).
5 Kelvin has written above the line and deleted, '⟨of same homogeneous dens^y as the liquid⟩'.

217 KELVIN to STOKES, [June or July 1862][1]
Stokes Collection, K146

valuable, and the explanations, descriptions, diagrams, and tables are such as to present the whole subject very clearly to the reader.

Mr. Stewart's, entitled 'On the Nature of the Forces concerned in producing Magnetic Disturbances'[2] is in my opinion a valuable contribution of data from observation, and of acute speculation, towards solving one of the most important and interesting problems of Terrestrial Physics.

The relation demonstrated in it, between the amounts of the sudden disturbances ⟨th⟩ [?] in horizontal and in vertical forces leading to the result that the direction of the disturbing forces to which they have have [*sic*] been due, has been generally inclined about 15° below the horizon is of the very highest importance.

It seems to me that a simpler and more natural view than that of the author supposing that two disturbing forces, nearly ⟨equal, &⟩ directly opposed, and sometimes nearly equal, are in operation, would be to suppose that there is a comparatively large single disturbing cause, varying gradually, & constituting the main disturbance of the magnetic storm: & that superimposed on this are the small disturbances constituting the 'peaks & hollows.' It is these small disturbances that depend on successive sudden actions of a force 15° below the horizon. They no doubt are connected with the large disturbance: & may very probably be induced results of it. An electrical analogy will make the idea I wish to convey clear enough. Suppose an electrical machine or two or three electrical machines, or an electrical machine carried gradually along in any direction, to be turned so as to keep it's [*sic*] prime conductor & secondary or negative conductor, both electrified, and suppose that, owing to it's [*sic*] action, a small conductor in some corner or fixed place of the room keeps acquiring, by a point or plane, a charge sometimes positive & sometimes negative & discharging by a spark but not at the same time discharging the prime or secondary conduction of the machine. Then a complete apparatus for observing & recording by three curves the resultant electric force, & its direction, at any point of the air in the neighbourhood would give three curves, with peaks & hollows, ⟨(⟩ like those described by Mr. Stewart, & indicating a sudden disturbing force in a nearly constant direction, which would be superimposed on the main disturbance curve, always similarly to one another whether themselves positive or negative, or whether superimposed on positive or on negative parts of the main disturbance curves.

I do not exactly understand what is meant when it is said (§32 for instance) that we have no means of ascertaining whether the disturbing force is negative or positive; or again whether it is due to a N pole or a S pole (or something to that effect.) The line of the disturbing force is known and the direction in that line in which the n. pole of the needle is drawn, is known in each case. What then is it that is unknown?

I have written a note[3] which I send in the book-packet, & which may, if sanctioned by the Council, be printed at the end of Mr Stewart's paper. It refers to another important result of Mr Stewart's to w^h therefore I need not refer more particularly here.

I enclose a diagram for Joule's & my last paper.[4] The diagram I gave you with the paper was along with it by mistake, & I should be obliged by your returning it. My address henceforth is Kilmichael, Brodick, by Ardrossan.

Yours truly
W THOMSON

1 The first part of this letter, which would have contained the date, is missing. However, the articles cited in notes 2 and 4 were received by the Royal Society on 14 and 19 June 1862,

respectively. Moreover, Kelvin's reference to his future address in the last sentence of the letter suggests that this letter preceded the next several by Kelvin which were written from that address during July and August 1862.

The surviving portion of this letter was mistakenly included with the incomplete first portion of a second letter and the two together numbered K146. The second letter, dated 18 December 1862, is letter 229.

2 B. Stewart, 'The nature of the forces concerned in producing the greater magnetic disturbances', *Phil. Trans.* (1862), 621–38.

3 Kelvin (129A).

4 Kelvin and Joule (10).

218 Kelvin to Stokes, 8 July 1862

Stokes Collection, K138

Kilmichael
Brodick, N.B.
July 8, 1862

My dear Stokes

I hope to have my paper[1] on elastic spheroids and shells, including demonstrations of the particular results used in my paper[2] on the Rigidity of the Earth, ready to send to you before the day you mention, the 17th.

I have got my solution of the general problem of the deformation of a shell, put in complete order now, and very nearly written out. The particular results are deducible from it; and I hope in the course of the next few days to get these details also written out & ready to send to you.

What do you think of Iceland and Teneriffe for observations on the fortnightly tides? We propose, Tait and I, to go one to the one place & the other to the other, and spend a month observing, before we write our chapter[3] on the Tides, but when that may be I do not know. I should choose Teneriffe for myself, and take the opportunity of finding whether Piazzi Smyth's result[4] that electricity on the top of the peak is always positive (i.e. air potential negative.), is correct. He says himself that it can not be depended on without verification, as he could not feel sure of his instrument. If it is true it must be fundamental in the theory of atmospheric electricity.

Yours very truly
William Thomson

1 Kelvin (133).

2 Kelvin (130).

3 Kelvin and Tait *TNP*.

4 Charles Piazzi Smyth, 'Astronomical experiment on the peak of Teneriffe', *Phil. Trans.* (1858), 527. Smyth (1819–1900) was director of the Edinburgh Observatory and, thus, Astronomer Royal for Scotland.

219 KELVIN to STOKES, 15 July 1862
Stokes Collection, K139

Kilmichael
July 15/62

MY DEAR STOKES

By this post (book post) I send you the substance of my paper[1] on elastic solids &c. Tomorrow I shall either write to you or send continuation or conclusion.

Prof Everett is here just now for electrometer drill. I propose sending the instrum^s^ along with him w^h^ will save freight & chances of breakage.

I enclose a letter he has written to me at my request.[2]

Yours very truly
W THOMSON

1 Kelvin (133).

2 This letter does not survive in the Stokes Collection, the Kelvin Collection, the Kelvin Papers at Glasgow, or the Royal Society.

220 KELVIN to STOKES, 16 July 1862
Stokes Collection, K140

Kilmichael, Brodick
July 16, 1862

MY DEAR STOKES

I am afraid I must ask for another week, or at the most 10 days, for finishing my paper.[1] What you have already (w^h^ I sent you by yesterday's post) might, by cutting off the last § or so, form a paper complete in itself, as it contains the complete solution of the problem for the shell, explicitly expressed, & the solution for the parallelepiped indicated enough to be completed explicitly by common Trigonom^l^ and algebraic work – each for the case in which surface displacements constitute the data.

On this account I hope it may be 'referred' at tomorrow's meeting; and perhaps also the concluding part (yet to come) might be included in the same resolution, on condition that it reaches you soon enough. It will contain the conclusion of the problem entered on in §26 (the shell, with given force ⟨of⟩ over its surface) provided I can get over a difficulty I feel at present; also, certainly the solution of the elastic shell or spheroid problem, when a given force (as centrifugal force, or tidal disturbing force) acts all through the mass, & either equilibrates itself between diff^t^ parts of the mass, or is equilibrated by surface pressure: also the particular case of tidal disturbing force, leading to the numbers used in my previous paper[2] on the rigidity of the earth: also the problem of the vibrations of a fluid globe, under mutual gravitation influence,

& either rotating or not rotating; and lastly the proof of the assertion in my former paper that rotation makes no sensible difference on the tidal deformation of an elastic solid globe, provided it is not so rapid as to itself cause any great deviation (spheroidal oblateness) from the spherical form.

Yours very truly
W THOMSON

P.S. Where ought I to address the conclusion of the paper in sending it to you?

1 Kelvin (133).
2 Kelvin (130).

221 KELVIN to STOKES, 22 July 1862
Stokes Collection, K141

Kilmichael
Brodick N.B.
July 22/62

MY DEAR STOKES

I fear every mention of the parallelepiped must be cancelled from my paper.[1] I was grievously in error in supposing that

$$\sin\frac{i\pi}{b}y \quad \sin\frac{i\pi}{c}z$$

as the expression for α, or β, or γ, at a surface $x=0$, can make each component of the displacement vanish at every point of the 4 planes $y=0$, $y=b$, $z=0$, $z=c$, and was thus led astray. If you have chanced to look into the matter you will have seen the mistake at once; and I am anxious to anticipate any announcement from you that you have found me out.

Nearly all I have written on the subject you will see stands, as being the solution for the plate, but the parallelepiped is not to be done so at all. I do not think however the problem is undoable, although I do not yet see how to do it. I am trying to rememember [*sic*] your method[2] for the hollow parallelepiped conductor of elec^y which I think may be useful.

Yours very truly
W. THOMSON

P.S. I hope almost immediately to send you the conclusion of my paper. I have just got off a 'Secular Cooling of the Earth'[3] which I have been owing to the R.S.E. & which has required to be pieced together &c & so has stopped me for three days.

1 Kelvin (133).
2 See Stokes (21) in *MPP*, I, 303–5.
3 Kelvin (128).

222 STOKES to KELVIN, 26 July 1862
Kelvin Collection, S400

Observatory Armagh Ireland[1]
26 July 1862

MY DEAR THOMSON,

I had not *yet* looked into your paper,[2] which I don't mind saying was referred to me. It is awaiting completion to go into the hands of the first referee.

My method[3] of the hollow ▱piped is the following, the problem being to find the potential due to an internal electrical point and the electricity induced on the surface.

Instead of taking the potential $V\left(=-\frac{\bar{m}}{r}\right)$ due to the point alone, having at the surface the value V_0 suppose, and completing it by a potential V' determined by the conditions $\nabla V'=0$ everywhere in the interior $\left(\nabla=\frac{d^2}{dx^2}+\frac{d^2}{dy^2}+\frac{d^2}{dz^2}\right)$ and $V'=V_0$ at the surface, which separation of the potential introduces a complexity not inherent in the problem, let us seek at once the complete potential supposing in the first instance the electricity diffused.[4] Then we must determine V by the general equation $\nabla V=-4\pi\varrho$ holding good everywhere in the interior and the condition $V=0$ at the surface. Consistently with this condition we may take

$$V=\sum\sum\sum A_{mnp}\sin\frac{m\pi x}{a}\sin\frac{n\pi y}{b}\sin\frac{p\pi z}{c}$$

and this condition being satisfied the development of $\frac{d^2V}{dx^2}$ &c will be got by differentiating under the sign ΣΣΣ so that

$$\nabla V=-\sum\sum\sum\left(\left(\frac{m\pi}{a}\right)^2+\left(\frac{n\pi}{b}\right)^2+\left(\frac{p\pi}{c}\right)^2\right)\mathrm{A}_{\mathrm{mnp}}$$

$$\times\sin(\ldots)\sin(\ldots)\sin(\ldots)$$

$$\text{and }-4\pi\varrho=-4\pi\frac{8}{abc}\int_0^a\int_0^b\int_0^c\varrho\sin\frac{m\pi x}{a}$$

$$\sin\frac{n\pi y}{b}\sin\frac{p\pi z}{c}\,dx\,dy\,dz$$

so that

$$A_{mnp} = \frac{\dfrac{2\pi}{abc}}{\left(\dfrac{m\pi}{a}\right)^2 + \left(\dfrac{n\pi}{b}\right)^2 + \left(\dfrac{p\pi}{c}\right)^2} I_{mnp}$$

where $$I_{mnp} = \int_0^a \int_0^b \int_0^c \varrho \sin \frac{m\pi x}{a} \,\&c\, dx\, dy\, dz$$

which on supposing the electricity collected into a point of w[h] the coord[s] are $x'y'z'$ becomes

$$I = \bar{m} \sin \frac{m\pi x'}{a} \sin \frac{n\pi y'}{b} \sin \frac{p\pi z'}{c}$$

One of the three summations may be effected by a known formula.

My address will be here for the next fortnight.

Yours very truly
G. G. STOKES

1 Kelvin has written on the envelope, 'Stokes. July 26 1852 [*sic*]. His method for parallelepiped'. The year *1852* is written in a different colour of ink than the rest and is no doubt a later addition. It was probably copied from the postmark on the envelope in which the *6* looks much like a *5*.
2 Kelvin (133).
3 See Stokes (21) in *MPP*, I, 303–5.
4 For Stokes's use of ▽, see letter 32, note 1.

223 KELVIN to STOKES, 12 August 1862
Stokes Collection, K142

Kilmichael, Brodick
N.B.
Aug 12, 1862

MY DEAR STOKES

Many thanks for your letter of the 26th. Before it reached me I had recovered my knowledge of your method[1] for the electrical problem of the parallelepiped. I believe it applies ⟨perfectly to⟩ to some cases of the elastic parallelepiped, but I have been obliged to attend to other work since then – and must set this question aside for some time. I have now finished the paper[2] up to the solution used in my previous paper[3] on the Rigidity of the Earth, but as the time is passed for your address at Armagh,[4] I send this there (and at the same time another line[5] to Cambridge) to ask⟨ing⟩ your proper address now. As soon as I receive it I shall send off the paper.

Yours very truly
W THOMSON

My address for ⟨letters posted several days after next Friday⟩ letters posted any time ⟨from⟩ on or before Tuesday the 19th is care of

A. Crum Esq[6]
Thornliebank
Glasgow

1 See Stokes (21) in *MPP*, I, 303–5.
2 Kelvin (133).
3 Kelvin (130).
4 See the previous letter.
5 See the next letter.
6 Alexander Crum (1828–93) was the brother of Kelvin's first wife.

224 KELVIN to STOKES, [12 August 1862][1]
Stokes Collection, K370

Kilmichael
Brodick
N.B.

MY DEAR STOKES

I shall be obliged by your writing me a line to say how I should address the concluding part of my paper,[2] which I have now in my hands ready to post. I shall send it off as soon as I hear from you.

I write to the same effect to Armagh, but I do not send the paper there as your fortnight is passed.

Yours very truly
W. THOMSON

1 This is clearly the 'line to Cambridge' referred to in the previous letter dated 12 August 1862.
2 Kelvin (133).

225 Kelvin to Stokes, 16 October 1862
Stokes Collection, K143

Steamer to Arran[1]
Oct 16/62

My dear Stokes

I hear that the copy for the plate for the curve, of Joule's & my last paper[2] is not forthcoming. You may remember perhaps that I sent the correct copy for this plate from Arran some little time after the paper had been read, & requested to have sent back to me the imperfect copy which by mistake I had given in along with the paper. The imperfect copy came back to me, and therefore I infer that the correct copy came into your hands du⟨t⟩ly. I am almost sure it was to you, not to Mr White,[3] that the correct copy was sent. If so, you will probably have it by you & I should be obliged by your forwarding it for the engraver. If not, will you remind Mr White that it is in his hands (as it would probably be, in such [a] case.) If any how it has gone amissing, I believe the imperfect copy in my hands can easily be made right, so do not let much trouble be taken in the matter.

Prof Bolzani[4] told me a few weeks ago that Lamé, in Liouville's Journal, had solved the elastic problem for spherical shells. I find that it is so (assuming his solution to be correct). The title & date are

Lamé ['] Équilibre d'Élasticité des Enveloppes
sphériques' Liouville's Journal 1854.[5]

This reference will have to be added to my paper if accepted for publication. Lamé professes to solve the problem for arbitraily given dist^n of force over outer & inner surfaces, & I see he sets about the problem with all generality. He works out no results, & leaves it in the clouds of 'Laplace's Functions'. If correct however his general solution must be tantamount to mine, when the superficial data are force.

So far as I see he does not touch the [easier][6] problem when the superficial data are displacements.

Yours truly
W Thomson

I shall tell this to Hopkins, from whom I have a letter[7] asking some explanations on certain points of the paper.

My address for letters posted Camb. or London on or before Monday Oct 20, will be Kilmichael Brodick N.B.

later, address Natural Philosophy Laboratory College, Glasgow

I have at last, for the first time got a students' working laboratory. It will be ready for the beginning of this session & will accommodate about 24 students.

1 Stokes has written at the top of the letter, 'An[d] 18/10/62'.
2 Kelvin and Joule (10).
3 Walter White was assistant secretary of the Royal Society from 1861 to 1885.
4 Apparently, J. Bolzani, author of 'Mathematische Untersuchungen über die Verbreitung des elektrischen Stromes in Körpern von gegebener Gestalt', *Archiv für wissenschaftliche Kunde von Russland*, **16** (1857), 45–110.
5 G. Lamé, 'Mémoire sur l'équilibre d'élasticité des enveloppes sphériques', *Comptes Rendus*, **37** (1853), 145–9, and Liouville's *Journal de mathématiques pures et appliquées*, **19** (1854), 51–87. Gabriel Lamé (1795–1870) was professor of physics at the École Polytechnique from 1832 to 1844 when he went to the University of Paris, first as graduate examiner and then as professor of mathematical physics and probability.
6 Kelvin's brackets.
7 W. Hopkins to Kelvin, 10 [October] 1862, Kelvin Collection, H127. Hopkins mistakenly dated the letter 'September'. The postmark, however, is 'October', and Kelvin's notation on the letter indicates that he received it on 13 October 1862.

226 KELVIN to STOKES, 26 November 1862

Stokes Collection, K144

Natural Philosophy Laboratory[1]
College, Glasgow
Nov 26/62

DEAR STOKES

I enclose a very short abstract according to your request. Not that the *very short* was by your request, but I hope you will think this enough, considering that the paper[2] is a mathematical one & must speak for itself.

I hope to follow it up with more important mathematical work, which I have already gone through, regarding rotatory motion, vibrations of elastic spheroids, waves at the surface of water of various depths, taking into account the compressibility of water &c. I do not know when I shall get all this written out as my whole time is taken up now between my College duties, & work for the book with Tait.[3]

Yours very truly
W THOMSON

P.S. I intended to write an Appendix giving shortly the analytical principles (equivalent to Laplace's functions: but practically simpl[if]ied & made symmetrical) actually used: also a new development of Laplace's functions: also, for the convenience of the reader, proofs of the equations of elastic solids &c. If you think this desirable I could do it yet on short notice.

1 Kelvin has added a footnote reading:
A new institution of a most satisfactory character, which I have commenced with this session, owing to University Commissioners' grant of assistant, and the College, proper rooms &c.
2 Kelvin (133).
3 Kelvin and Tait *TNP*.

227 KELVIN to STOKES, 8 December 1862

Stokes Collection, K145

Glasgow College
Dec 8/62

MY DEAR STOKES

Rankine some time ago told me of his investigation showing the trochoid as being rigorously the form of deep sea waves, motion in two dimensions. I did not learn his investigation thoroughly, but so far as I saw of it I saw no fallacy. Now that I think of it, he finds no translation in the upper parts, which I think must be wrong. This I think I see independently of your investigation. It is curious that your investigation should lead to the trochoid as a second (or third) approximation to the true form; but, from the fact of his finding no translation I suspect Rankine must have fallen into some error. I have had several conversations with him about it, but not since this occurred to me. I understand he has sent you a paper[1] for the R.S. on the subject and I should feel much obliged if you would let me know what you think of it: that is to say if, as I suppose is the case, you come immediately to a conclusion on its merits.

Will you also tell me if Earnshaw[2] on Russell's two great Solitary waves[3] is right, on the whole, or altogether? I shall have to make out something about waves for the book of Tait's & mine)[4] which is now in process of printing. I know nearly all about ⟨infinitely small⟩ waves of infinitely small height in deep water: also about ⟨infinitely⟩ waves of infinitely small height, in water infinitely shallow in proportion to the wave length. I have not worked out the transition cases when the wave is not infinitely long in comparison with the depth. This I presume is Earnshaw's.

Yours very truly
W. THOMSON

P.S. ⟨(This⟩ [?] Here is how I see there must be translation when the height is not infinitely small. Superimpose a velocity V, equal to that of the waves, on the whole mass of fluid. This will reduce to a case of steady motion. AB, and $A'B'$, any two verticals through crests (or hollows) are clearly surfaces for which $\phi = \text{const}$. Let q denote veloc. at any point. Hence

$\int q\,ds$ for a line of motion from A to A'
$= \int q\,ds$ " " " " " " B " B'.

(the motion being of course such that $udx + vdy$ is a complete differ^l).

But the path $\int ds$ is longer from A to A' than [from] B to B'. Hence the velocity is on the whole less from A to A' than from B to B'. Again, the path is

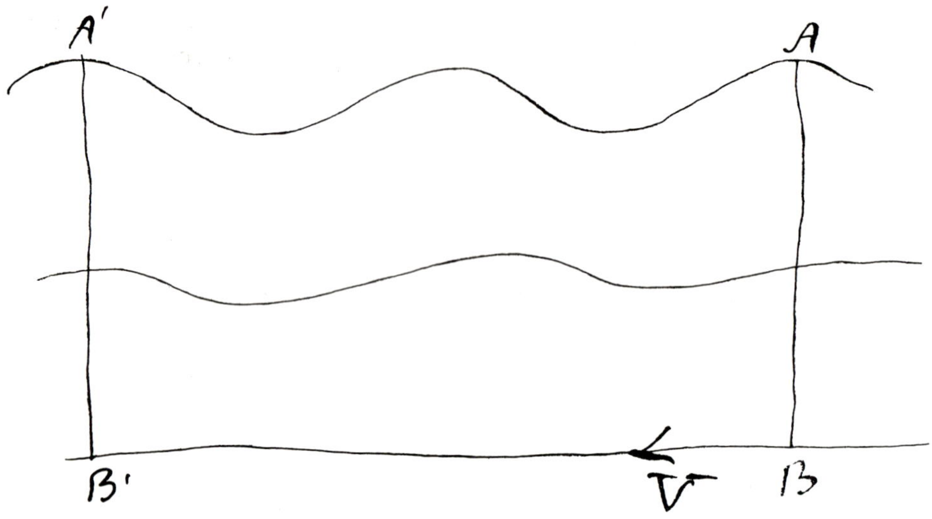

longer from A to A' than from B to B'. Therefore a fortiori a particle from A to A' must take more time than from B to B'. This is equivalent to your 'translation'.

1 W. J. M. Rankine, 'On the exact form and motion of waves near the surface of deep water', *Phil. Trans.* (1863), 127–38. Pages 137–8 contain additions dated October 1862 and June 1863 which explain the difference between Rankine's investigation and that of Stokes, as contained in Stokes (29). Giving no reference, Rankine credited J. S. Russell with first stating 'the trochoidal form of waves'. (P. 127.)

2 S. Earnshaw, 'The mathematical theory of the two great solitary waves of the first order', *Camb. Phil. Soc. Trans.*, **8** (1849), 326–41.

3 J. S. Russell, 'Report on waves', *Brit. Assoc. Rep.* (1844), 311–90. In the papers mentioned in the two previous notes, both Stokes and Earnshaw relied on Russell's report. John Scott Russell (1808–82) was a civil engineer in London.

4 Kelvin and Tait *TNP*.

228 STOKES to KELVIN, 9 December 1862

Kelvin Collection, S401

Lensfield Cottage Cambridge[1]
⟨8⟩9th Decr 1862

MY DEAR THOMSON,

I have not yet examined Rankine's paper[2] but I shall shortly do so as I am one of the referees. The work lookes [*sic*] so simple that it probably will be easily verified or disproved.

I see no necessary opposition between his results and mine[3] but as the single assumption (as a means of selection a particular kind of motion not as a substitute for any of the necessary equations) that I make besides that which is common to both investigations that we are dealing with a regular succession of waves is that $udx + [vdy + wdz]$ is a perfect differential and as I am confident of my own investigation it follows that in his (assumed to be correct) $udx + [vdy + wdz]$ must *not* be a perfect differential and consequently the case he contemplates must be applicable to a fluid which independently of the waves moves horizontally with a motion decreasing rapidly from the surface downwards.

Earnshaw's investigation[4] gives a rigorous result *for the portion of fluid comprised within planes at the beginning and end of the wave* but the condition, at the connexion of the wave and still water are violated as I pointed out to him when the paper was referred to me and as he admitted and altered his original MS accordingly. In fact the result as a whole would be true only on the supposition of the existence of impulsive forces acting for a moment on an infinitely thin layer of particles and changing at the next moment to another layer and so on.

The investigation of oscillatory waves for a finite depth is not more difficult than for an infinite only the expressions are somewhat longer.

Yours very truly
G. G. STOKES

1 Kelvin has written on the envelope, 'Dec 1862. Stokes on Earnshaw & Rankine Waves'.
2 See the previous letter, note 1.
3 Stokes (29). See the previous letter, note 1.
4 See the previous letter, note 2.

229 KELVIN to STOKES, 18 December 1862[1]

Stokes Collection, K146

Glasgow College
Dec 18, 1862

MY DEAR STOKES

Many thanks for your letter on hydrokinetics. What you tell me in it about Earnshaw's paper[2] will save me a great deal of trouble. I fancied he might have taken up, on strict analytical principles, the transition from the case of the depth [of] an infinitely small fraction of the wave length, (in which case clearly a solitary wave, either positive or negative, according to the law of a vibrating string, can be propagated, and all wave lengths go at same speed) to the case of the depth comparable with the wave length or infinitely great in comparison with the wave length, in neither of which can a solitary wave be propagated without change. But from what you say it appears that he throws no light on this question. Both from what you say, & from what I see in the opening of his paper, I conclude that his solution is not practically applicable to any real case. He opens his paper in a most unsatisfactory manner, proposing conditions as from observation, which, if true, must come out as mathematical propositions; and seems to have a vague idea of [The rest of the letter is missing.]

1 See letter 217, note 1.
2 See letter 227, note 2.

230 KELVIN to STOKES, 24 January 1863
Stokes Collection, K147

College, Glasgow
Jan 24, 1863

MY DEAR STOKES

I wrote you a letter on hydrodynamics in Jan 1858. If you chance to have preserved it & to be able to lay your hands on it without trouble (the latter condition essential) will you send it to me, as I have not preserved more than some slight memoranda as to its contents?

I commenced some weeks ago a letter thanking you for your answers about Rankine & Earnshaw,[1] and asking some more questions regarding the latter, but I was forced to leave the subject for a time, and I need not trouble you again in the meantime. Rankine is working at his solution, having extended it to finite depth. He is looking after the rotational quality.

Everett was here in summer and after a good deal of *drill*, got away [with] his former station electrometer (which met with the rough usage formerly) and a new portable one. He has carried them both safely to Windsor, Nova Scotia, and has commenced observing with them. I have had three letters from him recounting observations & I have engaged him to send me for communication to the R.S. a paper[2] of observation before the end of this session. A Mr Binney[3] there has bought the station electrometer & water collector for £20. The whole amount of the account, which I enclose, is £20–4–5. If as I understood was formerly the case, the Royal Society is still disposed to entertain my application for funds, will you lay these accounts before the⟨m⟩ Grant Committee?

The construction of the instruments for Mr Del[l]mann[4] was on the point of being commenced last August, when a new plan of electrometer occurred to me, which promised first to supersede the old form of portable electrometer, and which now seems likely to be most notable also for a stationary atmospheric electrometer. I have tested the new plan carefully and am now only waiting to get ⟨a⟩ good micrometer screws, to commence making instruments for Mr Del[l]mann accordingly. In the mean time, the sum granted by the Royal Society remains in White's[5] hands to the credit of the instruments, which I hope soon to be able to report as completed.

If you see Tyndall, will you tell him that I can give him a little portable electrometer $3\frac{1}{2}$ inches diameter by $3\frac{1}{2}$ i[nches] high, which can be carried with ease, even to the top of the Matterhorn, if he goes there.

I am sending to Troughton and Simms[6] for micrometer screws, of aluminium bronze. If you chance to know of any thing better, that I should do, you would oblige me much by telling me.

Yours very truly
W THOMSON

P.S. I am afraid I have neglected Moon's paper[7] too long, but I shall try to return it immediately (giving Rankine first an opportunity of seeing it). It will not do at all for publication I believe, but I must make sure that I know it sufficiently before committing myself to such an opinion.

Tyndall's paper[8] is with me & I hope soon to return it.

1 See letter 227, notes 1 and 2.
2 J. D. Everett, 'Account of observations of atmospheric electricity taken at Windsor, Nova Scotia', *Proc. Roy. Soc.*, **12** (1862–3), 683–707.
3 Edward Binney, mentioned in J. D. Everett to Kelvin, 13 October 1862, Kelvin Collection, E101.
4 Johann Friedrich Georg Dellmann (1805–70) taught at the gymnasium at Kreuznach.
5 James White (d. 1884) was a Glasgow instrument maker on whom Kelvin relied heavily and whose business was taken over after his death by Kelvin and others, the name of the firm being changed in 1901 to Kelvin and James White, Ltd.
6 Troughton and Simms were mathematical instrument makers in London.
7 R. Moon, 'On the true theory of pressure as applied to elastic fluids', *Proc. Roy. Soc.*, **12** (1862–3), 242–6. Robert Moon (1817–89) was eighth wrangler in 1838 and called to the bar in 1844. Kelvin's report on Moon's paper is in the Royal Society, RR.5.158 and 159.
8 J. Tyndall, 'On the relation of radiant heat to aqueous vapour. Third memoir', *Phil. Trans.* (1863), 1–12. Kelvin's referee's report is in the Royal Society, RR.5.273.

231 KELVIN to STOKES, 28 March 1863

Stokes Collection, K148

MY DEAR STOKES

Many thanks for your letter of yesterday. I meant decidedly what you thought me to mean, as to the two computers.

Yours very truly
WILLIAM THOMSON

Glasgow College
March 28/63

232 Kelvin to Stokes, 29 October 1863

Stokes Collection, RS419

University of London,
Burlington House, W.
Oct 29 1863

Dear Stokes

Hearing from Maxwell that it was to him, not me, that it had been intended to send Chambers' paper[1] I have given it to him.

I received it at the Royal Institution last June, with a note from you, so I suppose I was one of the referees, and I therefore leave a report[2] for you upon it: also a note[3] I have written w^h might be added to it if you think proper.

I send also an addition and Appendix to my paper[4] on Dynamical Problems regarding Elastic spheroids and Liquid globes. The addition is referred to in the *title*, and therefore I let it take the form of addition, not appendix but I have written on it 'added since the paper was read'. The whole is rather longer than I estimated when I wrote to you last but I hope the council will sanction its being printed along with the paper. I have prepared both the addition and Appendix very carefully for the press and although they look very rough I believe the printers will find no difficulty with them. They are quite ready for them as I have just returned the proofs up to the end of the matter already in type, for revise, and I told them I would try to get your sanction for having them put in hands immediately.

The §§ on the plane plate were in the original paper, and the only addition there was a connecting § of about six lines with no results. I therefore cancelled your footnote but told the printers I should write to you explaining it. The real change in the paper was cancelling 4 or 5 leaves which were wrong, and which you will remember I asked leave to withdraw before the whole paper was in your hands.

I am sorry not to have seen you in London, as I must leave tomorrow afternoon.

Yours truly
W Thomson

1 C. Chambers, 'The nature of the sun's magnetic action on the earth', *Phil. Trans.* (1863), 503–16. Charles Chambers (1834–96) was an assistant at Kew Observatory from 1856 to 1863 and later director of the government observatory near Bombay.

2 Kelvin's report is in the Royal Society, RR.5.47.

3 Kelvin (132A).

4 Kelvin (133).

233 Kelvin to Stokes, 30 October 1863
Stokes Collection, K149

Glasgow College
Oct 30/63

My dear Stokes

I cannot spare time to write out a fresh copy of my Appendix;[1] and my assistant is so very much pressed with other work at present that it would be a long time before I could get it copied by him. I cannot get it copied by a professional copyist, because of the 'Algebra'. But you shall have a printed copy from Edinburgh better and sooner than I could in any way get you a written copy.

I took extreme care to make the copy *correct* and *clear* for the compositor; although it is in an intolerable form for a *reader*, until it gets into type. I thought that (as you would probably not have it committed to referees) you would excuse my letting it into your hands in so rough a shape. But so far as the compositor is concerned I feel quite sure he would have no trouble. The only difficulty ever complained of *or felt* with reference to pencil copy is I find, on the score of light, and when ⟨the⟩ his light is good the compositor can set up from pencil copy as easily as from ink copy. I feel confident that you will find the printed copy you will get from the Edinburgh printers quite as correct as a first proof generally is. I shall see no proof of it before it goes to you; but I shall send, either to you or to the London printers, a corrected proof about two days later than you receive the proof from Edinburgh.

I should feel much obliged by your letting those who agreed with you in thinking the pencil scroll unfit for the compositors, know the substance of the above when you can do so without trouble.

Yours very truly
William Thomson

1 Kelvin (133). See the previous letter.

234 Kelvin to Stokes, 2 November 1863
Stokes Collection, K150

2 College, Glasgow
Nov 2/63

Dear Stokes

Here is the affair for forces given, and displacements infinitely small.

Let X, Y, Z, be forces applied at any point (x, y, z) of an infinite heterogen[e]ous solid, reckoned per unit of its volume. X, Y, Z may be regarded as each an absolutely arbitrary function of x, y, z.

Regard X, Y, Z as acting each constant, when the particle of the body acted upon moves through any infinitely small displacement. Then $\langle x\alpha+\rangle$

$$\int_{-\infty}^{+\infty}\iint(X\alpha+Y\beta+Z\gamma)\,dx\,dy\,dz$$

is the work they do if the whole solid is first held unstrained, and then gradually allowed or caused to take any arbitrary displacement $\langle x, y\rangle$ (α, β, γ): and of this work a portion $\iiint w\,dx\,dy\,dz$ goes to generate potential energy in the solid (w being the same quad[rati]c fu^{n} as in my 'App^{x}').[1] Hence for equilibrium

$$\int_{-\infty}^{+\infty}\iint\{X\alpha+Y\beta+Z\gamma-w\}\,dx\,dy\,dz \tag{1}$$

must be a max^{m} or minimum. For stable equil^{m} it must be a max^{m}. (All this may be stated in a somewhat different order, more clearly). Hence, taking the variation, & dealing with $\iiint \delta w\,dx\,dy\,dz$ as in the App^{x},

$$\iiint\left\{\left(X\langle\delta\rangle+\frac{d}{dx}\frac{dw}{de}+\frac{d}{dy}\frac{dw}{dc}+\frac{d}{dz}\frac{dw}{db}\right)\delta\alpha+\ldots\right\}dx\,dy\,dz=0$$

Hence the equ^{ns} of equil^{m} are

$$(2)\quad\left\{\begin{array}{l} X+\dfrac{d}{dx}\dfrac{dw}{de}+\dfrac{d}{dy}\dfrac{dw}{dc}+\dfrac{d}{dz}\dfrac{dw}{db}=0\\ \qquad\qquad\&\text{c}\end{array}\right.$$

Now let α', β', γ' be any $\langle$other$\rangle$ displacement not fulfilling these equ^{ns}: and let W & W' denote the values of W for α, β, γ & for α', β', γ'. Then

$$W'-W=\iiint dx\,dy\,dz\left\{X(\alpha'-\alpha)-\frac{dw}{de}(e'-e)\right.$$
$$\left.-\frac{dw}{dc}(c'-c)-\frac{dw}{db}(b'-b)\ \&\text{c}-H\right\}$$

where H denotes the same homog^{s} quad^{c} function of $e'-e$ &c that w is of e &c.

Integrating by parts and attending to (2) we see that this becomes $\langle W'-W-\iiint\rangle$

$$W'-W=-\iiint H\,dx\,dy\,dz$$

& therefore is essentially positive. Hence W is greater for (α, β, γ) which satisfy (2) than for any diff^{t} disp^{t} $(\alpha', \beta', \gamma')$. Hence no other disp^{t} can satisfy (2).

Now let the coefficients in the quadratic function be ⟨nothing⟩ each zero outside a certain surface, S. They become finite as rapidly as you please in entering across S, and they may be either constant or variable within S. And let X, Y, Z be each very large for a very thin shell just within S. They may vanish or be what you please for the space within this shell. This shell being supposed infinitely thin you have the case of superficial forces. Of course by attending to limits.

Thanks for your reminder about Lamé.[2] I had not forgotten it but was planning where to put in the statement. I thought of putting it in at the end of the part already in type, as there was a leaf missing (it had been lost somehow, ⟨which required⟩ not in my hands, *after* the paper was first given in; but I remember seeing it loose, torn off, when the M.SS. was returned to me) and I have therefore to write in something in any case, to make a connection. But perhaps as you say it would be better to make the statement at the outset of the paper, and I daresay there it will trouble the printers as little, as they have space to spare on the first page.

I had kept Tyndall's paper[3] carefully in the envelope in which I received it, and when I took it out for the last time, before sending it to you, I missed the drawing. I do not remember either seeing or missing the drawing before and as I cannot find it, and had no receptacles in which it can have got buried, I think I cannot have received it with the paper. I hope it will be found in the R.S.

Yours very truly
WILLIAM THOMSON

P.S. I find Lamé indicates the solution ⟨for⟩ for the spherical shell when there are internal forces in certain cases viz (1) constant gravity in parallel lines (2) centrif force (3) attraction in lines through the centre of the shell.

1 Kelvin (133).
2 See letter 225, note 5.
3 See letter 230, note 8.

235 KELVIN to STOKES, 3 November 1863

Stokes Collection, K151

2 College Glasgow
Nov 3/63

DEAR STOKES

When I said yesterday that X, Y, Z are absolutely arb[y] fu[ns] of x, y, z I should have added of course subject only to the cond[ns]

$$\iiint X\,dx\,dy\,dz = 0 \quad \&c$$

$$\& \quad \iiint (Yz - Zx)\,dx\,dy\,dz = 0.$$

These conditions secure that

$$\iiint (X\alpha + Y\beta + Z\gamma)\,dx\,dy\,dz = 0$$

for every displacement that produces no alteration of form or volume in any part of the solid.

Hence we can show analytically that

$$\iiint \{(X\alpha + Y\beta + Z\gamma) - w\}\,dx\,dy\,dz$$

has a maximum (which what I wrote yesterday demonstrates if we assume that there is some configuration of equilibrium, ⟨)⟩ but did not demonstrate without thus relying on dynamical knowledge) ⟨For, X, Y, Z being of course infinitely small⟩ For, change α, β, γ into $n\alpha$, $n\beta$, $n\gamma$. The integral becomes

$$\iiint \{n(X\alpha + Y\beta + Z\gamma) - n^2 w\}\,dx\,dy\,dz$$

⟨Hence⟩ By making n large enough the value of this may be made negative: & if α, β, γ are any thing which makes $X\alpha + Y\beta + Z\gamma$ positive we may by making n small enough, render the integral positive. When $n = 0$ it vanishes. Hence it has a maximum. And since X, Y, Z are infinitely small, this maximum is attained with infinitely small values of α, β, γ.

Perhaps if I can get this & yesterday's statement put into shape in time I may introduce it into the Appendix.[1]

Yours very truly
WILLIAM THOMSON

1 Kelvin (133).

236 STOKES to KELVIN, 3 November 1863
Glasgow University Library, Kelvin Papers, S84

Lensfield Cottage Cambridge
3^d Novr 1863

MY DEAR THOMSON,

I ⟨hardly⟩ never doubted that the compositors could ⟨compose⟩ set up from the MS you gave me almost free from error, that is, ⟨is⟩ as well as from a neatly written MS, but did not think it fair to impose such a task upon them, especially as the MS. was almost entirely ⟨spoken⟩ written in pencil, which must be very trying to them, working as they must a good deal at this time of year very much by ⟨sun-⟩ gas-light. I was often urged before to ⟨reject⟩ refuse to accept your[1] rough MSS, but hitherto I have taken them, seeing that though rum uns to look at they were good uns to go, cruelty to animals not being taken into consideration. I doubt not a compositor would set up the type, if my letter were to be printed, as correctly from page 1 as from the rest, yet I should think it hardly fair to put that page before him. Had your MS been written in ink I should have sent it to press, though I should probably have been urged to the contrary; but consideration for the compositors' eyes moved my compassion, so that the fact of the MS. being written in pencil was the last straw that broke the camel's back (poor patient animal!)

I heard today from Mr White that there is no trace of Tyndall's drawing[2] in our drawers, and if it were not sent to you I cannot think what has become of it. Will you have the goodness to engage so far in antiquarian research as to endeavour to make out something of its history at that remote period when the paper itself first came into your possession?

Yours very truly
G. G. STOKES

1 This is the end of the first page of Stokes's letter. In addition to the many deletions, Stokes has interlined several words to illustrate the roughness of Kelvin's own manuscript for the appendix to Kelvin (133).

2 See letter 230, note 8.

237 KELVIN to STOKES, 28 November 1863

Stokes Collection, K152

2 College, Glasgow
Nov 28/63

DEAR STOKES

I forget exactly what I wrote to you regarding the uniqueness of the solution for elastic solid, forces given, but I suspect there is a fallacy in it, obvious enough as to the *thing*; but a little puzzling as to the analysis expressing it; which later is to be rectified by taking into account terms of second degree in the variation of

$$\iiint (X\alpha + Y\beta + Z\gamma)\,dx\,dy\,dz$$

when $\delta\alpha$, $\delta\beta$, $\delta\gamma$ are such as to represent variation in position, of a rigid body, ⟨and if⟩ when X, Y, Z vary so that the same particle of the body experiences the same force in its different positions.

The fact is there are obviously two positions, one of stable, and the other of unstable, equilibrium if X, Y, Z remain the same for the same particle of the body in its different positions; I am not quite sure whether what I wrote to you was inconsistent with this or not: but I am just now trying to get the whole matter ⟨into⟩ written out in correct shape.

I have had nothing from the printers all this time, though I expected a revise[1] before leaving London.

The additional matter which you now have in type can, if you please, go into their hands as it is. There are only two corrections to be made on it, so far as I have yet been able to find. One of them I enclose. The other is to dele a suffix 2 in equ^{n} (112) and in the line following it. I have made besides one or two slight changes by way of improvement, but I keep my own copy to see if I can get in the case of *force given*, in time.

Yours very truly
W THOMSON

P.S. On farther consideration I see that the internal equilibrium of a heterogeneous elastic solid may be unstable & consequently the solution multiple, when force acts on the particles. Thus for instance NS a magnetic bar enclosed in a shape of jelly N', S' fixed external magnetic poles.

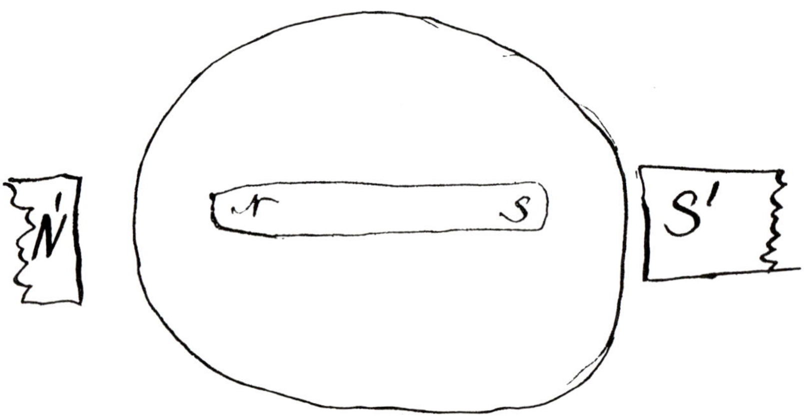

The upshot is I suppose that you were right in what you wrote last, & I wrong in my answer to it.

I must think about it a little longer before [I][2] can decide whether to offer you anything more in the way of Appendix to Appendix to Appendix to my present paper,[3] but in the meantime I hope the printers are pleasing themselves or or and [*sic*] you, and not waiting for more from me.

1 Of Kelvin (133).
2 The corner of the letter is torn away here, but *I* seems to be the missing word.
3 Kelvin (133).

238 KELVIN to STOKES, 30 November 1863
Stokes Collection, K153

2 College
Glasgow
Nov 30/63

DEAR STOKES

Merely to find the equations of equilibrium when the data are forces is, as said in §64, very easy. But to investigate the criterion of singleness & discrimination of stability and instability is not to be thought of for the present paper.[1] I have added therefore a §68 referring to the question, and if you approve the whole may go to press as it is.

Yours truly
W THOMSON

1 Kelvin (133).

239 KELVIN to STOKES, 3 December 1863
Stokes Collection, K154

2 College, Glasgow[1]
Dec 3/63

MY DEAR STOKES

Jenkin wishes to become a candidate for the Royal Society and I have asked M[r] White to send him a blank form accordingly.[2] The enclosed slip is his own proposal for filling up the blank, before signatures. A good deal more might be said about his ability, & his zeal for science. Do you think such should be added to his own ⟨sugg⟩ statement, or is [it] not as well to let the latter speak for itself. He certainly is very zealous for science, & has already done good work besides having proved remarkable ability. I sh[d] feel much obliged by a line from you to say what you think advisable. Or if you think proper to send Jenkin's draft statement back to himself, with any suggestion that may occur to you, you need not be at the trouble of writing to me.[3] I hope my App[x] reached you a few days ago, with the additional § as to data of force.[4]

Yours truly
W THOMSON

I have nothing yet from Taylor & Francis.[5]

Rankine's paper[6] rec[d] & will be attended to & returned I hope within the month.

1 Stokes has written at the top of the letter, 'ans[d] 4/12/63'.
2 H. C. F. Jenkin became a fellow of the Royal Society in 1865.

3 Kelvin has written in the margin, 'His address is F Jenkin Esq 6 Duke Street Adelphi London'.
4 Kelvin (133).
5 Taylor and Francis were printers and publishers in London who published the Royal Society's *Proceedings* and *Philosophical Transactions*. Richard Taylor (1781–1858) had also edited Taylor's *Scientific Memoirs*, the fifth and final volume of which appeared in 1852. William Francis (1817–1904) also served as editor of the *Philosophical Magazine*.
6 W. J. M. Rankine, 'On plane water-lines in two dimensions', *Phil. Trans.* (1864), 369–91.

240 Kelvin to Stokes, 19 December 1863

Stokes Collection, K155

2 College Glasgow
Dec 19/63

Dear Stokes

Not till this morning I have received from the printers revise of my paper,[1] with App^{x}. I see that the latter does *not* contain the addition, in reference to the problem of *force given*, which I sent you about a fortnight ago. I think it important that the question of stability or instability sh^{d} be noticed, & therefore I would like to insert something about it before returning the proof. If you have no objections to the section I added (in w^{h} I quote you) will you send me as soon as you can conveniently do so, the proof on which I wrote it. Or if ⟨not⟩ you have any objection I should be glad to hear from you. In any case I should be glad to have the proof, ⟨as⟩ I last sent you, as it contains one or two corrections.

Yours very truly
W Thomson

1 Kelvin (133).

241 Kelvin to Stokes, 27 December 1863

Stokes Collection, K156

The Rouken,
Thornliebank,
Glasgow Dec 27/63

My dear Stokes

I am very sorry to hear of the loss[1] you have had and I feel much concerned about the danger you have yourself suffered. I hope you are still improving steadily, and that you will ⟨still⟩ soon be quite strong again.

I am sorry that you should have had trouble looking for my proof.[2] Do not waste another thought upon it. I can easily supply the corrections &c on the R.S. proof.

Yours always truly
W Thomson

I hope the others of your family have perfectly recovered, if not escaped the scarlet fever.

1 Stokes's infant child died.
2 Kelvin (133).

242 Kelvin to Stokes, 17 February 1864
Royal Society of London, RR.5.214

2 College, Glasgow, Feb 17/64

My dear Stokes

I am ordered to take as little fatigue of either standing or walking about as possible, having done rather too much at Christmas between curling & walking, and been (apparently in consequence) much lamer since. I had scarcely any sense whatever of lameness before Christmas & am told that I have no reason for apprehension if I give myself rest.

And keep my leg up which brings me to the point i.e. that I must ask you to excuse my reporting on Rankine in pencil. I sent it to you book post two days ago.

I have gone carefully through his paper,[1] and find several decided novelties in it, besides the practical application. The most important of these seem to me to be the lines trajectories[2] of minimum & of max^{m} velocity of water relatively to ⟨solid⟩ solid, & the appli[c]ation to show the crests & hollows of waves produced by the motion; the 'lissenoid' [spelled *lissoneoïd* in Rankine's paper] core is very remarkable.

The practical application would I believe by *very approximate* for the case

of an infinitely deep ship, of prismatic hor^{l} sec^{n}; & all the author's principles & practical deductions would hold for this case; also approximately for a pier such as that of a bridge. But the application is not, as the author supposes, even approximate for a real ship. His statements on this subject would require *very decided correction* but this would leave the main substance of the paper untouched.

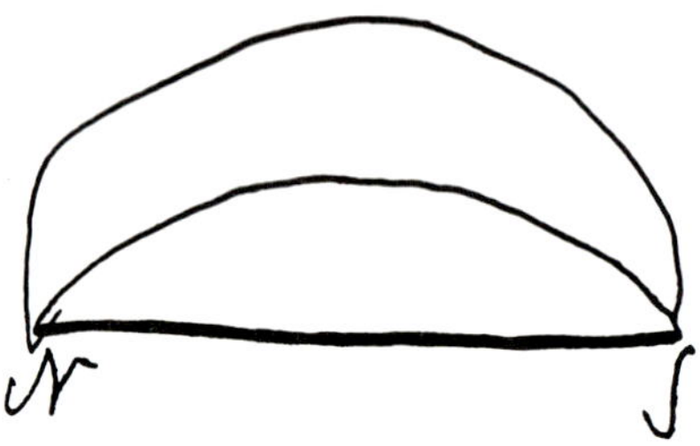

A much closer approx^{n} to real ship water lines would be had by taking a solid of revolution moving through a liquid instead of a case in which the motion is entirely in parallel planes. Thus by starting with the known lines of force between the poles of an inf^{y} thin unif^{y} & long. magnetised bar (instead of the circles through *N* & *S*, which he takes, & which are the lines of force

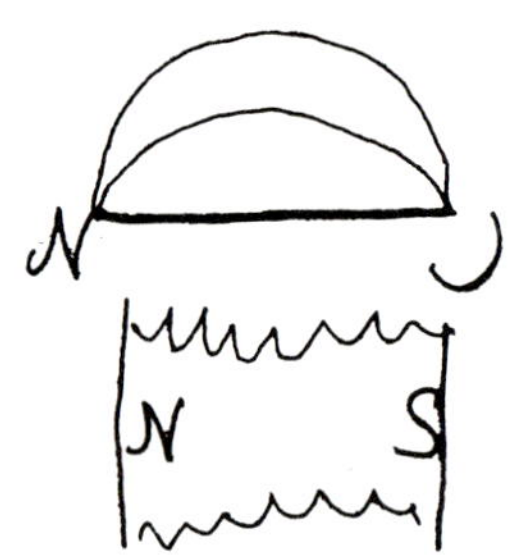

between the two parallel edges of an infinitely long transversely magnetised thin lamina) he would get in a similar way a set of 'neoids' more 'neoidal' than his own, and also 'oogenous' although from a differently formed oval. These would apply very approximately to the case of a ship whose draft of water is equal to ⟨its⟩ half her breadth, but even this is a good deal too deep for a practical ship.

The general character of the deductions would still hold, & it would be both theoretically & practically interesting to compare the forms of lines, the proportions for the lissenoid, &^{c} &^{c} with those of Rankine's present paper.

Maxwell I believe was the first to use the diagonal method of drawing, for lines of force & I think should be referred to.[3] I am not sure whether Lamé should be referred to or not; but perhaps your paper, actual[l]y referred to, contains the whole matter & really anticipates Lamé, w^{h} I did not remark before.[4]

Yours very truly
WT

I can see Rankine & talk over the matter with him if desired.

1 See letter 239, note 6.
2 Kelvin has written *trajectories* immediately above *lines* but has not deleted *lines*.
3 Rankine cited Maxwell in a brief appendix.
4 Rankine did cite Stokes (1) but not Lamé.

243 STOKES to KELVIN, 18 February 1864
Royal Society of London, RR.5.215

College, Glasgow
Feb 18/64

DEAR STOKES

Water lines deduced from an oval of revolution correspond to a ship drawing water equal to *half* its breadth (clearly) not its breadth, as I said yesterday inadvertently. This is not so far out for a real ship: but even still is too great a proportion of depth.

Yours truly
W THOMSON

244 KELVIN to STOKES, 20 April 1864
Stokes Collection, K157

2 College, Glasgow
April 20/64

MY DEAR STOKES

Besides thanking you for your last letter and for the trouble you have taken with my paper,[1] (which I hope will after all, notwithstanding the stiffness of the subject be, thanks to you and to the good 'reader' of the proofs, tolerably accurate) I have for the last three or four weeks, an accumulation of things to write but have been continually prevented by the incessant engagements of this period of our Glasgow session with *the book*[2] (w^{h} goes on very slowly being only now at the 400th page) in addition. One of these will now bear no farther delay, and that is to remind you of Jenkin's qualifications as a candidate for the R.S.,[3] of which you said it might be well I should write you something about this time. He is remarkably clever and eager in the pursuit of *science*, both in acquiring knowledge himself, and in endeavouring to obtain new results by experiment. He has already communicated several valuable papers to the Royal Society. His first[4] (which was not published in the Transactions, but in the Government Telegraph Committee's Blue Book) is in one respect the most valuable; as it contains an accurate determination of a property of matter (the specific resistance of the insulator of the cables tested) in approximate absolute measure. His last[5] is very interesting and valuable with reference to practical signalling, and contains remarkable verifications of the mathematical theory, besides supplying the numbers required for reducing the results to absolute measure for speed of signalling. This result, when the resistance is known in absolute electromagnetic measure, leads to the determination of the capacity for charge, in absolute electromagnetic measure, according to the mathematical theory; and a measurement of the ⟨electrostatic⟩ capacity for charge in electrostatic measure is all that has to be added to complete the data for a comparison between electrostatic and electromagnetic measure, quite independent of Weber's. Jenkin has worked hard for the units' committee[6] in this most important & difficult matter, & along with Maxwell, and B. Stewart, have already gone through the work ⟨fro⟩ [?] of measuring resistance in absolute electrostatic measure, with a degree of accuracy much superior to that attained by Weber. Altogether Jenkin is a most useful man for science, and a remarkably sound worker both in pure science and in engineering applications. He is in my opinion eminently deserving of a place in the Royal Society.

I wished also to write to you that I have now completely solved the question of finding the vibrations of elastic solid or hollow spheres; which even after the solution of the equilibrium problem, has difficulties of its own. I see every prospect of being able to write it out completely for the R.S. before the end of this session. If I do not succeed in doing this, I may perhaps at least be allowed to make a communication of some results which could be published in the Proceedings.[7]

I suppose it is too late to apply this year for any ⟨grant⟩ allowance out of the Government Grant for scientific purposes; but if it is not, or if the application could stand over till next year, I should wish to make one for assistance in pendulum observations. I have, in the course of the last two years, formed and partly tested a new plan which has two objects.

1. A very accurate absolute determination at head quarters.

2. Observations in different places by means of a transportable pendulum, adapted to be extremely invariable.

The absolute pendulum I propose should be shaped as in the sketch; of flat

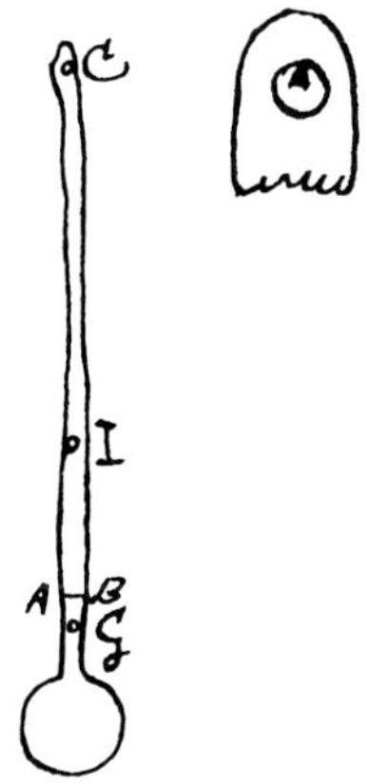

stiff metal (I have at present one cut of sheet brass about $\frac{1}{8}$ i[nch] thick but it is too flexible. A casting of bell-metal will no doubt be the best thing). It is to have three holes bored through it: one as near as possible to the end of the bar: another (*G*) through its centre of gravity: and the third *I*, at a distance from the second approximately equal to the radius of gyration, as nearly as may be. I swing it on a fixed knife edge (instead of as in Kater's method[8] by a knife edge attached to itself). By certain statical (& perhaps ⟨dynam⟩ kinetic) experiments with the pendulum supported on a knife edge through the hole *G*, the distance of the centre of gravity from a mark *AB* across the bar, may be determined with extreme accuracy. Then by oscillating it successively on a

knife edge through *C* and through *I*, the values of *k* and *g* may be determined, with far less probable error as is easily proved than by Kater's method. (Of course *k* is *approximately* known, before the hole *I* is bored).

I have a method of observing which will give the determination of the oscillations about *G* with the same accuracy as is attainable by the common method of coincidences in cases in which the time is approximately 1 sec. This consists in attaching ⟨(on Helmhol[t]z's principle) a lense [*sic*] (or a⟩ a mirror (on Lissajou[s]'s principle)[9] or an object lense (on Helmhol[t]z's)[10] to a pendulum vibrating accurately once per second, & by aid of it observing a bright point on the gravity pendulum, and noting the points in which it cuts two rectangular axes in the field of view. The invariable pendulum is shaped as in the second sketch and is vibrated on a knife edge through either aperture the positions of these being such as to give the shortest possible time of vibration, and therefore to make the error due to abrasion in the course of use, insensible. The greatest error I anticipate is from the curvature of the knife edges but some preliminary trials seem to show that this will be very small.

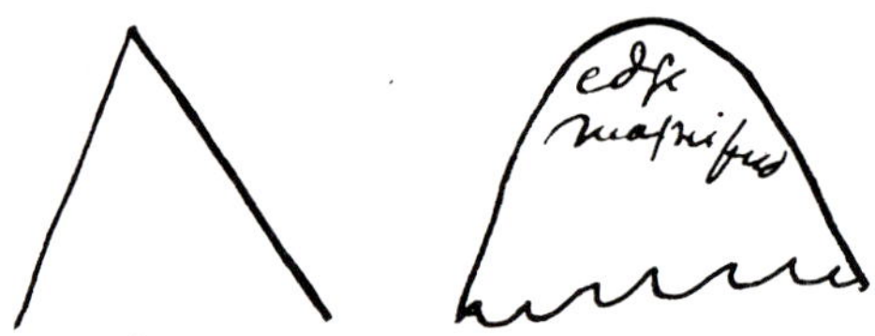

We have now great facilities for this kind of observation in my laboratory, as I have two electric clocks within reach of it kept constantly regulated within a fraction of a second of Greenwich mean time, by Prof Grant[11] at the other end of a telegraph wire three miles long; & by the regulating current alone I ⟨can⟩ keep an auxiliary pendulum, carrying the lense, vibrating with a simple harmonic motion, not disturbed as a clock pendulum is by the shock of the escapements. If I had funds (or a fair prospect of having funds next year) for a vacuum apparatus, &c I think I could get a good determination this summer.

The invariable pendulum will be very useful for a gravity survey of England & Scotland which I hope will some time be undertaken by the Royal Society or British Association. All along the lines of telegraph it can be carried out with the greatest ease, because either from Greenwich or from Glasgow, the pendulum-regulating current could be had any night for three or four hours, as I know the telegraph companies will be ready to arrange for.

I hope soon to send a report o⟨n⟩f my experiments on contact electricity and on thermal conductivity.[12] I have now got trustworthy results in both subjects and am pushing the latter on as much as possible at present.

I shall be in London a few days at the beginning of May & again for a week about the twentieth; which I hope will give me some opportunities of seeing you.

[not signed]

1 Kelvin (133).
2 Kelvin and Tait *TNP*.
3 See letter 239, note 2.
4 F. Jenkin, 'On the insulating properties of gutta-percha', in *Report of the joint committee appointed ... to inquire into the construction of submarine telegraph cables ..., Parliamentary Papers*, **62** (1860), 464–81.
5 F. Jenkin, 'Experimental researches on the transmission of electric signals through submarine cables, Part 1. Laws of transmission through various lengths of one cable', *Phil. Trans.* (1862), 987–1017.
6 See the 'Report of the committee on standards of electrical resistance', *Brit. Assoc. Rep.* (1864), 345–67.
7 He did not do so.
8 Henry Kater (1777–1835) served as a surveyor with the British Army in India, became a fellow of the Royal Society in 1815, and received the society's Copley Medal in 1817 for his pendulum experiments, the results of which were published in the *Philosophical Transactions* for 1818 and 1819.
9 Jules Antoine Lissajous (1822–80) was professor of physics at the Lycée Saint-Louis from 1847 to 1874. See his 'Mémoire sur l'étude optique des mouvements vibratoires', *Annales de Chimie*, **51** (1857), 147–231.
10 See H. Helmholtz, 'On the motion of the strings of a violin', *Phil. Mag.*, **21** (1861), 393–6, in which Helmholtz reported on attaching the object lens of a microscope to one branch of a tuning fork in order to study a vibrating violin string. The paper was originally delivered to the Royal Philosophical Society of Glasgow.
11 Robert Grant (1814–92) was professor of astronomy at Glasgow from 1859 until his death.
12 He did not do so.

245 STOKES to KELVIN, [late 1865 or early 1866][1]
Stokes Collection, RS507A

The Royal Society
Burlington House, London, W.
186 [*sic*]

SIR,

I am directed to send you the accompanying Paper, – 'On the viscosity of gases By Mr J. C. Maxwell'[2] and to request that you will favour the Committee of Papers with your opinion as regards its eligibility for publication in the Philosophical Transactions.

The Committee hope that you will be able to examine the Paper sufficiently for the purpose within a month from the present date; if, however, your engagements prevent, I should feel obliged by an intimation to that effect.

I remain,
Sir,
Your obedient Servant,
'G. G. Stokes'
Secretary R.S.[3]

Prof[r] W. Thomson

P.S. As you are already familiar with the object and results of this short paper, about which you can doubtless at once express an opinion I hope you will not keep the paper, since it has now been some while in our hands, having been kept for the Bakerian Lecture which Maxwell cannot deliver till February being engaged in the Senate House Exam[n] at Cambridge.[4]

1 The paper mentioned in note 2 was received on 23 November 1865 and read on 8 February 1866. The first part of this letter is a form letter, and the portions in Stokes's own hand are enclosed in inverted commas.

2 J. C. Maxwell, 'On the viscosity or internal friction of air and other gases', *Phil. Trans.* (1866), 249–68.

3 The rest of the letter is in Stokes's hand.

4 Maxwell examined in the mathematical tripos in 1866, 1867, 1869, and 1870, before becoming Cavendish professor at Cambridge in 1871.

246 Kelvin to Stokes, 11 April 1866
Royal Society of London, RR.6.178

Western Club Glasgow.[1]
April 11, 1866

My dear Stokes

I return Maxwell's paper[2] on the Viscosity of air and other gases by book post. I am very glad to have had an opportunity of reading it before its publication in the Transactions. It is most interesting and valuable. The evidence it contains as to the accuracy of the results in absolute measure is very satisfactory.

I hope the author will be induced to continue the investigation for other gases, and for liquids.

The plan he adopts is quite the same as one I had long contemplated, and intended applying myself, for liquids (except that I intended to have only one vibrating disc.) It seems to me ⟨best⟩ better than any other plan adapted both for determining the viscosity, and the coefficient of slip if any, between the fluid & the disc. I hope Maxwell will *soon* be able to settle the very important question of slip for liquids. Investigations of this kind are of national importance, and any thing that money can do to promote them, whether by supplying a convenient experimental laboratory, and a sufficient number of thoroughly qualified operators to carry out the work, or in any other way ought to be done by the government. If the government knew its own interest even ⟨in⟩ on a[3] strictly and simply economical grounds, it would do every thing that money can do to promote the *execution* of good experiments. You of all others know how much there is between the planning & the execution of experiments for most important objects. Could not the Royal Society move government for the establishment of laboratories for *investigation*, in which teaching would be thoroughly subordinate to the search for new knowledge of properties of matter.

Suppose such a laboratory was established in London with you or Maxwell, or both, in charge, the results would be wo[r]th hundreds of fold the annual cost, even in material economies w^h^ would arise from the knowledge gained. Besides, the directors of such institution or institutions would naturally form an advisory council for the gov^t^ (admiralty, army, customs &^c^) which would save hundreds of thousands wasted in useless experiments on a large scale, for every thousand spent in keeping up the proposed laboratory. I do think the Royal Society might move in this matter with effect.

Yours always truly
W Thomson

1 Stokes has written at the top of the page, '1866 Maxwell by Thomson'.
2 See the previous letter, note 2.
3 Apparently, after writing 'in a', Kelvin changed *in* to *on* but forgot to delete *a*.

247 Stokes to Kelvin, 1 September 1866
Kelvin Collection, S402

4 Windsor Terrace Malahide[1]
1st Septr 1866

My dear Thomson,

There is a paper[2] of Maxwell's which has been a long time in your hands, and we should be glad of a report.

I have taken several series of measures with my Iceland spar prism, using the bright line D. The result is that I can detect no error in Hygens's construction. The error, if any, in the index at 45° to the axis and thereabouts, compared with the index calculated from the constants obtained at 90° to the axis (i.e. the principal indices) appears not to go beyond the *5th* place of decimals, which place I cannot answer for, especially as the cut faces are not so truly plane as I could wish.

The supposition that the difference of velocity arises from a difference of inertia instead of a difference of elasticity which would lead to

$$\frac{1}{v^2} = \frac{\cos^2\theta}{a^2} + \frac{\sin^2\theta}{b^2}$$

instead of Fresnel's

$$v^2 = a^2\cos^2\theta + b^2\sin^2\theta$$

would give an index $\left(\frac{1}{v}\right)$ at 45° differing from that resulting from Hygens's construction by .00940, nearly 1 in the *second* place of decimals. This supposition therefore must be absolutely rejected. The section of the extraordinary wave-surface by a plane through the axis is no mere oval, symmetrical with respect to two rectangular axes, and accordingly an *approximate* ellipse, the approximation depending upon the smallness of $a - b$ compared with a or b, but a veritable ellipse. Of course I don't mean to say but that the terms on which dispersion depends may modify the exact elliptic form, or that variation of inertia might have some minute and altogether subordinate influence. Whether the disturbing (if it does disturb) influence of dispersion is such as might be expected to be sensible in observation is a point which I mean to take into consideration.

I heartily congratulate you on the successful laying of the Atlantic Cable. From our not hearing anything about it I fear the fishing attempt has proved a failure. I think you told me that you did not mean to wait for this if the cable were laid successfully, but return to Valentia, and accordingly I think it possible you may have arrived there by this.

Yours very truly
G. G. STOKES[3]

1 Kelvin has written on the envelope:

Sep 1, 1866
Stokes: finds extraordinary ray in Iceland spar *verifies Huyghens*
Result published first in Phil Mag Oct 1872

The paper is Stokes (87).

2 J. C. Maxwell, 'On the dynamical theory of gases', *Phil. Trans.* (1867), 49–88.

3 Kelvin has written below Stokes's signature, 'The result is published in a 21 line communication to Phil Mag. of Oct 1872, which I procured by personal attacks on Stokes every time I was in Cambridge from Sep 1866 till 1872. K[elvin] Dec 1, 1903'.

248 KELVIN to STOKES, 1 October 1866
Stokes Collection, K158

The Rouken
Thornliebank
Glasgow
Oct 1, 1866

DEAR STOKES,

Please say where you would like me to send Maxwell's paper[1] if not to Lensfield Cottage.[2] I shall be in London on Tuesday & Wed^y till Thursday morning where my address will be care of the Rev Dr King.[3] 25 Colville Square Nottinghill. After that I return here. I hope to be able to send you the paper before I leave London. I only got your letter 10 days ago on returning in the Great Eastern. I am sorry for the delay of the paper in my hands, but nearly all the time has been during my absence, so that even if I had got it read through I could not have returned it to you till the return of the Great Eastern. I was much interested in what you told me of your proof of Huyghens' construction.

I shall return your Airy's Tides[4] very soon. It was of great use to me almost every day during a large part of the time of my absence.

I met with very sad news on my return, the loss of one of my wife's sisters which has been a great blow to us all here. She had been very ill since last January, but it was only by a telegram before going in to Heart's Content the second time that I had intelligence preparing me for what might happen.

Mrs Thomson joins with me in kind regards to you and Mrs Stokes.

Yours always truly
W THOMSON

1 See the previous letter, note 2.
2 Kelvin has added a footnote here reading, 'where I shall send it if I do not hear from you'.
3 David King (1806–83) was the husband of Kelvin's sister, Elizabeth (1819–96).
4 G. B. Airy, 'Tides and waves', in Edward Smedley, Hugh James Rose, and Henry John Rose, eds., *Encyclopaedia Metropolitana*, 20 vols. (London, 1845), v, 241*–396*. (Asterisks are in the original.)

249 KELVIN to STOKES, 13 October 1866
Royal Society of London, RR.6.179

Blackdales Largs
by Greenock
Oct 13/66

MY DEAR STOKES

You have no doubt by this time got my letter addressed to Lensfield

Cottage, but if not no matter now. I hope my delay in returning Maxwell's paper[1] (which I posted, addressed to you at Malahide, two days after my return from London, being just 18 days after the Great Eastern's arrival) has [not] caused delay in its publication, although I fear it must have been inconvenient to you. It was with great difficulty I could find time to read it and think about it sufficiently and I had to use my last railway journey to learn a little more of it otherwise I sh[d] have returned it several days sooner.

There can be no doubt as to its suitability for the Transactions, in my opinion. But it might I think be considerably improved by guarding & modifying some what some of the statements. For instance throughout the idea, explicitly expressed in the beginning, is adhered to, of regarding the molecules as mere centres of force, repelling as D^{-5}. Now this is inconsistent with giving them kinetic energy of rotation round their centres of inertia and if they were really mere centres of force we *must* have $\beta=0$, and so the proportions of the spec. heats, &c could not be made to fit the reality of gases. It should I think be explained in a sentence or two early in the paper that the molecules are regarded *not* as centres of force but as really (according with their name) little heap[s] of matter, acting on one another with forces *not* in lines through their centres of inertia (in as much as ⟨increase⟩ change in the energy of motions of their centres is accompanied with increase of their rotatory energies keeping the average proportion $\beta:1$ constant, and that the average value of the component of the mutual force, between each pair in the direction of the line joining their centres is prop[l] to D^{-5}.

⟨This [illegible word] conditions analogous to this could even be realized for instance by making each molecule a group of magnets (arranged to constitute what⟩

This supposition differs from those previously worked on by Herapath, Clausius, & Maxwell only in making D^{-5} the law of average repulsion, instead of $\left(\frac{D}{a}\right)^{-12}$ where a is the radius of a hard spherical atom, and n is a very great (infinitely great) number.[2]

There is one very important investigation, against the conclusion drawn from which I think the author should warn the reader (as he warned me in correspondence 3/4 of a year ago).[3] It is that in which the law of temperature of air in a vertical column is sought. The conclusion, as Maxwell wrote to me, violates Carnot's principle. He put it to me thus, and I thought of it a long time, without being able to see through it. Neither can I now when I have had his paper in my hands. But yet I feel convinced that the conclusion must be wrong.

Imagine a closed vessel of impermeable material – a vertical column (circular or rectangular if you please) with a ⟨symmetri⟩ plane partition *EF* also impermeable, dividing it into two symmetrical parts, communicating freely at

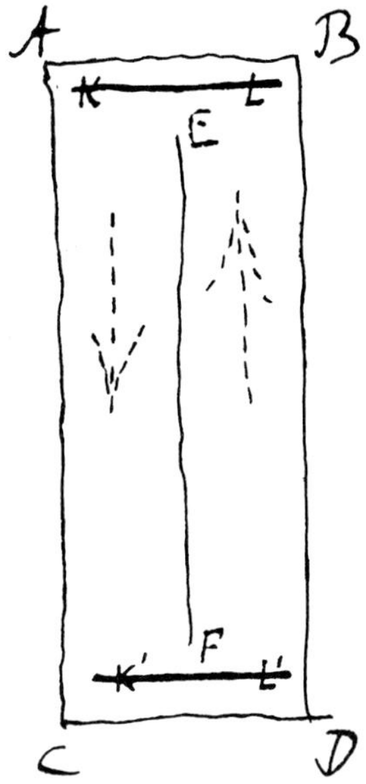

the top and bottom of the partition but perfectly separated elsewhere. Fix plates of ⟨metal⟩ polished silver *KL*, *K′ L′* sym[m]etrically in the positions shown & let the halves of these towards *K*, *K′* be smoked. Leave the whole to itself and and [*sic*] the Maxwell equilibrium will be unstable. For let a ⟨a⟩ current ⟨will⟩ ever so slight be established in the direction shown by the arrow heads. Then in virtue of the air being (as Maxwell's conclusion makes it) hotter at the top, and the consequent radiation from *K* to *K′*, the air moving down from *K* will become cooled below its equilibrium temperature and that moving horizontally from *K′* to ⟨*K′* an⟩ *L′* and upwards from *L′* will become hotter for the same reason. Thus the current will be kept flowing, & by putting a wind mill in any proper position work will be done till all the kinetic energy of the particles is gone: i.e. the whole heat of the gas will convert itself into weights raised. Surely there must be some flaw in a kinetic investigation that leads to such a result.[4] What the flaw may be in Maxwell's investigation if any I have not been able to see. But I cannot see it as a rigorous mathematical working out from the premises and therefore it seems to me probable that some of the assumptions used as stepping stones must be in-valid. The only definite suggestion that has occurred to me is that Maxwell does not seem to have introduced the change of energy in the free motions of each particle between successive collisions due to gravity and that possibly by taking this into account a different conclusion may be had without giving up any of the assumptions.

I have made a few pencil marks on the first page or two, where it appeared to me that the author had stated too unguardedly fallacious conclusions of the older molecular attempts.

It is obvious (and easily proved by experiment as I have done in lecturing on the subject) that for cork Poisson's ratio is nearer o than $\frac{1}{4}$. Everett in experiments made this summer on glass & brass which will soon be communicated to the R.S.[5] has found for glass something decidedly less than $\frac{1}{4}$ (something about .22 or .23 if I remember right). For brass he finds a considerably larger value than Kirch[h]of[f] had found for his brass. Besides, a long time ago I asserted (Phil. Trans. somewhere about 1855)[6] that a solid violating Poisson's condition could be ⟨constructed⟩ built up of small parts (molecules) each fulfilling it. Here for instance is one way of doing this. Make ordinary spiral springs of steel wire suppose, (w[h] ⟨very nearly⟩ some what approximately does really fulfil Poisson's condition & may be supposed to do so rigorously for the present) and build up a solid of such molecules, ⟨conne⟩ ⟨cast into⟩ connected in any way you please. In as much as when you compress or elongate a single spiral spring in the direction of its axis, the lateral expansion or contraction is very small, the Poisson ratio for ⟨such⟩ a ⟨com⟩ solid so composed will be nearer o than $\frac{1}{4}$. It no doubt is so for a ball of worsted or for a cube of sponge & no doubt cork is composed, so for as first grade of molecular structure, some what like sponge.

This reminds me of one of Maxwell's initial remarks w[h] requires correc[n] – viz that ⟨molecular⟩ theories of matter are either atomic, or the one that supposes all matter continuous *and homogeneous*. The only views that have ever appeared to me true or natural as to the constitution of matter are those that suppose all space to be full but the properties of known bodies to be due to or necessarily associated with molecular structure or of a sponge or other organic tissues or brick work, i.e. that there are vast variations of density from point to point within spaces of dimensions some small fraction of a wave length (though not inappreciably small.)

I am at a loss where to address this but as you say nothing about leaving Malahide I let it take its chance there.

Yours always truly
WILLIAM THOMSON[7]

1 See letter 247, note 2.

2 John Herapath (1790–1868) was a self-educated man who ran a private school and in 1836 became editor of the *Railway Magazine and Annals of Science*. His 'A mathematical inquiry into the causes, laws, and principal phenomena of heat, gases, gravitation, &c', *Annals of Philosophy*, 2nd series, 1 (1821), 273–93, 340–51, 401–16, became well-known as a long-neglected treatment of kinetic theory. For an account of nineteenth-century kinetic theory, see Stephen G. Brush, *The Kind of Motion We Call Heat: A History of the Kinetic Theory of Gases in the 19th Century* (Amsterdam, Oxford, New York, 1976).

3 See Maxwell to Kelvin, 27 February 1866, Glasgow University Library, Kelvin Papers, M19.

4 Kelvin has added a footnote here reading, 'I am very confident that the bodily convective equilibrium of temperature in a vertical column must agree with the molecular equil[m] of temperature in the true molecular theory'.

5 J. D. Everett, 'Account of experiments on the flexural and torsional rigidity of a glass rod, leading to the determination of the rigidity of glass', *Phil. Trans.* (1866), 185–92.

6 See Kelvin (91) in *MPP*, III, 36–8.

7 Stokes has written below Kelvin's signature, '1866 Maxwell by Thomson'.

250 KELVIN to STOKES, [February? 1867][1]

Stokes Collection, K372

mistakes on which it is founded.

Do you know of any theorem to the following effect. It is impossible for a continuous bulk, *V*, of a perfect fluid ever to become divided: i.e. to become separated into isolated parts so that it would be impossible to pass from some point *A* to some other point, *B*, of the matter of *V*, without passing through some fluid not belonging to *V*: provided either there is no solid, in the fluid (the fluid infinite); or no solid with a hole or holes through it like a ring, or any more complex shape involving 'multiple continuity' as Riemann[2] calls it. It is clearly possible to divide a Volume *V* into parts by a solid with a properly shaped trough through it; but I think not otherwise. What I am particularly interested in is an infinite fluid with no solids through it, but vortex motion in parts of it (i.e. $\frac{dv}{dz}-\frac{dw}{dy}$ &c not $=0$, except in certain parts). You know Helmhol[t]z I suppose, on 'Wirbelbewegung'.[3] This and observations on smoke rings have set me to a very promising atomic theory regarding wh I shall soon I hope send you something in print from the Proceedings RSE.[4]

Two days ago I despatched the last regular MSS. for Vol I of our book[5] and in the course of a few weeks I hope to get it all through the press.

I have been making a commencement of reducing Tide guage [*sic*] observations (Ramsgate for one year being my first supply).[6] I find that on the average of the whole year the level is very sensibly higher at 10.30 am and lower [at] 10.30 pm (mean solar time). This no doubt is because the solar diurnal tide is greater in winter than in summer on acct of the excentricity of the earth's orbit. I find that the level is very sensibly higher on the average of the 24^h in autumn than in spring. (? Melting of snow & ice in north polar regions). I intend to work out by strict harmonic analysis the solar semiannual, diurnal, & semidiurnal tides, & the corresponding lunar ⟨corre⟩ tides: also the variations of each one to excentricities of orbits, &c &c. There will be great labour, but after I get the Ramsgate series well worked, which I think I can do, I hope to be able to suggest a methodical plan, by which tide guage [*sic*] observations might be regularly reduced & good results found. Nothing could well be worse than the rules for reduction hitherto followed.

I hope your family are all well. My wife unhappily is obliged to winter at Torquay but has got through the bad season well.

Yours very truly
W THOMSON

The smoke ⟨rig⟩ rings give exquisite subjects for problems especially regarding their vibrations.

1 The first part of the letter, which would have contained the date, is missing. However, the article cited in note 4 was read to the Royal Society of Edinburgh on 18 February 1867. Also, Kelvin's work with the Ramsgate tides appears to be in a more preliminary state in this letter than in the next.
2 Georg Friedrich Bernhard Riemann (1826–66) became extraordinary professor of mathematics at the University of Göttingen in 1857 and full professor in 1859.
3 H. Helmholtz, 'Ueber Integrale der hydrodynamischen Gleichungen welche den Wirbelbewegungen entsprechen', Crelle's *Journal für die reine und angewandte Mathematik*, 55 (1858), 25–55.
4 Kelvin (161).
5 Kelvin and Tait *TNP*.
6 Kelvin was the principal member of the British Association's Committee for the Purpose of Promoting the Extension, Improvement, and Harmonic Analysis of Tidal Observations. See the committee's series of reports in *Brit. Assoc. Rep.* (1868), 489–510; (1870), 120–51; (1872), 355–95; and (1876), 275–307.

251 KELVIN to STOKES, 21 March 1867
Stokes Collection, K159

Glasgow College
March 21/67

MY DEAR STOKES

I enclose figures showing what I said to you yesterday as to higher water at Ramsgate in autumn than in spring. The general concurrence of the two sets of means seems to prove that the result is genuine, and you see the deviations on each side of the mean of all the 350 × 8 heights amounts to as much as $1\frac{1}{2}$ inch at times. The earth must be a very bad time keeper if ⟨all⟩ the melting of Polar ice has anything like this effect.

The 70 days' period was merely to get 10 & 14 as a factor for some matters of convenience as to a rough preliminary.

Yours truly
W THOMSON

P.S. I intended to send you the diurnal tide on average of year but my calculator (a student, amateur of reductions of observations) has locked it up with his papers.

252 KELVIN to STOKES, 29 March 1867

Stokes Collection, K160

Glasgow College
March 29/67

DEAR STOKES

I find that even a solid ring can never cause the separation into isolated portions of any volume, once one, of a fluid. I don't quite yet see a general proof of the indivisibility.

I see that a bullet does not pierce a plane of air in front of it but draws it forward & stretches it, indefinitely the farther it (the bullet) goes.

Yours very truly
W THOMSON

253 STOKES to KELVIN, 24 April 1867

Kelvin Collection, S403

Lensfield Cottage Cambridge[1]
24th April 1867

MY DEAR THOMSON,

I had a good deal on hands when I got your letter containing the question, and though I did not see my way to getting the separation about the possibility of which you wrote to me, ⟨but⟩ I kept the matter over for further consideration, and a day or two afterwards I got your letter in which you told me you had convinced yourself the thing was not possible.

In the course of a letter to Pratt lately (who by the bye has been writing some heterodoxy in the Phil. Mag.)[2] I saw the solution of a problem which when I thought of it some 20 years ago in connexion with Clairaut's Theorem &c.[3] I did not see, namely, how to get the most general expression for the density of matter within a closed surface which shall produce a given potential at the surface. I have written a short paper[4] on the subject for the Proceedings R.S. but the thing is absurdly simple when once pointed out, and if it so seems to you without this pointing out I should be glad of a line, as in that case I might like to withdraw my paper.

Yours very truly
G. G. STOKES

1 Kelvin has written on the envelope, 'Stokes Ap 24 1867. small paper correcting error of Pratt re Clairaut's theorem & density to produce given potential. (K[elvin] Nov 30 1903)'.

2 J. H. Pratt published several articles in the *Philosophical Magazine* during 1866 and 1867, but the April 1867 number contained one of a series entitled, 'To find what changes may be made in the arrangement of the mass of a body, without altering its outward form, so as not to affect the attraction of the whole on an external point', *Phil. Mag.*, **32** (August 1866), 132–5, and **33**

(April 1867), 261–4; (May 1867), 332–5; and (June 1867), 445–6. See also Stokes's two letters to Pratt, both dated 9 March 1867, in the Stokes Collection, P618 and P619. In Pratt's 'On Professor Stokes's proof of Clairaut's theorem', *Phil. Mag.*, **34** (July 1867), 25–6, he apologized for misrepresenting Stokes's research.

3 See Stokes (27) and (31).

4 Stokes (79).

254 KELVIN to STOKES, 12 June 1867
Stokes Collection, K161

Glasgow College
June 12, 1867

DEAR STOKES

I shall send you a paper[1] (very short) which will be despatched from here either on Friday or Saturday next 'On a simple apparatus for the augmentation and maintenance of an electric charge', for the R.S. which I presume will meet for the last time this session, ⟨this⟩ tomorrow week.

I hoped to see you in London about the beginning of this month but we were called away by news of the sudden illness of my father-in-law. I intended to speak to you about the interior distribution of matter, but afterwards did not write, as I did not see anything to make your communication[2] unnecessary as you suggested. I don't see through your title but whether I see through your solution as a known thing when I see your paper, or not, I am sure it will do good to publish it.

Yours always truly
WILLIAM THOMSON

1 Kelvin (162).

2 Stokes (79).

255 KELVIN to STOKES, 16 June 1867
Stokes Collection, K162

The Rouken
Thornliebank
Glasgow
June 16/67

DEAR STOKES

I shall send you the paper[1] on the generation & maintenance of electric charges, addressed to you at the R.S., and posted in Glasgow not later than Tuesday. I left it yesterday, dictated & taken down complete, but I want to revise it tomorrow and also to see to copy for three simple wood cuts. It is intended only for the Proceedings. I presume there will be no objection to the wood cuts.

Yours truly
WILLIAM THOMSON

1 Kelvin (162).

256 KELVIN to STOKES, 8 May 1868
Royal Society of London, R.6.8

Nice May 8, 1868

MY DEAR STOKES

Before leaving Glasgow I sent you back Airy's paper on Earth currents,[1] but I did not succeed in finding time to write along with it. I have however little to add to the verbal report I gave you at Cambridge. I think the paper ought decidedly to be published in the Transactions. It contains I believe the first step towards such a comparison between earth-currents, and magnetic disturbances as is required from observation before any hypothesis to explain the connection can be properly weighed.

The diagrams referred to in your memorandum attached to the paper had not arrived before I sent it off. I should think however, though I have not seen them, that it will be quite worth the cost, to give a good deal of them with the paper.

Yours very truly
WILLIAM THOMSON

P.S. If you know of any thing more regarding the numerical calculation of the solutions of the equation

$$\frac{d^2u}{dr^2}+\frac{1}{r}\frac{du}{dr}+pu=0$$

than is to be learned from your papers in the Camb. Trans (Pendulums, & Numerical calculations of certain definite integrals)[2] I should be much obliged by your informing me. I mean to have each solution for u, and $\frac{du}{dr}$, calculated, both for p pos. & p neg., through some such range as from $r=1$ to $r=10$, for $p=\pm 1$. This is of course very easy (by the diverging series you give) to a fair degree of approx^n, for all values of r exceeding 5 or 6; but there will be a good deal [of] labour to complete the ⟨four⟩ 8 tables of values, for sufficiently close intervals as to r, to be useful for practical purposes, and if tables are already published I should be very glad to be spared it. If you have anything to tell me will you address it to M[r] D. Macfarlane, Nat. Phil. Laboratory College Glasgow, who has been making the calculations for me, and has your papers at present in his hands. He writes to me once a week when I can give my address which just now I cannot do.

1 G. B. Airy, 'Comparison of magnetic disturbances recorded by the self-registering magnetometers at the Royal Observatory, Greenwich, with magnetic disturbances deduced from the corresponding terrestrial galvanic currents recorded by the self-registering galvanometers of the Royal Observatory', *Phil. Trans.* (1868), 465–72.

2 Stokes (47) and (41).

257 Kelvin to Stokes, 19 October 1868

Royal Society of London, RR.6.181

Burnmouth
Largs by Greenock
Oct 19, 1868

My dear Stokes

I return you by book post along with this Maxwell's paper[1] on a Method of making a direct comparison between electrostatic and electromagnetic ⟨units⟩ force, which I think ought to be published by all means in the Transactions.

According to your request in your letter of Sep 17 I sent without more than two days delay to M[r] White in a registered pacquet, the paper on Solar Physics by De la Rue, ⟨Stuar⟩ Stewart and Loewy.[2] It too I think should be printed with however only such of the tables as are suited to be useful to persons working independently on the same subject, or proper to allow the reader to judge of the evidence afforded by observation for any inferences that may afterwards be drawn. It seemed to me that the *auxiliary* tables (about three of the five if I remember right) which are comparatively short, might be generally useful but that, at all events at this stage of the investigation, the long tables of observations might be with advantage replaced with short abstracts, or deferred until a ⟨later⟩ continuation of the commun[n] to the R.S. containing a discussion of the results is made. I was in London lately and attempted to see Stewart ⟨at Kew⟩ on the subject but found he was away on a tour of observatories. I think the authors might be consulted, and if they wish the tables to be published complete at present for which if they do wish it they no doubt would have good reason you might properly determine to publish them.

Yours always truly
William Thomson[3]

1 J. C. Maxwell, 'On a method of making a direct comparison of electrostatic with electromagnetic force; with a note on the electromagnetic theory of light', *Phil. Trans.* (1868), 643–58.

2 W. De la Rue, B. Stewart, and B. Loewy, 'Researches on solar physics. Heliographical positions and areas of sun-spots observed with the Kew photoheliograph during the years 1862 and 1863', *Phil. Trans.* (1869), 1–110. Warren De la Rue (1815–89) worked with the company established by his father, Thomas De la Rue (1793–1866), which produced various paper products, including cards, envelopes, and stamps. Warren made contributions to chemistry and astronomy, especially in developing new instruments for the photographic study of the moon and the sun. There is an extensive correspondence from Warren De la Rue to Stokes in the Stokes Collection. Benjamin Loewy (b. 1831) was an assistant at Melbourne Observatory from 1860 to 1864, an assistant at Kew Observatory from 1864 to 1869, and from 1870 professor of natural philosophy at International College in London.

3 Stokes has written at the bottom of the page, '1868 De la Rue, Stewart, Loewy [on] Solar physics, also Maxwell by Thomson' and '1868 Maxwell also De la Rue (Solar obs. Kew) by Thomson'.

258 Kelvin to Stokes, 2 December 1868

Stokes Collection, K164

Glasgow College
⟨Nov⟩ Dec 2/68

My dear Stokes

I have no objection whatever to your sending my report,[1] which I return herewith, to De la Rue & C°. I shall attend to what you say as to reports in future – ⟨to⟩ and shall write each on separate paper with no other matter.

As to Le Roux's[2] thermoelectric researches, if he sends an account of all he does to the Royal Society, or to me & I undertake to present a translation to the Royal Society, before it appears elsewhere would it not be considered that the object of the Gov[t] Grant would be as properly carried out in case of assistance being given to him, as if he were to make the experiments in England or Scotland? It would be a continuation of my former work, described in my paper 'Electrodynamic Qualities of Metals'[3] for which assistance was formerly given to me, and would tend to answer, and I trust have effect in answering, some important questions which I put as rising out of those researches, but which work in other directions has hitherto prevented me from attempting to answer myself by exp[t].

Yours always truly
William Thomson

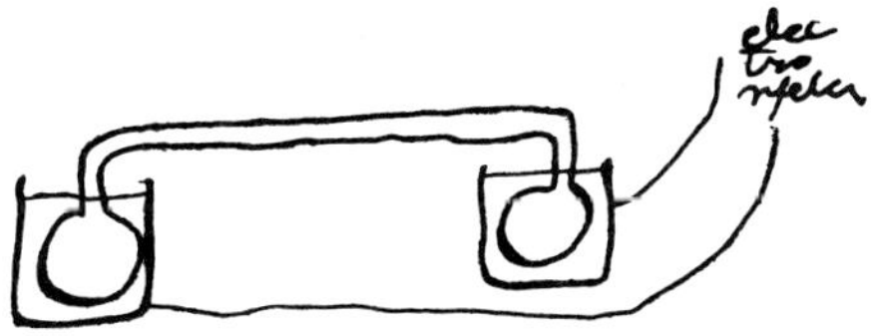

P.S. The specific inductive (electrostatic) capacity of glass increases when heated and diminishes when cooled. I use two glass balls connected with glass tube charged internally with elec^{y}, & hermetically sealed. A little moisture inside forms inside coating. Difference of temperature between the balls, connected outside by insulated water cans, affects the electrometer.

I hope very soon to send a short paper for the Proceedings on spec. ind. capac.[4]

1 Apparently, the previous letter, which Kelvin presumably meant could be sent to Warren De la Rue and his co-authors, not to Thomas De la Rue & Co., the London firm. See the previous letter, note 2.

2 François Pierre Le Roux (1832–1907) was professor of physics in the school of pharmacy at the University of Paris.

3 Kelvin (92).

4 Kelvin (171A).

259 KELVIN to STOKES, 14 December 1868

Stokes Collection, K165

Glasgow College
Dec 14, 1868

MY DEAR STOKES

I am more inclined to wish for a reprint of Fourier in French, than a translation.[1] The language should be no barrier to any one who could read such a book at all, and I think it would be more interesting to read it in Fourier's own words than in a translation. A very condensed exposition of all that Fourier has, to make room for a great deal more, is a want that will soon no doubt be much felt at Cambridge, and no doubt also soon supplied, but Fourier's own book in his own language will I think always be interesting; and for many years at all events would probably be useful 'for Tripos purposes'.

Yours truly
WILLIAM THOMSON

P.S. I find *very roughly* about $\frac{1}{100}$ per cent of its own amount as the augmentation of spec ind. capac of glass per 1° Cent elevation of temperature. I shall soon have a fair measurement.[2]

1 An English translation of Fourier's 1822 book appeared in 1878. (J. B. J. Fourier, *The Analytical Theory of Heat*, trans. A. Freeman (Cambridge, 1878).) The question of a translation no doubt occurred in 1868 in regard to the movement to reform the mathematical tripos, a movement which led to the introduction of heat, electricity, and magnetism into the tripos in 1873. There was not a reprint in Britain of Fourier's book in French during this period, although there were editions in Breslau in 1883 and Paris in 1888. Alexander Freeman

(1838–97) was fifth wrangler in 1861, fellow of St John's College from 1862 to 1882, and rector of Murston, Kent, from 1882 to 1897.

2 See Kelvin (171A).

260 Kelvin to Stokes, 2 March 1869

Stokes Collection, K166

Glasgow College
March 2, 1869

Dear Stokes

Is there any chance that anything can be done to promote thermoelectric measurement & discovery through Leroux's work, by assistance from the R.S. grant, in any of the ways suggested in my last letter to you. If I remember right General Sabine in writing to me on the subject in summer said that the thing might possibly be done by letting the grant be given in my name (as to me) and my becoming responsible for its proper application. This I would consent to if the Council preferred it to voting the grant to Le Roux, and consented to doing it at all. I shall look for General Sabine's letter & enclose it if I can find it.[1]

What is to be done to correct the attributing of Joule's result about the heating of copper by motion through the magnetic field, to Foucault in the last Proc^{gs}?[2] You remember that when it was proposed to attribute it to Foucault at the time the medal was given to him you wrote to me[3] and found that I agreed with you that it was wrongly attributed to Foucault by others: and it was not done so (I suppose in consequence of your objection) by the R.S.

Yours truly
William Thomson

1 Sabine's letter is not in the Kelvin Collection, the Glasgow Kelvin Papers, the Stokes Collection, or the Royal Society. However, see Kelvin to Sabine, 4 June 1868, Stokes Collection, RS646. Kelvin wrote to Sabine from Italy saying he had recently seen Le Roux in Paris. They had agreed on experiments necessary to resolve differences between them in thermo-electricity, and Kelvin wanted the Royal Society to support Le Roux's conduction of the experiments because 'there seems to be no source from which such assistance can be had in France'.

2 I have not found this reference.

3 See letters 134 to 137.

261 KELVIN to STOKES, 30 March 1869
Royal Society of London, RR.6.11

March 30/69[1]

DEAR STOKES

I send by Book-post along with this, Airy 'On the Diurnal and Annual ⟨Irregularities⟩ Inequalities of Terrestrial Magnetism'.[2] I am of opinion that it ought to be published in the Transactions. The graphic constructions are most interesting. It is very desirable that a thorough comparison between the Greenwich and Kew observations should be made. It is also desirable I think that they should all be reduced harmonically. This would give a more precisely definite character to any such result as those very interesting ones of the Ast. Royal in the present paper in which he finds a single varying force to constitute the lunar disturbance, & a corresponding (though very different) result for the solar.

I am sorry to have kept the paper so long. I intended to return it immediately but had let it stand over much longer than I thought, when I received your reminder.

I am anxious to write to LeRoux on the subject of my former letter to you and should be glad to hear from you what (if any) prospect there is that something could be done.

Yours very truly
WILLIAM THOMSON

1 Stokes has written at the top of the letter, '1869 Airy by Thomson'.
2 G. B. Airy, 'On the diurnal and annual inequalities of terrestrial magnetism, as deduced from observations made at the Royal Observatory, Greenwich, from 1858 to 1863; with a note on the lunodirunal and other lunar inequalities, as deduced from observations extending from 1848 to 1863', *Phil. Trans.* (1869), 413–24.

262 KELVIN to STOKES, 4 April 1869
Stokes Collection, K167

Glasgow College
April 4, 1869

DEAR STOKES

I never thought of 'the Society's funds being used for separate illustrations' of my private copies; but I thought that perhaps engravings of some kind or lithographs had been or were to be executed for the Proceedings on a separate page, of which I ⟨e⟩should have copies to accompany the printed slip containing the letter press which the printers had sent me. The thing is a matter of no consequence, but if anything had been done from the photographs for the Proceedings I thought I might as well have it to accompany the the [*sic*] printed article.[1]

Yours very truly
WILLIAM THOMSON

1 Kelvin (170), which contains only two simple figures.

263 KELVIN to STOKES, 3 August 1869
Stokes Collection, K168

Largs, by Greenock
Aug. 3, 1869

DEAR STOKES

I could not answer your letter until I had an opportunity of telegraphing through the cable to members of the units committee[1] on the other side. You know the result. I am sorry none of them is to be at Exeter. I tried through the cable to persuade them, urging that all three should be at Exeter to tell about the cable, but could only get a negative reply. Unless Maxwell is to be at the meeting I am afraid 'electric measurement' will not be personally represented this time. But a good deal of work has been going on quietly, and I hope that by next year there will be some results worth reporting. I am sorry not to be at the meeting myself, and I should certainly in other circumstances have made an effort to attend, under your presidency. I hope you will have a pleasant and in all respects successful and satisfactory meeting. I shall look for reports of the proceedings with much interest.

I am now settled here, having got back from Brest last week, where I was on account of duty undertaken this time last year. I do not intend to go from home again if I can help it, and I am refusing every proposal as to cables or anything else that could take me away.

Since I did not manage to get an account of experiments on specific inductive capacity prepared at the right time for printing, I think it will be better to delay it a little, until some farther experiments, for which preparations are in progress in my laboratory just now, shall have been made. In the mean time I have a short article[2] almost ready, and if there should be much delay in getting the new exp[ts] done I shall send what I have.

Yours truly
W THOMSON

1 The 'Report of the committee on standards of electrical resistance', *Brit. Assoc. Rep.* (1869), 434–8, gives a list of the committee's members.
2 Kelvin (171A).

264 KELVIN to STOKES, 17 August 1869

Stokes Collection, K169

Aug 17, 1869

DEAR STOKES

I am sending Report ⟨for⟩ of Tides Committee's work[1] to be in the hands of the secretary of Section A., and along with it accounts of expenditure, from the commencement of the grants. The expense has been almost wholly for calculations in the Nautical almanac office. It has exceeded by about £41 the whole amount of the grants which I have received (£200, in two years.) We are now able to make far more rapid progress than ever, having matured the plans of calculation, E. Roberts,[2] recommended to me by Mr Farley[3] of the Nautical Almanac Office, according to Airy's advice, has proved most satisfactory. He is very trustworthy, and very intelligent. He enters into the principle of the calculation in such a way as to give much more value to his work than if he were a mere calculator.

I hope the Admiralty will relieve us of the expense of continuing the investigation after this year. But for this year it is very important that the work should not be allowed to languish and I hope a grant of £200 may be had from the Association.[4] Wishing you *therefore* a *numerous*, and for all reasons a pleasant meeting I remain

Your always truly
WILLIAM THOMSON

1 Kelvin's report for 1869 was not published until 1870. See the report of the Committee for the Purpose of Promoting the Extension, Improvement, and Harmonic Analysis of Tidal Observations in *Brit. Assoc. Rep.* (1870), 120–51. Kelvin's 1869 report is on pages 120–1.

2 Edward Roberts, of the Nautical Almanac Office, headed the team of calculators in that office who reduced tidal observations for the British Association committee on tides.

3 Richard Farley (1810–79) was a calculator for the Nautical Almanac Office before becoming in 1837 a first assistant.

4 In 1869 Kelvin received £100 for the committee on tides and £50 for the committee on underground temperature. (*Brit. Assoc. Rep.* (1869), p. lxxx.)